Spectral Theory and Differential Equations

V. A. Marchenko's 90th Anniversary Collection

American Mathematical Society

TRANSLATIONS

Series 2 • Volume 233

Advances in the Mathematical Sciences—66

(*Formerly Advances in Soviet Mathematics*)

Spectral Theory and Differential Equations

V. A. Marchenko's 90th Anniversary Collection

E. Khruslov
L. Pastur
D. Shepelsky
Editors

American Mathematical Society
Providence, Rhode Island

2010 *Mathematics Subject Classification.* Primary 34B20, 34L25, 34L40, 34M60, 35B27, 47B36, 60B20, 60J67, 81Q20, 81U40.

Library of Congress Card Number 91-640741
ISBN-13: 978-1-4704-1683-6
ISBN-10: 1-4704-1683-2
ISSN 0065-9290

∞ The paper used in this book is acid-free and falls within the guidelines
established to ensure permanence and durability.
Visit the AMS home page at http://www.ams.org/

10 9 8 7 6 5 4 3 2 1   19 18 17 16 15 14

This volume is dedicated to Professor V. A. Marchenko in honor of his 90th birthday. It contains original papers written by his colleagues and former students and focuses on the areas to which he made important contributions: spectral theory of differential operators and related topics of mathematical physics, including inverse problems of spectral theory, homogenization theory, random matrices, and the theory of integrable systems.

Contents

Preface

Prominent mathematician Vladimir Aleksandrovich Marchenko was born on July 7, 1922, in Kharkov (the Soviet Union, now Ukraine). With this volume the authors and editors celebrate his 90th birthday.

Like many people of his generation, Marchenko was strongly affected by World War II during his youth. He was a student at the Department of Physics and the Department of Mathematics (the latter by correspondence) at Leningrad (now St. Petersburg) University in 1939. He experienced World War II while on a summer vacation with his mother and sister, who lived in Kharkov. They were not able to leave Kharkov because of health problems in the family and had to stay there during the whole occupation, October 1941–August 1943. This was a very hard time, requiring enormous efforts just to survive, including the production and sale of hand-made safety matches.

After the liberation of Kharkov in 1943, Marchenko continued his education at Kharkov University, this time in the Department of Physics and Mathematics. He graduated in 1945, received his Ph.D. in 1948 (in two years instead of the usual three), his D.Sc. in 1951, and became a full professor of the department in 1952. In 1959–1960 he was strongly involved in the organization of the Institute for Low Temperature Physics and Engineering in Kharkov and became the head of the Department of Mathematical Physics of the Institute in 1960. Since then Marchenko's scientific life has been entirely linked with the Institute. He is now an Emeritus Professor of the Mathematical Division of the Institute, continuing his research work, writing books, generously sharing with colleagues his erudition and vision of mathematics and, as always, keeping a lively interest in advances in modern science.

Marchenko is the author of more than 130 publications, including 12 monographs. He obtained fundamental and well-known results in harmonic analysis, the theory of almost periodic functions, the spectral theory of differential and finite-difference operators, the theory of inverse problems of spectral analysis and scattering theory, the spectral theory of large random matrices, the theory of diffraction of electromagnetic waves on periodic structures, homogenization theory for boundary-value problems of mathematical physics in domains with complicated microstructure, and the theory of completely integrable nonlinear evolution equations. One can mention at least three quite active fields, inverse problem of spectral theory, homogenization theory, and random matrix theory, where Marchenko is among the founding fathers and where his seminal works are well known and regularly cited. This shows that together with his virtuoso analytical techniques, he has a deep understanding of the trends and challenges of various branches of mathematics, setting their standards and perspectives sometimes well ahead of their time. We refer the reader to the recent paper by Yu. M. Berezansky et al. (Russian Math. Surveys **67**:5 (2012), 186–191) for more details on his biography and results.

For more than half a century Marchenko has been the leader of the Kharkov school of mathematical physics and spectral theory of operators, which dates back to the 19th century and includes A. M. Lyapunov and V. A. Steklov among its founders. A brilliant researcher and outstanding teacher and personality, he possesses the gift to make deep and difficult topics of analysis, spectral theory and mathematical physics attractive for young people via his lectures, personal discussions, and his well-known seminar. For many years, the seminar has been the center of activity of an informal and large group who conduct research on a wide range of subjects of modern mathematics and mathematical physics.

Marchenko's outstanding scientific achievements made his name well known to mathematicians all over the world. He is a member of the National Academy of Sciences of Ukraine and the Russian Academy of Sciences, a Doctor Honoris Causa of the University Denis Diderot (Paris) and Kharkov University, and is a member of the Royal Norwegian Society of Sciences and Letters. He is a recipient of the Vernadsky Gold Medal of the National Academy of Sciences of Ukraine, the highest academic distinction of the country.

Marchenko's colleagues, friends, and students congratulate him on his 90th birthday and wish him good health and all the best in the coming years.

The Editors

Amer. Math. Soc. Transl.
Volume **233**, 2014

Local Inverse Scattering Problem as a Tool of Perturbation Analysis for Resonance Systems

V. M. Adamyan, G. J. Martin, and B. S. Pavlov

ABSTRACT. The unperturbed resonance scattering systems $A \oplus H \equiv \mathbf{A}_0$ are characterized by eigenvalues $\{a\}$ of the insulated system A embedded into the absolutely continuous spectrum of the complementary component H. These embedded eigenvalues give rise to resonances of the perturbed system $[A \oplus H]_\varepsilon \equiv \mathbf{A}_\varepsilon$ with an interaction switched on. The resonances and the resonance states of the perturbed system can't be calculated by the standard analytic perturbation procedure, but may be obtained, following the celebrated H. Poincaré idea, via a modified two-step procedure, which consists of construction of a solvable model $\mathbf{A}_0^\varepsilon$ in the first step, followed by the standard analytic perturbation procedure $\mathbf{A}_0^\varepsilon \longrightarrow \mathbf{A}$ in the second step, which yields the full scattering matrix $S(\mathbf{A}_\varepsilon, H) = S(\mathbf{A}_\varepsilon, \mathbf{A}_0^\varepsilon) \times S(\mathbf{A}_0^\varepsilon, H)$ of the perturbed system as a product of two factors serving the scattering matrices of the pairs $\mathbf{A}_\varepsilon, \mathbf{A}_0^\varepsilon$ and $\mathbf{A}_0^\varepsilon, H$ respectively. Here the left factor $S(\mathbf{A}_\varepsilon, \mathbf{A}_0^\varepsilon$ depends analytically on the perturbation parameter and the right factor $S(\mathbf{A}_0^\varepsilon, H)$, playing the role of the local scattering matrix, is not analytic with respect to ε, but takes into account exactly the "dangerous resonances " approaching the embedded eigenvalues of the unperturbed system on the selected spectral interval when $\varepsilon \to 0$. In this paper we aim to calculate the local scattering matrix and to restore the intermediate Hamiltonian $\mathbf{A}_0^\varepsilon$ based on the factor $S(\mathbf{A}_0^\varepsilon, H)$, which corresponds to the local inverse scattering problem.

1. Dedication

The remarkably rich area of analysis motivated by inverse problems remains—for decades—a centre of attraction for a large community of mathematical physicists. The basic works of Vladimir Alexandrovich Marchenko have shaped the field, and are considered the "bible" of the community, where the enlightened find rules and advice. His philosophy of inverse problems helped us to find a persuasive—we hope—pretext for the substitution of the standard analytic perturbation procedure in resonance scattering by the *quantum* analog of the modified two-step version of the analytic perturbation procedure, suggested by H. Poincaré for problems of celestial mechanics with small denominators.

2010 *Mathematics Subject Classification.* Primary 47-XX.

B. S. Pavlov acknowledges support from a grant of the Russian Academy of Sciences, RFFI 97-01-01149. G. J. Martin and B. S. Pavlov were partially supported by the NZ Marsden Fund. V. M. Adamyan was supported by the Swiss National Science Foundation through SCOPES Grant IZ73Z0-128135.

2. Introduction

A typical resonance system is formed as a union of a basic region such as a resonator, or quantum dot or a quantum well surrounded by barriers, and a "reservoir" attached to this basic region by thin channels, narrow necks, or—in the quantum situation—by tunnelling.

This decomposition corresponds to a coherent process on the basic region which is fed by the exterior reservoirs; see for instance [**30**]. On the reservoirs; which are assumed to be homogeneous and neutral, the dynamics is assumed to be free. The resonance properties of the basic region and the coupling on the contacts will define the dynamics of the resonance system. Despite the weak coupling of the components of a quantum resonance system, the standard analytic perturbation theory cannot be used to describe the dynamics because of the presence of the embedded eigenvalues in the original unperturbed system (where there is vanishing coupling).

The dynamics of similar classical systems in celestial mechanics, with small denominators, were considered by H. Poincaré, [**29**], who suggested a two-step version of the analytic perturbation procedure which consists of choosing a modified unperturbed Hamiltonian as an "intermediate operator" and taking into account only "dangerous resonances" and then subsequently constructing the analytic perturbation series on that base. An extended discussion of this can be found in [**23**], where the resonance scattering on a quantum junction is considered.

For a junction, see Figure 1 below; the role of the basic region is played by a compact domain Ω_{int}, with a piecewise smooth boundary, and the role of the reservoir is played by the union $\omega = \cup_m \omega^m$ of straight leads, width δ, attached to it.

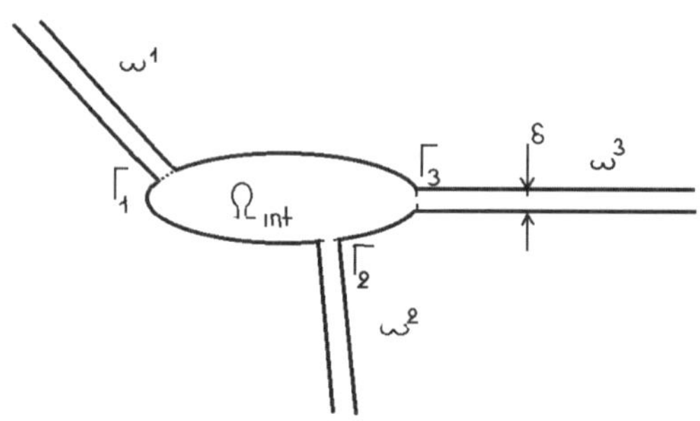

FIGURE 1. A junction $\Omega \equiv \Omega_{in} \cup \omega^1 \cup \omega^2 \ldots \omega^M$, $M = 3$. The bottom sections Γ_m of the leads ω^m, width δ, are shown by the dashed lines.

2.1. Local spectral analysis of the resonance system. The one-body quantum scattering problem on the junction can be reduced to the spectral problem for the relevant Schrödinger equation with real piecewise smooth, compactly supported, potential vanishing on the leads and with the Dirichlet boundary condition on $\partial\Omega$, complemented by the Meixner condition of square integrability of the gradient of the solution of the equation on the junction Ω. In the relevant acoustic scattering problem the potential is zero and the Neumann boundary condition is imposed on $\partial\Omega$. The spectrum of the corresponding selfadjoint spectral problem for the Laplace equation $-\Delta u - \lambda u \equiv \mathcal{L}u - \lambda u = 0$ on the junction $\Omega \equiv \Omega_{int} \cup \omega$ consists of a finite or, possibly, a countable set of eigenvalues accumulating at $+\infty$ and a countable set of absolutely continuous branches $\sigma_l \equiv [\frac{\pi^2 l^2}{\delta^2}, \infty), l = 0, 1, 2, \ldots$ equivalent to ones on the set of leads $\omega^m : (0 < x^m < \infty)$, $\omega = \cup_{1\leq m\leq M}\omega^m$ separated from the basic region by the bottom sections $\Gamma_m : x^m|_{\Gamma_m} = 0$. See the extended discussion of spectral properties of junction in [**23**] and references therein. The operator $\mathcal{L}$ is generally considered as a perturbation of the "simpler" selfadjoint operators defined by an additional Dirichlet or Neumann boundary condition on $\cup_{m=1}^{M}\Gamma_m =: \Gamma$,

$$\mathcal{L} \longrightarrow L^\omega \oplus L_{int} \;=\; \mathcal{L}_0 \;.$$

The spectrum $\sigma(L_{int})$ of the operator L_{int} is discrete and the spectrum σ^ω of the operator L^ω is absolutely continuous, consisting of a countable set of branches $\sigma^\omega = \cup_{m=1}^{M} \cup_{l=1}^{\infty} \sigma_l^m$ corresponding to the parts L_l^m of L^ω

$$L_l^m \;=\; -\frac{d^2}{dx^2} + \frac{\pi^2 l^2}{\delta^2}, \quad l \geq 0 \;,$$

with the homogeneous Neumann boundary condition $\frac{du}{dx}\big|_\Gamma = 0$ at the bottom sections $x^m\big|_{\Gamma_m} = 0$. The operators L_l^m on the leads are obtained from L^ω via separation of variables with the basis of cross-section eigenfunctions

$$\{e_l^m\} = \left\{\sqrt{\frac{2-\delta_{0l}}{\delta}} \cos\frac{\pi l y}{\delta}\right\}, \; l = 0, 1, 2, \ldots, \; m = 1, 2, \ldots M$$

with the local transversal coordinate on ω_m denoted by y.

The component eigenfunctions of L^ω on each lead ω^m and is represented by the scattering Ansatz, i.e., the linear combinations

$$\psi_l^m(x) \;=\; \chi_+^l(x) + S_l\, \chi_-^l(x), \quad x = x_m \geq 0 \;,$$

of oscillating modes with reflection coefficients $S_l = 1$:

$$\chi_\pm^{m,l} = e^{\pm i\sqrt{\lambda-\lambda_l}\; x}\, e_l^m(y) := e^{\pm iK_+^{m,l}\, x}\, e_l^m(y), \quad \lambda > \lambda_l = \pi^2\, l^2 \delta^{-2}. \tag{2.1}$$

The perturbed operator $\mathcal{L}$ is obtained from $\mathcal{L}_0$ by replacing the homogeneous Neumann boundary condition on the bottom sections Γ by a smooth matching condition with the solution of the relevant homogeneous equation in Ω_{int}. The corresponding scattered waves are obtained by matching, on Γ, a solution of the Schrödinger equation on Ω_{int} with the scattering Ansatz (see for instance [**24**]):

(2.2)

$$\psi_l^m(x) = \begin{cases} \chi_+^l(x) + \sum_{\pi^2 r^2/\delta^2<\lambda} S_{l,r}^{m,m}\chi_-^r(x) + \sum_{\pi^2 r^2/\delta^2>\lambda} s_{l,r}^{m,m}\xi^r(x), & x \in \omega^m \\ \sum_{\pi^2 r^2/\delta^2<\lambda} S_{l,r}^{m,n}\chi_-^r(x) + \sum_{\pi^2 r^2/\delta^2>\lambda} s_{l,r}^{m,n}\xi^r(x), & x \in \omega^n, \\ & n \neq m, \end{cases}$$

composed of the oscillating modes in the open channels and the decreasing (evanescent) modes in the closed channels

$$\xi_-^{m,s} \;=\; e_s^m(y)\, e^{-\sqrt{\lambda_s-\lambda}\; x} := e^{-K_-^{m,s}\, x}\, e_s^m(y), \quad \lambda < \lambda_s \;, \tag{2.3}$$

associated with the thresholds $\lambda_s^m = \pi^2\, s^2\, \delta^{-2}$ of the closed channels in the leads; see [**23**] for details. In this paper we study the low–energy scattering on the first spectral band,

$$0 \leq \lambda \leq \pi^2 \delta^{-2} \equiv \Delta_0$$

in the open channel, representing the cross-section space $L_2(\Gamma) =: E = E^1 \oplus E^2 \oplus E^3 \dots E^M$ as an orthogonal sum of the entrance subspaces $E_\pm$ of the open and closed spectral channels respectively.

(2.4)

$$L_2(\Gamma) =: E = E_+ \oplus E_- = \left[E_+^1 \oplus E_+^2 \oplus E_+^3 \dots E_+^M\right] \oplus \left[E_-^1 \oplus E_-^2 \oplus E_-^3 \dots E_-^M\right].$$

Consider the operators L_Γ^D and L_Γ^N defined on $W_2^2(\Omega_{int})$ with the Meixner conditions at the inner corners and Neumann boundary conditions on $\partial\Omega_{int}\backslash\Gamma$ and, respectively, Neumann or Dirichlet boundary conditions on Γ. The corresponding Green functions are obtained from the following equations:

(2.5)

$$L_\Gamma^D G_\Gamma^D = -\triangle G_\Gamma^D + V G_\Gamma^D = \lambda G_\Gamma^D + \delta(x-y), \quad \frac{\partial G_\Gamma^D}{\partial n}\Big|_{\partial\Omega_{int}\backslash\Gamma} = 0, \quad G_\Gamma^D\Big|_{x\in\Gamma} = 0,$$

(2.6)

$$L_\Gamma^N G_\Gamma^N = -\triangle G_\Gamma^N + V G_\Gamma^N = \lambda G_\Gamma^N + \delta(x-y), \quad \frac{\partial G_\Gamma^N}{\partial n}\Big|_{\partial\Omega_{int}\backslash\Gamma} = 0, \quad \frac{\partial G_\Gamma^N}{\partial n}\Big|_{\Gamma} = 0.$$

Denote by σ^N and σ^D respectively the spectra of the above operators L_Γ^D and notice that $L_\Gamma^N = L_{int}^N$. According to the general theory of second–order elliptic equations, the solution u to the homogeneous equation

$$L_{int} u = \lambda u,$$

with the boundary data $u\big|_\Gamma = u_\gamma$, $\frac{\partial u}{\partial n}|_{\partial\Omega_{int}\backslash\Gamma} = 0$, is represented by the relevant Poisson integral

$$u(x) = \int_\Gamma \mathcal{P}_\Gamma(x, \gamma, \lambda) u_\gamma(\gamma)\, d\gamma\,,$$

involving the kernel $\mathcal{P}_\Gamma(x,\gamma) = -\partial G_\Gamma^D(x,\gamma)/\partial n$.

The corresponding boundary current is

$$\frac{\partial u}{\partial n}\Big|_{x\in\Gamma} = -\int_\Gamma \frac{\partial^2 G_\Gamma^D(x,\gamma,\lambda)}{\partial n_x \partial n_\gamma} u_\gamma(\gamma)\, d\gamma =: \mathcal{DN}_\Gamma(\lambda) u_\gamma\,.$$

The formal integral operator $\mathcal{DN}_\Gamma$ is "the relative Dirichlet–to–Neumann mapping". See [**25, 23**]. The properties of the relative DN–mapping are similar to those of the classical DN–mapping in [**31, 14, 15**]. It is a Nevanlinna class function $\mathcal{DN}(\lambda)$ for Im $\lambda \leq 0$, with poles at the eigenvalues of the relative Schrödinger operator L_Γ^D. The relative DN–map is a pseudo–differential operator of order 1: for $W_2^2(\Omega)$ solutions u the DN–map acts from $W_2^{3/2}(\Gamma)$ to $W_2^{1/2}(\Gamma)$, whereas for $W_2^{3/2}(\Omega)$ generalized solutions the DN–map acts from $W_2^1(\Gamma)$ to $L_2(\Gamma)$.

Henceforth, we shall use both versions of this theory in parallel. The mapping

$$u(x) = \int_\Gamma G_\Gamma^N(x,\gamma,\lambda) \frac{\partial\psi}{\partial n}\Big|_{\partial\Omega_{int}} d\,\gamma =: Q_1^\Gamma \frac{\partial\psi}{\partial n}\Big|_\Gamma, \quad x \in \Omega_{int}\,,$$

gives a solution of the Neumann boundary problem. The trace of the solution on Γ

$$u(x)\big|_\Gamma = \int_\Gamma G_\Gamma^N(x,\gamma) \frac{\partial\psi}{\partial n}\Big|_{\partial\Omega_{int}} d\,\gamma =: \mathcal{ND}\frac{\partial\psi}{\partial n}\Big|_\Gamma, \; x \in \Gamma\,,$$

defines the relative Neumann–to–Dirichlet mapping which is inverse to the relative Dirichlet–to–Neumann mapping defined above, while λ is not an eigenvalue of either problem:

$$\mathcal{ND}_\Gamma \, \mathcal{DN}_\Gamma \;=\; I_\Gamma.$$

Here $\mathcal{ND}_\Gamma$ is the restriction of the whole ND–map of Ω_{int} onto Γ. In particular for W_2^2- solutions u the corresponding ND–mapping acts on the set of all regular spectral points λ of the Neumann Schrödinger Operator, from $W_2^{1/2}(\Gamma)$ onto $W_2^{3/2}(\Gamma)$. For $W_2^{3/2}$ solutions the ND–mapping acts acts from $L_2(\Gamma)$ onto $W_2^1(\Gamma)$.

2.2. Scattering matrix via intermediate DN and ND maps. The coefficients $S_{l,r}$, $s_{l,r}$ of the scattering Ansatz (2.2) can be found, at least in principle, from the infinite linear system obtained by substitution of the scattering Ansatz into the matching condition. See for instance [**23**]. Selecting $E_\pm$ as indicated above in (2.4) will represent the ND–mapping of L_Γ^N by a 2×2 operator matrix with elements $\mathcal{ND}_{\pm,\pm} \;=\; P_\pm \mathcal{ND} P_\pm$

$$\mathcal{ND}_{int}^{\Gamma} = \begin{pmatrix} \mathcal{ND}_{++} & \mathcal{ND}_{+-} \\ \mathcal{ND}_{-+} & \mathcal{ND}_{--} \end{pmatrix}. \tag{2.7}$$

A similar decomposition of the DN–mapping of the Schrödinger operator L_Γ^D on Ω_{int}

$$\mathcal{DN}_{int}^{\Gamma} \;=\; \begin{pmatrix} \mathcal{DN}_{++} & \mathcal{DN}_{+-} \\ \mathcal{DN}_{-+} & \mathcal{DN}_{--} \end{pmatrix} \tag{2.8}$$

was used in [**23**] in the course of construction of a convenient representation for the scattering matrix on the first spectral band $\Delta_0 = [0, \pi^2\, \delta^{-2}) =: [0,\, \lambda_{\min})$. We denote

$$e_0^m = \begin{cases} 1/\sqrt{\delta}, & \text{if } y \in \Gamma_m\ , \\ \quad 0, & \text{if } y \in \partial\Omega_{int} \backslash \Gamma_m\ . \end{cases}$$

In accordance with the general notations $E_+ = \bigvee_{open} e_l$ (2.1,2.2,2.3), we set on the open spectral band Δ_0 :

$$\begin{aligned} K_+ \;&\equiv\; \textstyle\sum_m \sum_{open} \sqrt{\lambda}\, e_l^m\rangle\langle e_l^m \;&=\;& \textstyle\sum_m \sqrt{\lambda}\, e_0^m\rangle\langle e_0^m\ , \\ K_- \;&\equiv\; \textstyle\sum_m \sum_{closed, l\geq 1} \sqrt{\lambda_l - \lambda}\, e_l^m\rangle\langle e_l^m \;&=\;& \textstyle\sum_m \sum_{l \geq 1} \sqrt{\lambda_l - \lambda}\, e_l^m\rangle\langle e_l^m\ . \end{aligned}$$

Hereafter we will use the standard bra/ket notations, $e >< e' : u \to e < e', u >$, with the bar on the first factor of the dot–product in $E = L_2(\Gamma)$. The exponents of oscillating and decreasing modes on the first spectral band $\left[0, \pi^2 \delta^{-2}\right)$ spanned by the vectors $e_\pm \in E_\pm$ are $\chi_\pm c_+ = e^{\pm i K_+ x} c_+$ and $\chi_- c_- = e^{-K_- x} c_-$ respectively. The matrices $S_{l,r}^{m,n}$ and $s_{l,r}^{m,n}$, which are defined by matching the scattering Ansatz to the solution of the homogeneous equation on Ω_{int}, constitute respectively the scattering matrix (the square table of amplitudes in front of the oscillating modes)

$$S \;=\; \sum_{m,n=1}^{M} \quad \sum_{l\, open,\ r\, open} \quad S_{l,r}^{m,n} e_l^m\rangle\ \langle e_r^n\ ,$$

and the amplitudes of the evanescent modes

$$s \;=\; \sum_{m,n=1}^{M} \quad \sum_{l\ closed,\, r\ open} \quad s_{l,r}^{m,n} e_l^m\rangle\ \langle e_r^n.$$

The scattering matrix of the junction is represented (see [**23**] and Theorem 2.5) in terms of the matrix elements $\mathcal{DN}$ as

$$S(\lambda) = (iK_+ + \mathcal{M})^{-1}(iK_- - \mathcal{M}) , \tag{2.9}$$

with $\mathcal{M} = \mathcal{M}(\lambda)$ defined by the corresponding "Krein formula":

$$\mathcal{M} = \mathcal{DN}_{++} - \mathcal{DN}_{+-}\frac{I}{\mathcal{DN}_{--} + K_-}\mathcal{DN}_{-+} . \tag{2.10}$$

Here $\mathcal{M}$ is an *Intermediate DN-map*, associated with the *Intermediate Hamiltonian*; see below and [**23**].

In this paper we aim to compare the transport properties of the junctions with the same basic region and leads of various width δ attached orthogonally to the basic region (vertex domain). In fact we consider the shrinking of the junction, keeping the vertex domain unchanged, but the width δ of the leads serves as a small parameter in the course of calculation of the scattering matrix. Thin networks, with small δ, are characterized by a large distance between the neighboring spectral thresholds, $\pi^2(l+1)^2\delta^{-2} - \pi^2 l^2\,\delta^{-2} = (2l+1)\pi^2\,\delta^{-2}$. As proved in [**23, 3**], for a thin junction, the denominator $\mathcal{DN}_{--} + K_-$ is invertible on a "major part of the essential spectral interval" Δ; see the definition below. The zeros of the denominator $\mathcal{DN}_{--} + K_-$ have an important operator–theoretic meaning: they are eigenvalues of the *intermediate Hamiltonian* below the threshold $\lambda_{min} = \pi^2\delta^{-2}$ of the corresponding continuous spectrum. The intermediate Hamiltonian is defined as the splitting of the original operator $\mathcal{L}$ on the junction by imposing an additional *partial zero boundary condition* in the open channel on the bottom sections Γ of the leads:

$$P_+ u\big|_\Gamma = 0 , \tag{2.11}$$

complemented by the usual smooth matching condition on Γ in the closed channels:

$$\mathcal{L} \longrightarrow L^0 \oplus l^0 = \mathcal{L}_\Gamma .$$

Here $l^0 = -\dfrac{d^2}{dx^2}$ in $L^2(0,\infty)\times E_+) =: \mathcal{H}_+$, with zero boundary condition at the origin $u(0) = 0$, and L^0 is defined in the orthogonal sum of the channel space $L^2(0,\infty)\times E_- =: \mathcal{H}_-$ of the closed channels and $L_2(\Omega_{int})$ on W_2^2 - smooth functions, subject to the Meixner condition and the matching condition on Γ in closed channels:

$$L^0 : D(L^0) \longrightarrow L_2(\Omega_{int}) \oplus \mathcal{H}_- .$$

We now recall the following theorem whose proof follows more or less the pattern of proof given for Theorems 2.1 and 3.1 in [**23**]. We leave the details to the reader.

THEOREM 2.1. *The operators L^0, l^0 are self-adjoint. The spectrum of l^0 is absolutely continuous and coincides with*

$$\sigma_a(l^0) = [0,\,\infty), \qquad \textit{with multiplicity } M.$$

The absolutely continuous spectrum of L^0 consists of a countable set of branches

$$\sigma_a(L^0) = \bigcup_{l\geq 1}\sigma_a^l \equiv [\lambda_{min},\infty),$$

where each branch $\sigma_a^l \equiv \left[\frac{\pi^2 l^2}{\delta^2},\infty\right)$, $l\geq 1$ has multiplicity M, and the total multiplicity is growing stepwise on the thresholds $\frac{\pi^2 l^2}{\delta^2} =: \lambda_l$ separating the spectral bands

$\Delta_l = [\lambda_l, \lambda_{l+1}]$. *The discrete spectrum of* L^0 *consists of a countable set of eigenvalues* λ_s^0 *accumulating at infinity. The singular spectrum of* L^0 *is empty. The solution of the inhomogeneous equation*

$$L^0 u - \lambda u = f \in L_2(\Omega_{int})$$

is represented, for complex λ, *by the integral*

$$u(x) = \int_{\Omega_{int}} G^0(x, s, \lambda) f(s)\, ds\ .$$

The countable set of eigenvalues of L^0 *is found from the equation* $[DN_\Gamma + K_-]\, e = 0$ *or* $[I + ND_\Gamma K_-]\, e = 0$, *and the corresponding eigenfunctions are constructed by matching the exponentials* $e^{-K_- x}e$ *in the closed channels,* $e \in E_-$, *with solutions of the corresponding inner relative boundary problem represented by the integral* $-\int_\Gamma G_\Gamma^N K_- e(\gamma) d\Gamma$.

We now make a definition.

DEFINITION 2.2. Denote by $Z^D =: Z$ the set of poles of $\mathcal{DN}_{--}$ on $\Delta \subset \Delta_0 = [0, \pi^2\delta^{-2})$, and consider an open neighborhood U_Z^D of Z, and the major part of the corresponding complement $\Delta \backslash U_Z^D =: \hat{\Delta}_Z^D$, which plays the role of the "major part of the essential spectral interval" Δ.

For any U_Z^D there exists δ such that K_- dominates $\mathcal{DN}_{--}$ on $\Delta \backslash U_Z^D$:

$$\| K_-^{-1}\mathcal{DN}_{--} \|_{W_2^{1/2}(\Gamma)} < 1\ , \qquad \| K_-^{-1}\mathcal{DN}_{--} \|_{W_2^{3/2}(\Gamma)} < 1\ .$$

This implies the following lemma for which the proof follows as in [**23**]:

LEMMA 2.3. *The denominator* $\mathcal{DN}_{--} + K_-$ *is invertible from* $W_2^{1/2}(\Gamma)$ *to* $W_2^{3/2}(\Gamma)$ *and from* $L_2(\Gamma)$ *to* $W_2^1(\Gamma)$ *for* δ *small enough, on the corresponding major part of the essential spectral interval.*

The above expression (2.10) for $\mathcal{M}$ contains singularities in the first and second summands, inherited from the unperturbed Schrödinger operator L_Γ^D on the quantum well. This expression can be transformed to a more convenient form, where these singularities compensate each other; see [**3**]. Then the scattering matrix of $\mathcal{L}$ is calculated in terms of $\mathcal{M}$ (see (2.13)), while only non–compensated singularities remain on Δ, at the eigenvalues of the intermediate Hamiltonian, below the first threshold $\frac{\pi^2}{\delta^2} \equiv \lambda_{min}$.

As in (2.10) we will need a similar formula for the intermediate Neumann-to-Dirichlet map $\mathcal{N}$ to use for the scattering matrix. It is also represented as a "Schur complement" (see [**23**] and [**22**]):

$$\mathcal{N} = \mathcal{ND}_{++} - \mathcal{ND}_{+-}K_- \frac{I}{I_- + \mathcal{ND}_{--}K_-}\mathcal{ND}_{-+}\ , \tag{2.12}$$

where $I_- := P_-$ plays the role of unity in E_-. The operator-function $\mathcal{N}$ plays the role of the Neumann-to-Dirichlet map of the intermediate Hamiltonian and we call (2.12) the *Krein formula for* $\mathcal{N}$.

We aim to find the derivation of a corresponding formula for the scattering matrix of the junction in terms of $\mathcal{N}$, and we must make sure that the denominator $I_- + \mathcal{N}_{--}K_-$ is invertible, for thin junctions, on the corresponding major part of Δ.

To calculate $\mathcal{N}$ based on (2.12) we have to find the inverse of $I_- + \mathcal{N}_{--}K_- = \left(K_-^{-1} + \mathcal{N}_{--}\right) K_-$, which is equivalent to solving the equation:

$$\left(K_-^{-1} + \mathcal{N}_{--}\right)\ K_- u \ = \ f \in W_2^{3/2}(\Gamma)\ ,$$

or the similar equation with $f \in W_2^1(\Gamma)$.

In fact there is a finite set, which we denote by Z^N, of (vector) zeros of the denominator $I_- + \mathcal{N}_{--}K_-$ on the essential spectral interval $\Delta \subset \Delta_0$. These zeros are in one-to-one correspondence with the vector-zeros of a finite matrix function; see [**23**]. Moreover, the following statement is true and can be found in [**21**].

LEMMA 2.4. *For each neighborhood U_Z of the set Z^N of zeros, and all δ sufficiently small, the corresponding major part $\hat{\Delta} = \Delta \backslash U_Z$ the inverse*

$$\left[I_- + \mathcal{N}\mathcal{D}_{--}K_-\right]^{-1}$$

of the denominator exists and is a bounded operator in $E_- = L_2(\Gamma) \ominus E_+$.

The matrix function $\mathcal{N}$ may also have singularities at the eigenvalues of L_{int}^N, that is, the poles of G_{int}^N. Hence the inverse $[I + \mathcal{N}\mathcal{D}_{--}K_-]^{-1}$ exists on the complement $\Delta \backslash \left\{\sigma^N \cup Z^N\right\} =: \hat{\Delta}$ of the union of these two sets as is noted in [**21**].

THEOREM 2.5. *Upon substitution of the scattering Ansatz (2.2) into the matching conditions on Γ we obtain the following formula for the scattering matrix on $\hat{\Delta}$*

$$S = \left[iK_+ + \mathcal{M}\right]^{-1}\left[iK_+ - \mathcal{M}\right]\ , \tag{2.13}$$

$$S = \left[\mathcal{N}iK_+ + 1\right]^{-1}\left[I - \mathcal{N}iK_+\right]. \tag{2.14}$$

Proof. The scattering Ansatz generated by the entrance vector $e \in E_+$ is formed from the incoming wave $e^{iK_+x}e$, the transmitted/reflected wave $e^{-iK_+x}Se$ and the evanescent wave $e^{-K_-x}se$:

$$\Psi_e \ = \ e^{iK_+x}e + e^{-iK_+x}Se + e^{-K_-x}se\ .$$

The boundary data of the scattering Ansatz at the bottom sections Γ should match on Γ the boundary data of the solution of the homogeneous Schrödinger equation inside Ω_{int}:

$$L_{int}\psi \ = \ \lambda\psi, \qquad \frac{\partial\psi}{\partial n}\Big|_{\partial\Omega_{int}\backslash\Gamma} \ = \ 0\ ,$$

$$\Psi\big|_\Gamma = \Psi_e(0) = e + Se + se\ , \qquad \frac{\partial\psi}{\partial n}\Big|_\Gamma \ = \ \Psi_e'(0) \ = \ iK_+e - iK_+Se - K_-se\ . \tag{2.15}$$

Using the matrix representations (2.8 and 2.7) for $\mathcal{D}\mathcal{N}$, $\mathcal{N}\mathcal{D}$, we obtain from (2.15) two equivalent linear systems which describe matching conditions on Γ

$$\begin{aligned} iK_+(1-S)e &= \mathcal{D}\mathcal{N}_{++}(1+S)e + \mathcal{D}\mathcal{N}_{+-}se\ , \\ -K_-se &= \mathcal{D}\mathcal{N}_{-+}(1+S)e + \mathcal{D}\mathcal{N}_{--}se\ , \end{aligned} \tag{2.16}$$

and

$$\begin{aligned} (I+S)e &= \mathcal{N}\mathcal{D}_{++}\,iK_+(I-S)e - \mathcal{N}\mathcal{D}_{+-}K_-se, \\ se &= \mathcal{N}\mathcal{D}_{-+}\,iK_+(I-S)e - \mathcal{N}\mathcal{D}_{--}K_-e. \end{aligned} \tag{2.17}$$

Eliminating the component se from these equations and using the expressions for $\mathcal{M}$, $\mathcal{N}$ we obtain the desired representation for the scattering matrix (2.13,2.14). Note that the first formula (2.13) was proved in [**23**]. □

The relation $\mathcal{M}\mathcal{N} = I$ observed from comparison of the formulas (2.14, 2.13) has an important operator–theoretic meaning. This derives from the fact that $\mathcal{M}$, $\mathcal{N}$ are respectively DN and ND–maps of the intermediate Hamiltonian.

3. Krein formula for the intermediate ND–map with compensated singularities.

Equation (2.14) shows that, for the scattering matrix the dynamics on the junction are defined by the asymptotic properties of $\mathcal{N}$ near the singularities. On the other hand, the set of singularities of $\mathcal{N}$ on the essential spectral interval $\Delta \subset \Delta_0$ is formed by the zeros of the denominator $I + \mathcal{N}\mathcal{D}_{--}K_-$ and possibly the poles of $\mathcal{N}\mathcal{D}$ (the eigenvalues of L^N_{int}) inherited from the L^N_Γ. In fact the latter ones in the first and second addenda compensate each other, so that only the zeros of the denominator $I + \mathcal{N}\mathcal{D}_{--}K_-$ survive in the final expression (2.12). In this paper, following [**21**], we reveal the compensation of singularities in the Krein formula for (2.12). The proof of a similar compensation result for (2.10) can be found in [**28**].

The compensation of singularities also enables us to obtain a more convenient representation for $\mathcal{N}$. This new representation implies the corresponding new formula (2.14) for the scattering matrix of $\mathcal{L}$, and, eventually, various approximate expressions for it. These approximate expressions may be used as a base for construction of corresponding "fitted" solvable models of the junction in the form of star-shaped 1D quantum graphs and for derivation of the boundary condition at the vertex.

Our analysis in this section is based on the matrix decomposition of the ND–map (2.7) with respect to the basis of subspaces $E = E_+ \oplus E_-$. The corresponding spectral problem for L^N_Γ is a standard inner Neumann problem. Denote by φ_l, λ^n_l the eigen-pairs of the Neumann spectral problem (2.6), with $\lambda^n_l \in \Delta \subset [0, \pi^2\delta^{-2}]$, $E^\Delta \equiv \bigvee_{\lambda^n_\Delta} \varphi_l$ and set $\varphi_l \stackrel{\Gamma}{\longrightarrow} \varphi_l\big|_\Gamma$, $E^\Delta \stackrel{\Gamma}{\longrightarrow} \Gamma E^\Delta \equiv E^\Delta_\Gamma$.

Assumption 1: *The mapping Γ_Δ is an isomorphism of E^Δ onto E^Δ_Γ and the orthogonal projection P_+ is an isomorphism $\Gamma_\Delta E^\Delta \stackrel{P_+}{\longrightarrow} P_+ E^\Delta_\Gamma$; hence $\Gamma_\Delta E^\Delta$ and $P_+E^\Delta_\Gamma$ are skew-connected.*

Hereafter we also need the corresponding adjoint operator $\Gamma^+_\Delta = \sum_l \varphi_l\rangle\langle\Gamma\varphi_l =: \mathcal{T}$. The operators Γ_Δ and $\Gamma_\Delta \equiv \mathcal{T}$ will be used when comparing the singularities of matrix elements of $\mathcal{N}\mathcal{D}_{\pm,\pm}$ at the Neumann eigenvalues in the proof of Theorem 3.1.

In the course of the proof we also need to estimate the displacement of the eigenvalues of the corresponding operators when replacing the partial Dirichlet boundary conditions (2.11) on Γ by the matching in E_+. To focus on this problem we shall represent the compact in $L_2(\Gamma)$) operator–function $\mathcal{N}\mathcal{D}$ via an appropriate rational approximation:

$$\mathcal{N}\mathcal{D}^\Gamma_{int} =: \mathcal{N}\mathcal{D} = \sum_{\lambda^n_1 \in \Delta} \frac{\varphi_l\big|_\Gamma\rangle\langle\varphi_l\big|_\Gamma}{\lambda^n_l - \lambda} + \mathcal{K}^\Delta \ =: \mathcal{N}\mathcal{D}^\Delta + \mathcal{K}^\Delta,$$

with an error $\mathcal{K}^\Delta$ corresponding to the contribution from the Neumann spectrum outside Δ.

Assumption 2: *The width δ of the leads, the essential spectral interval $\Delta \subset [0, \pi^2 \delta^2)$ and the rational approximation can be selected such that on Δ the error of the rational approximation is relatively small:*

$$\sup_\Delta \| \mathcal{K}^\Delta K_- \|_{W_2^1(\Gamma)} < 1 . \tag{3.1}$$

The above assumption is fulfilled, in particular, if the spacing between the selected group of eigenvalues on the given interval Δ and the complementary set of eigenvalues is sufficiently large. We will not formulate an exact condition which guarantees (3.1) for given Δ; we will just note that, due to the compactness of the resolvent, for any Δ there exist δ, n_Δ, μ such that the error of some finite rational approximation of the resolvent is small:

(3.2)

$$G^N(x,s,\lambda) = G^N(x,s,\mu) + (\lambda-\mu)Q_2^\Delta = G^N(x,s,\mu) +$$

$$+ (\lambda-\mu)\sum_{l=1}^{n_\Delta} \frac{\varphi_l(x)\rangle\,\langle\varphi_l(s)}{(\lambda_l^n-\lambda)(\lambda_l^n-\mu)} + (\lambda-\mu)\sum_{l=n_\Delta+1}^{\infty} \frac{\varphi_l(x)\rangle\,\langle\varphi_l(s)}{(\lambda_l^n-\mu)(\lambda_l^n-\lambda)}$$

$$= \sum_{l=1}^{n_\Delta} \frac{\varphi_l(x)\rangle\,\langle\varphi_l(s)}{\lambda_l^n-\lambda}$$

$$+ \left[G^N(x,s,\mu) - \sum_{l=1}^{n_\Delta} \frac{\varphi_l(x)\rangle\,\langle\varphi_l(s)}{\lambda_l^n-\mu} + (\lambda-\mu)\sum_{l=n_\Delta+1}^{\infty} \frac{\varphi_l(x)\rangle\,\langle\varphi_l(s)}{(\lambda_l^n-\mu)(\lambda_l^n-\lambda)} \right].$$

The condition (3.1) can be verified on the basis of the above representation, if $\lambda \in \Delta$, μ is large negative, and n_Δ is sufficiently large. The estimate (3.1) is also valid for the error framed by the projections onto $E_\pm$: $P_\pm \mathcal{K}_2^n P_- =: \mathcal{K}_\pm$.

To represent $\mathcal{N}$ in a form with compensated singularities, we have to solve

$$(I + \mathcal{N}\mathcal{D}_{--}K_-)\,u = \mathcal{N}\mathcal{D}_{-+}f . \tag{3.3}$$

We will do this with the help of the Banach principle by considering the matrices (2.7 and 2.8) in the basis $E_+ \oplus E_-$, with elements $\mathcal{N}\mathcal{D}_\pm^\Delta$ and using the corresponding rational approximations with errors $\{\mathcal{K}_\pm^\Delta\}$.

Denote the following by

$$\begin{aligned}
\mathcal{K}_{++} - \mathcal{K}_{+-}K_-(I+\mathcal{K}_{--}^\Delta K_-)^{-1}\mathcal{K}_{-+} &=: \mathcal{N}_{reg} ,\\
\{P_+ - \mathcal{K}_{+-}^\Delta K_-(I+\mathcal{K}_{--}^\Delta K_-)^{-1}\}\mathcal{T}^+ &=: \mathcal{J}\mathcal{T}^+,\\
\mathcal{T}K_-(I+\mathcal{K}_{--}^\Delta K_-)^{-1}\mathcal{T}^+ &=: V(\lambda),\\
L^\Delta - \lambda I^\Delta + V(\lambda) &=: L^\Delta(\lambda),\\
\mathcal{T}\{P_+ - K_-(I+\mathcal{K}_{--}^\Delta K_-)^{-1}\mathcal{K}_{-+}^\Delta\} &=: \mathcal{T}\mathcal{J}^+.
\end{aligned} \tag{3.4}$$

Hereafter we call $V(\lambda)$ an "auxiliary" potential. It measures the difference between the part L^Δ of L_{int}^N in the spectral subspace E_Δ^n, and the auxiliary operator–function $L^\Delta(\lambda)$.

THEOREM 3.1 (Compensation of singularities N). *The Krein formula (2.12) for the intermediate ND–map can be rewritten in a more convenient form, with compensated singularities at the eigenvalues λ_s^n of the operator L_{int}^N, so that only* *the zeros of the denominator of the second (polar) term remain as singularities of the whole expression:*

$$\mathcal{N} = \mathcal{N}_{reg} + \left\{P_+\mathcal{T}^+ - \mathcal{K}^\Delta_{+-}K_-(I+\mathcal{K}^\Delta_{--}K_-)^{-1}\mathcal{T}^+\right\}\frac{I}{L^\Delta(\lambda)}\{\mathcal{T}P_+ \\ -\mathcal{T}K_-(I+\mathcal{K}^\Delta_{--}K_-)^{-1}\mathcal{K}^\Delta_{-+}\}. \tag{3.5}$$

PROOF. We treat equation (3.3) by using the Banach principle, under the assumption (3.1). This implies:

$$K_-u + K_-(I+\mathcal{K}^\Delta_{--}K_-+)^{-1}\mathcal{N}\mathcal{D}^\Delta_{--}K_-u \;=\; K_-(I+\mathcal{K}^\Delta_{--}K_-)^{-1}\mathcal{N}\mathcal{D}^\Delta_{-+}f\,. \tag{3.6}$$

Notice that $\mathcal{N}\mathcal{D}^\Delta$ is connected with the part L^Δ of L^N in E^Δ, by

$$\mathcal{N}\mathcal{D}^\Delta = \mathcal{T}^+\frac{I}{L^\Delta-\lambda I^\Delta}\mathcal{T}\,.$$

Then, denoting

$$\frac{I}{L^\Delta-\lambda I^\Delta}\mathcal{T}K_-u =: v\,,$$

we rewrite the above equation (3.6) as an equation for v, and obtain the solution of it in terms of the inverse of the finite matrix $L^\Delta(\lambda)$:

$$v \;=\; [L^\Delta(\lambda)]^{-1}V\frac{I}{L^\Delta-\lambda I^\Delta}\mathcal{T}P_+f\,.$$

Substituting the result into (3.6), we get:

$$K_-u \;=\; K_-(I+\mathcal{K}^\Delta_{--}K_-)^{-1}\left[\mathcal{T}^+\frac{I}{L^\Delta-\lambda I^\Delta}\mathcal{T}+\mathcal{K}^\Delta_{-+}\right]P_+f\; -$$

$$K_-(I+\mathcal{K}^\Delta_{--}K_-)^{-1}[L^\Delta(\lambda)]^{-1}\mathcal{T}K_-(I+\mathcal{K}^\Delta_{--}K_-)^{-1}\left[\mathcal{T}^+\frac{I}{L^\Delta-\lambda I^\Delta}\mathcal{T}+\mathcal{K}_{-+}\right]P_+f\,.$$

Then

$$\begin{aligned}\mathcal{N}f \;=\;& P_+\mathcal{T}^+\frac{I}{L^\Delta-\lambda I^\Delta}\mathcal{T}P_+f - \mathcal{K}^\Delta_{++}f \;-\; P_+\left[\mathcal{T}^+\frac{I}{L^\Delta-\lambda I^\Delta}\mathcal{T}P_- + \mathcal{K}^\Delta_{+-}\right]\\ &\times\Big\{K_-(I+\mathcal{K}^\Delta_{--}K_-)^{-1}\left[\mathcal{T}^+\frac{I}{L^\Delta-\lambda I^\Delta}\mathcal{T}+\mathcal{K}^\Delta_{-+}\right]P_+f-\\ &K_-(I+\mathcal{K}^\Delta_{--}K_-)^{-1}\mathcal{T}^+[L^\Delta(\lambda)]^{-1}\mathcal{T}K_-(I+\mathcal{K}^\Delta_{--}K_-)^{-1}\\ &\times\left[\mathcal{T}^+\frac{I}{L^\Delta-\lambda I^\Delta}\mathcal{T}+\mathcal{K}^\Delta_{-+}\right]P_+f\Big\}.\end{aligned}$$

The leading terms inside the parentheses give:

$$K_-(I+\mathcal{K}^\Delta_{--}K_-)^{-1}\mathcal{T}^+\left[\frac{I}{L^\Delta-\lambda I^\Delta}-\frac{I}{L^\Delta(\lambda)}V\frac{I}{L^\Delta-\lambda I^\Delta}\right] =$$

$$K_-(I+\mathcal{K}^\Delta_{--}K_-)^{-1}\mathcal{T}^+\frac{I}{L^\Delta(\lambda)}\,. \tag{3.7}$$

Taking into account the leading term of the first addendum, we obtain the compensation of polar terms $\approx (L^\Delta-\lambda I^\Delta)^{-2}$ order -2:

$$P_+\mathcal{T}^+\frac{I}{L^\Delta-\lambda I^\Delta}\mathcal{T}P_+f - \mathcal{T}^+\frac{I}{L^\Delta-\lambda I^\Delta}V\frac{I}{L^\Delta(\lambda)}\frac{I}{L^\Delta-\lambda I^\Delta}\mathcal{T}P_+f \\ = P_+\mathcal{T}^+\frac{I}{L^\Delta(\lambda)}\mathcal{T}P_+f. \tag{3.8}$$

The lower–order terms in parentheses containing $\approx (L^\Delta - \lambda I^\Delta)^{-1}$ of order -1 give:
(3.9)
$$-\mathcal{K}_{+-}^\Delta K_-(I+\mathcal{K}_{--}^\Delta K_-)^{-1}\mathcal{T}^+ \left[\frac{I}{L^\Delta - \lambda I^\Delta} - \frac{I}{L^\Delta(\lambda)}V\right]\mathcal{T}P_+ f$$
$$= -\mathcal{K}_{+-}^\Delta K_-(I+\mathcal{K}_{--}^\Delta K_-)^{-1}\mathcal{T}^+ \frac{I}{L^\Delta(\lambda)}\mathcal{T}P_+ f\ ,$$

and the adjoint expression. Thus, the whole singular part of the expression (2.12) is reduced to a term which contains only the main singularity $[L^\Delta(\lambda)]^{-1}$:

$$\mathcal{K}_{+-}^\Delta K_-(I+\mathcal{K}_{--}^\Delta K_-)^{-1}\ \mathcal{T}^+ \frac{I}{L^\Delta(\lambda)}\mathcal{T}\ K_-(I+\mathcal{K}_{--}^\Delta K_-)^{-1}\mathcal{K}_{-+}^\Delta\ .$$

The remaining terms, which do not have singularities on Δ, yield:

(3.10) $$\mathcal{K}_{++}^\Delta - \mathcal{K}_{+-}^\Delta K_-(I+\mathcal{K}_{--}^\Delta K_-)^{-1}\mathcal{K}_{-+}^\Delta\ =: \mathcal{N}_{reg}\ .$$

Note that the operator $K_-(I+\mathcal{K}_{--}^\Delta K_-)^{-1}$ is self–adjoint on the first spectral band. Then, collecting all regular and singular terms, we replace (2.12) by the corresponding modified Krein formula with compensated singularities inherited from L_{int}^N:

$$\mathcal{N} = \mathcal{N}_{reg} + \left\{P_+\mathcal{T}^+ - \mathcal{K}_{+-}^\Delta K_-(I+\mathcal{K}_{--}^\Delta K_-)^{-1}\mathcal{T}^+\right\}\frac{I}{L^\Delta(\lambda)}\left\{\mathcal{T}P_+ - \right.$$
$$\left. - \mathcal{T}K_-(I+\mathcal{K}_{--}^\Delta K_-)^{-1}\mathcal{K}_{-+}^\Delta\right\}\ .$$

This completes the proof. □

Observation. Consider the component of the scattered wave in the open channel $L_2(0,\infty)\times E_+ = L_2(0,\infty)\times\left[E_+^1 \oplus E_+^2 \oplus E_+^3 \oplus \dots E_+^M\right]$:

$$\Psi_e^+ = e^{ipx}\,e + e^{-ipx}Se.$$

This vector-function satisfies the homogeneous equation $-\left[\Psi_e^+\right]'' = \lambda\,\Psi_e^+$ in
$\left[L_2(0,\infty)\times E_+^1\right] \oplus \left[L_2(0,\infty)\times E_+^2\right] \oplus \left[L_2(0,\infty)\times E_+^3\right] \oplus \dots \left[L_2(0,\infty)\times E_+^M\right] \oplus$
on the corresponding one-dimensional star-junction with the boundary condition at the vertex:

(3.11) $$\mathcal{N}(\lambda)\left[\Psi_e^+\right]'(0) - \Psi_e(0) = 0 \ \text{ or } \ \left[\Psi_e\right]'(0) - \mathcal{M}\left[\Psi_e\right](0) = 0\,,$$

which do not yet define an operator, contrary to the classical situation described in [**13, 9, 6**], because they are energy–dependent (λ - dependent). The boundary condition (3.11) at the vertex give for Ψ_e^+, similarly to [**32**], the correct expressions (2.14, 2.13) for the scattering matrix [**26**]. But neither of the above conditions defines a self–adjoint operator on the 1D star-graph $\omega^1 \cup \omega^2 \cup \omega^3 \cdots \cup \omega^M \equiv \omega$; hence the corresponding dynamics can't be modeled by means of the traditional quantum mechanics, with use of orthogonality of eigenfunctions, via construction of the unitary evolution, based on the theory of self–adjoint operators. A similar situation, for an approximate scattering matrix represented in terms of a Dirichlet–to–Neumann map, was resolved in [**26**] via construction of the corresponding zero-range fitted solvable model, whose scattering matrix coincides with the approximate scattering matrix, obtained via rational approximation of the original scattering matrix. The Hamiltonian of the solvable model in [**26**] is a selfadjoint operator; its eigenfunctions are orthogonal and the evolution is unitary, according to axioms of the standard quantum mechanics.

In fact the above expression (3.5) for the intermediate ND-map may provide numerous rational approximations, and each of them may be selected as a base for the corresponding solvable model, whose dynamics may be described in terms of quantum mechanics. In the next section we will review several rational approximations for the junction and after that we construct a relevant solvable model, reducing the problem to inverse scattering.

4. Approximate scattering matrix and the boundary conditions at the vertex

It is hard to resist the natural temptation to substitute an analysis of the dynamics on a thin 2D network by the much easier problem for a model but credible 1D network; see for instance [**1, 19**] and numerous papers quoted therein. Though the description of the general selfadjoint condition at the junction of the 1D quantum graph can be easily found, the selection of realistic boundary conditions from the wide list of selfadjoint boundary conditions at the vertex of a 1D junction presents difficultties. Yet physicists prefer to experiment theoretically and practically; see for instance [**7**], with the Datta-Das Sarma boundary condition for the T- junction, suggested in [**12**]:

$$\psi_1 = \beta^{-1}\psi_2 = \psi_3\,, \quad \psi_1' + \bar\beta\psi_2' + \psi_3' = 0. \tag{4.1}$$

Here $\{\psi_i\}_{i=1}^3 := \vec\psi$ is the vector of limit values of the wave function on the wires and $\{\psi_i'\}_{i=1}^3 := \vec\psi'$ is the vector of the values of the corresponding outward derivative at the vertex. Introducing the corresponding orthogonal projection

$$P_0 = \frac{1}{\beta^2+2}\begin{pmatrix} 1 & \beta & 1 \\ \beta & \beta^2 & \beta \\ 1 & \beta & 1 \end{pmatrix},$$

one can present the Datta-Das Sarma boundary condition in the form

$$P_0^\perp\vec\psi = 0\,, \quad P_0\vec\psi' = 0. \tag{4.2}$$

The scattering matrix of such a junction is constant $S = I - 2P_0 = P_0^\perp - P_0$. In [**12**] the real β is interpreted as a free parameter describing the strength of coupling between the leg and the bar of the T-junction. The fitting of the Datta-Das Sarma model has never been discussed. The simplest non-symmetric T-junction (see Figure 2(c)) was first seen in [**17**]. The fitting requires a deeper understanding of the junction as a *resonance scattering system*, involving spectral properties of the corresponding spectral problem on the basic region; see [**28**].

In this section we construct several rational approximations of the intermediate ND-map and suggest, based on them, several types of boundary conditions at the vertex of a model 1D junctions, similar to the Datta-Das Sarma condition (4.2). In particular, we recover the quantum-mechanical meaning of the phenomenological parameter β. Generally, our formulas are aimed to suit a wider class of junctions and at higher temperatures.

Notice first that the main formula (3.5) of the previous section is exact, subject to the standard analytic perturbation procedure for its elements on an essential spectral interval $\Delta \subset \Delta_0$. In this section we calculate, based on (3.5), an approximate scattering matrix on the corresponding major part $\hat\Delta \subset \Delta$; see the analogous

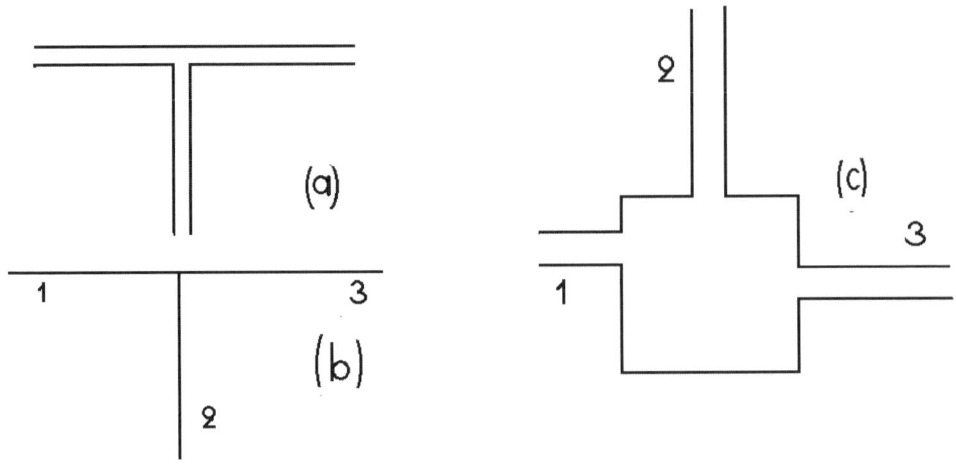

FIGURE 2. Standard orthogonal T-junctions: (a) symmetric 2D T-junction, (b) symmetric model 1D T-junction, and (c) asymmetric 2D T-junction

definition (2.2) above. Theorem 4.4 below opens up a way to construct a solvable model for the Neumann junction in terms of quantum mechanics, with orthogonal eigenfunctions and unitary evolution, and even fitted in a low–energy region on the major part of the corresponding essential spectral interval $\hat{\Delta}$. Moreover, the dynamics of the solvable model may emulate realistic single-body dynamics on the junction because the resonances of the model calculated from the rational approximation of (3.5) approximate exact resonances. The main tool to be used in the program of calculation of the resonances is the matrix Rouché theorem [**16**].

In the sketch below we do not develop any general analysis, but just calculate the zero–order and the first-order terms in the corresponding approximate expressions for the ND–map and for the scattering matrix, on *a major part of the essential spectral interval* Δ. In the course of calculating the approximate scattering matrix we assume that

Assumption 3: *The rational approximation of the intermediate ND map is selected so that the corresponding error is small as an operator* $W_2^1(\Gamma) \to L_2(\Gamma)$, *that is*

$$\sup_{\Delta} \| \mathcal{K}^{\Delta} \|_{W_2^1(\Gamma)\times L_2(\Gamma)} = o_1 \,, \tag{4.3}$$

with a scalar positive function o_1, *which is uniformly small on* Δ.

Then at first-order with respect to δ, we can extract the following corollaries from Theorem 3.1. These corollaries serve as the building blocks for Theorem 4.4.

COROLLARY 4.1. *Under Assumptions 2 and 3, the functions* $\mathcal{K}^{\Delta}_{+-}K_-$, *on* Δ *and* $\mathcal{K}^{\Delta}_{++}$ *can be neglected in the regular term*

$$\mathcal{N}_{reg} \approx \mathcal{K}^{\Delta}_{++} - \mathcal{K}^{\Delta}_{+-} K_- \mathcal{K}^{\Delta}_{-+} =: \mathcal{N}_{reg,appr}$$

$$(4.4) \qquad = \mathcal{K}^{\Delta}_{++} - \mathcal{K}^{\Delta}_{+-} K_- [\frac{I}{K_-}] K_- \mathcal{K}^{\Delta}_{-+} \approx o_1 - o\left(\frac{I}{K_-}\right) o_1^2.$$

Thus, for thin leads, $\mathcal{N}_{reg}$ can be neglected in the formula (3.5), compared with the polar second term on the major part of the essential spectral interval.

COROLLARY 4.2. *The second summand $-\mathcal{K}^{\Delta}_{+-} K_- (I + \mathcal{K}^{\Delta}_{--} K_-)^{-1}$ of the left factor in the numerator of the second (polar) term of (3.5):*

$$P_+ - \mathcal{K}^{\Delta}_{+-} K_- (I + \mathcal{K}^{\Delta}_{--} K_-)^{-1} =: \mathcal{J}$$

and one of the adjoint, which is the right factor of the polar term, can be estimated for thin leads as

$$\mathcal{K}^{\Delta}_{+-} K_- (I + \mathcal{K}^{\Delta}_{--} K_-)^{-1} \approx \mathcal{K}^{\Delta}_{+-} K_- = o_1 ,$$

so that we can substitute $\mathcal{J}\mathcal{T}^+$, with a minor error o_1, on the corresponding major part of the essential spectral interval, by $P^+\mathcal{J}$ and $\mathcal{T}\mathcal{J}$ by $\mathcal{T}P_+$.

COROLLARY 4.3. *For a thin junction, on an essential spectral interval $\lambda \in \Delta = [a, b]$ situated in low-energy region $0 \le a < b << \pi^2 \delta^{-2}$,*

$$V(\lambda) = \mathcal{T}^+ K_- \frac{I}{I + \mathcal{K}_{--} K_-} \mathcal{T} \approx \mathcal{T}^+ K_- \mathcal{T}$$

$$= \sum_{r,s=1}^{n_\Delta} \sum_{l \ge 1} \sum_{m=1}^{M} e_l^m\rangle \langle e_l^m \sqrt{\frac{\pi^2 l^2}{\delta^2} - \lambda} \, \langle \varphi_s^n|_{\Gamma^m}, e_l^m\rangle \, \langle e_l^m, \varphi_r^n|_{\Gamma^m}\rangle$$

$$\approx \sum_{r,s=1}^{n_\Delta} \sum_{l \ge 1} \sum_{m=1}^{M} e_l^m\rangle \langle e_l^m \left[\frac{\pi l}{\delta} - \frac{\lambda \delta}{2\pi l}\right] \langle \varphi_s^n|_{\Gamma^m}, e_l^m\rangle \, \langle e_l^m, \varphi_r^n|_{\Gamma^m}\rangle =: V^+_{appr} - \lambda V^-_{appr}.$$

For thin junctions the potential $V(\lambda)$ can be obtained from $V^+_{appr} - \lambda V^-_{appr}$ by an analytic perturbation procedure. Denote by Z_V the set of zeros λ_s^V, e_s^V of the function $L^V(\lambda) := L^\Delta + V^+_{appr} - \lambda V^-_{appr} - \lambda I$, $\lambda_s^V \in \Delta$. Then we can develop an analytic perturbation procedure for $[L(\lambda)]^{-1}$, based on $[L^V(\lambda)]^{-1}$, on the corresponding major part $\hat{\Delta}^V \subset \hat{\Delta}$ of the essential spectral interval: on the complement in Δ of an open neighborhood U_Z^V of the vector–zeros Z_V of $L^V(\lambda)$ in the essential spectral interval. Based on that, one can also develop an analytic perturbation procedure for the full scattering matrix S based on substitution of $\mathcal{N}$ by the corresponding rational approximations.

Hereafter we substitute $L^\Delta(\lambda)$ by the approximate expression $L^V(\lambda)$. One can derive from the matrix Rouché theorem [**16**] that for any U_Z^Δ small enough, there exists a neighborhood $U_Z^V \supset U_Z^\Delta$, such that the total multiplicity of zeros Z_V in each connected component of U_Z^V is equal to the total multiplicity of zeros Z_Δ of $L^\Delta(\lambda)$ in the corresponding component of U_Z^Δ.

Denote by φ_s^V and λ_s^V the eigen-pairs of the formal spectral problem:

$$(4.5) \qquad [L^\Delta + V^+_{appr}] \, \varphi_s^V = \lambda [I^\Delta + V^-_{appr}] \varphi_s^V ,$$

and by U the matrix connecting two bases: $\varphi_s^V = \sum_{t=1}^{n_\Delta} U_{st} \varphi_t$ in E^Δ. Then

$$[L^\Delta_{appr}(\lambda)]^{-1} = \sum_{s=1}^{n_\Delta} \frac{1}{\lambda_s^V - \lambda} \sum_{t,r} U_{st} \bar{U}_{sr} \, \varphi_t\rangle \langle \varphi_r =: \sum_{s=1}^{n_\Delta} \frac{q_s^V}{\lambda_s^V - \lambda}$$

with $q_s^V := \sum_{t,r} U_{st}\bar{U}_{sr}\ \varphi_t\rangle\langle\varphi_r$. In view of the matrix Rouché theorem, the vector zeros e_s, λ_s of the denominator L^Δ on Δ are situated, for thin networks, near the eigen-pairs φ_s^V, λ_s^V of the spectral problem (4.5). Then,using the notations introduced above and neglecting $\mathcal{N}_{reg}$, due to Corollary 1, we summarize the above calculations in the form of the following statement:

THEOREM 4.4. *In the case of thin junction the ND–map (3.5) of the intermediate Hamiltonian can be substituted, with a minor error, on the major part $\hat{\Delta}$ of the essential spectral interval by the expression*

$$\mathcal{N} \approx P_+\mathcal{T}^+\sum_{s=1}^{n_\Delta}\frac{q_s^V}{\lambda_s^V-\lambda}\mathcal{T}P_+ =: \mathcal{N}^V(\lambda)\,. \tag{4.6}$$

The corresponding approximation of the scattering matrix on the major part $\hat{\Delta}$ of the essential spectral interval is represented as

$$S(\lambda)\approx\frac{I-i\mathcal{N}^VK_+}{I+i\mathcal{N}^VK_+}\equiv S^V(\lambda), \tag{4.7}$$

and $S(\lambda)$ can be obtained from $S^V(\lambda)$ on $\hat{\Delta}^V$ by an analytic perturbation procedure. The above approximate scattering matrix S^V can be obtained formally by matching the scattering Ansatz $P_+\Psi_V = e^{iK_+x}e + e^{-iK_+x}S^Ve$ into the boundary condition at the vertex:

$$\mathcal{N}^V(\lambda)P_+\Psi_V'(0) + P_+\Psi_V(0) = 0. \tag{4.8}$$

The constructed formal scattering matrix S^V serves an approximation, on the major part $\hat{\Delta}$ of the essential spectral interval, of the original scattering matrix of the operator $\mathcal{L}$. One can prove, following [**26**], that there exists a solvable model of the original scattering system in the form of a 1D quantum star–graph, which has exactly the scattering matrix (4.7). In this sense, the solvable model associated with the boundary condition (4.8) is "fitted". It follows from the Rouché theorem, quoted earlier, that for each small δ there exists a neighborhood U_V^Δ of the spectrum of the problem (4), such that the total multiplicity of eigenvalues of L_{appr}^Δ on each connected component of U_V^Δ coincides with the total multiplicity of the zeros of the denominator $L^\Delta(\lambda)$ on U_V^Δ. The example below contains a discussion of the simplest case, with a single simple eigenvalue of L_{int}^N on Δ.

COROLLARY 4.5. *Consider the case when the open spectral branch $[0,\infty)$ is spanned by the constant cross section functions, $E_+ = \bigvee_{m=1}^M e_0^m$, where $e_m(\gamma) = \delta_{sm}$ if $\gamma\in\Gamma=\cup_{s=1}^M\Gamma_s$. Then the role of the exponential in the open channel is played by a diagonal matrix in $E_+ = \bigvee_m e_m$*

$$K_+(\lambda) = \sqrt{\lambda}\delta^{-1}\lceil e_m\rangle\langle e_m\rfloor = \sqrt{\lambda}P_+ = pP_+\,. \tag{4.9}$$

The orthogonal projections of the traces of the eigenfunctions $\varphi_s^V\big|_\Gamma$ of the spectral problem (4) onto E_+ are obtained by averaging of each component $\varphi_s^V\big|_{\Gamma_m}$ on the *section* Γ_m*:*

$$P_+\mathcal{T}\varphi_s^V = \begin{pmatrix} \delta^{-1}\int_{\Gamma_1}\varphi_s^V(\gamma)\,d\gamma \\ \delta^{-1}\int_{\Gamma_2}\varphi_s^V(\gamma)\,d\gamma \\ \delta^{-1}\int_{\Gamma_3}\varphi_s^V(\gamma)\,d\gamma \\ \vdots \\ \delta^{-1}\int_{\Gamma_M}\varphi_s^V(\gamma)\,d\gamma \end{pmatrix} =: \begin{pmatrix} \varphi_s^1 \\ \varphi_s^2 \\ \varphi_s^3 \\ \vdots \\ \varphi_s^M \end{pmatrix} =: \vec{\varphi}_s^V . \tag{4.10}$$

Here the elements of the column $\vec{\varphi}_s^V$ are calculated for the normalized eigenvector φ_s^V of the spectral problem

$$\left[L^\Delta + V_{appr}^+\right]\varphi_s^V = \lambda[I^\Delta + V_{appr}^-]\varphi_s^V ,$$

with $\|\varphi_s^V\|_{L_2(\Omega_{int})} = 1$. The orthogonal projection P_s^V onto the 1–dimensional subspace spanned by $\vec{\varphi}_s^V$ is

$$P_s^V = \frac{\vec{\varphi}_s^V\rangle\,\langle\vec{\varphi}_s^V}{\delta\sum_{m=1}^M|\varphi_s^m|^2} , \tag{4.11}$$

and $q_s^V = \alpha_s^2\,P_s^V$, with $\alpha_s^2 = \delta\sum_{m=1}^M|\varphi_s^m|^2$. Then on a major part of the essential spectral interval

$$\mathcal{N}^V \approx \sum_{s=1}^{n_\Delta}\frac{\alpha_s^2\,P_s^V}{\lambda_s^V - \lambda} =: \mathcal{N}_a^V ,$$

and the corresponding approximate scattering on the major part of the essential spectral interval is given by Theorem 4.4 with the above $\mathcal{N}_a^V$, and $K_+ = pP_+$.

One can make one more step toward simpler approximate expressions for the scattering matrix. One can see that for a thin network that $V_{appr}^\pm$ are estimated by $\delta^{3/2}$. Neglecting both of them, we substitute φ_s^V by $\varphi_s + o(\delta^{3/2}) \approx \varphi_s$ and λ_s^V by $\lambda_s^n + o(\delta^{3/2}) \approx \lambda_s^n$. Then we obtain a simpler approximation of $\mathcal{N}$ in terms of the averaging $\vec{\varphi}_s$ of the eigenfunctions φ_s on Γ_m:

$$\mathcal{N}^V \approx \sum_{s=1}^{n_\Delta}\vec{\varphi}_s\Big\rangle\frac{1}{\lambda_s^n - \lambda}\Big\langle\vec{\varphi}_s =: \mathcal{N}_a .$$

Then, from equation (2.14) for the scattering matrix, we derive an approximate expression for the scattering matrix on the corresponding major part of the essential spectral interval:

$$S_a(\lambda) = [1 + ip\mathcal{N}_a]^{-1}\,[1 - ip\mathcal{N}_a] , \tag{4.12}$$

for thin junctions in terms of eigen-pairs of the Neumann Schrödinger operator L_{int}^N on Ω_{int}. This accomplishes our calculation of the scattering matrix in the zero-order approximation with respect to δ in the case when several eigenvalues of L_{int}^N are present on Δ.

EXAMPLE 4.6. Consider a thin star–shaped junction $\Omega = \Omega_{int}\cup\omega$ with the Neumann Schrödinger operator $\mathcal{L}$ on it. Assume that only one eigenvalue $\lambda_0 \in \Delta$ of L^Δ is present on an essential spectral interval, situated on the first spectral band $[0, \pi^2\delta^{-2})$, with constant cross–section eigenfunctions. For a thin junction under the conditions of Theorem 4.4, we consider the corresponding approximate scattering matrix substituting λ_0^V, φ_0^V by λ_0, φ_0, and assuming that $\| \varphi_0 \|= 1$ and using $\vec{\varphi_0} = P_+\varphi_0\big|_\Gamma$ with the square norm:

$$|\alpha|^2 = |\vec{\varphi_0}|^2_{L_2(\Gamma)} = \delta^{-1} \sum_{m=1}^{M} \left| \int_{\Gamma_m} \varphi_0(\gamma) d\gamma \right|^2 .$$

Then

$$\mathcal{N}_a = \frac{\vec{\varphi_0}\rangle \langle \vec{\varphi_0}}{\lambda_0 - \lambda} = \frac{|\alpha|^2 P_0}{\lambda_0 - \lambda},$$

and due to (4.12),

$$S_a(\lambda) = \left(1 - ip\frac{|\alpha|^2 P_0}{\lambda_0 - \lambda}\right) / \left(1 + ip\frac{|\alpha|^2 P_0}{\lambda_0 - \lambda}\right),$$

with the orthogonal projection P_0 onto the 1–dimensional subspace spanned by $\vec{\varphi_0}$. This means that for any $e \in E_+$ the corresponding component of the scattering Ansatz Ψ_α in the first (open) channel of the wires, $P_+\Psi_e(x) = e^{ipx}e + e^{-ipx}\, S_a e$, fulfills the following boundary conditions:

$$(4.13)\qquad \begin{aligned} (I + S_a)e &= \frac{2(\lambda_0 - \lambda)}{\lambda_0 - \lambda + ip|\alpha_0|^2 P_0} e = P_+\Psi(0) \,, \\ ip(I - S_a)e &= \frac{2ip\alpha_0^2}{\lambda_0 - \lambda + ip|\alpha_0|^2} P_0 e = P_+\Psi'(0) \,. \end{aligned}$$

This implies that the component $P_+\Psi_e$ of the scattering Ansatz satisfies the Schrödinger equation $-[P_+\Psi_e]$ on the quantum graph ω and the boundary condition at the vertex:

$$(4.14)\qquad P_0\left[P_+\Psi(0)\right] = 0, \qquad P_0^\perp\left[P_+\Psi'(0)\right] = 0,$$

which can be interpreted similarly to the Kirchhoff boundary condition. Indeed, since only the first spectral branch (with the constant cross-section functions) is open, we have, for the normalized eigenfunction φ_0 of L^N_{int}:

(4.15)

$$P_+\varphi_0 = \begin{pmatrix} \delta^{-1}\int_{\Gamma_1}\varphi_0(\gamma)\,d\gamma \\ \delta^{-1}\int_{\Gamma_2}\varphi_0(\gamma)\,d\gamma \\ \delta^{-1}\int_{\Gamma_3}\varphi_0(\gamma)\,d\gamma \\ \vdots \\ \delta^{-1}\int_{\Gamma_{M-1}}\varphi_0(\gamma)\,d\gamma \\ \delta^{-1}\int_{\Gamma_M}\varphi_0(\gamma)\,d\gamma \end{pmatrix} = \begin{pmatrix} \bar{\varphi}_0^1 \\ \bar{\varphi}_0^2 \\ \bar{\varphi}_0^3 \\ \vdots \\ \bar{\varphi}_0^{M-1} \\ \bar{\varphi}_0^M \end{pmatrix} = \vec{\varphi_0} \,, \qquad P_0 = \frac{\vec{\varphi_0}\rangle \langle \vec{\varphi_0}}{|\vec{\varphi_0}|^2} .$$

Since for $P_+\Psi(0)$ and $P_+\Psi'(0)$ we have:

$$(4.16)\qquad P_+\Psi(0) = \begin{pmatrix} \bar{\Psi}_1 \\ \bar{\Psi}_2 \\ \bar{\Psi}_3 \\ \vdots \\ \bar{\Psi}_{M-1} \\ \bar{\Psi}_M \end{pmatrix} = \vec{\Psi}_0 \,, \qquad P_+\Psi'(0) = \begin{pmatrix} \bar{\Psi}'_1 \\ \bar{\Psi}'_2 \\ \bar{\Psi}'_3 \\ \vdots \\ \bar{\Psi}'_{M-1} \\ \bar{\Psi}'_M \end{pmatrix} = \vec{\Psi'}_0.$$

Then for the first boundary condition (4.14) we have:

$$(4.17)\qquad P_0\Psi = \frac{\delta\sum_{s=1}^{M}\Psi_s\,\bar{\varphi}_0^s}{|\vec{\varphi_0}|^2} = 0,$$

and it can be interpreted as a condition of the weighted balance of the Ansatz Ψ at the vertex. Similarly the second boundary condition (4.14) can be interpreted as a condition of the weighted continuity of the current at the vertex:

$$(4.18)\qquad \frac{\bar{\Psi}'_m}{\bar{\varphi}_0^m} = \frac{\delta\sum_{s=1}^M \bar{\Psi}'_s\,\bar{\varphi}_0^s}{|\vec{\varphi}_0|^2} = \frac{\bar{\Psi}'_r}{\bar{\varphi}_0^r} =: \bar{\Psi}'(0)/\bar{\varphi}(0)\ ,\quad 1\le m,r\le M\ .$$

Here $\bar{\Psi}'(0)/\bar{\varphi}(0)$ plays the role of the weighted current at the vertex.

In the case of the symmetric T-junction the projections $P_0, P^{\perp}$ can can be substituted by the projections onto the orthogonal complements $Q_0^{\perp}, Q_0$ in E_+, which gives the Datta-Das Sarma boundary condition (4.1,4.2), thus enabling us to interpret the phenomenological parameter β and the relevant projection in spectral terms of the relevant differential equation on the basic region. With respect to projections $Q_0^{\perp}, Q_0$ the boundary conditions (4.14) take the form

$$(4.19)\qquad Q_0^{\perp}\left[P_+\Psi(0)\right] = 0,\qquad Q_0\left[P_+\Psi'(0)\right] = 0,$$

and can be interpreted, vice versa, as the condition of the weighted continuity and the weighted balance of the current respectively, similarly to the standard Kirchhoff boundary condition. See an extended discussion and more examples in [**28**].

5. Local scattering matrix and the analytic perturbation procedure

The one-body Hamiltonians of resonance scattering systems are obtained from a small perturbation $\beta^{-1}\approx 0$ of selfadjoint operators with eigenvalues embedded into the continuous spectrum. The standard analytic perturbation procedure can't be applied to these systems because of zero spacing of the unperturbed spectrum containing the embedded eigenvalues. H. Poincaré suggested in [**29**] a modified two-step analytic perturbation procedure for classical analogs of the resonance systems: the systems with small denominators in celestial mechanics. Following H. Poincaré, one should first select a new ("intermediate") unperturbed Hamiltonian which has the same dangerous resonances preventing the convergence of the analytic analytic perturbation series for the original problem. Then the second step is to use the standard analytic perturbation procedure based on the new unperturbed Hamiltonian to obtain convergence.

There are, of course, no hints in [**29**] concerning the general construction of the intermediate Hamiltonian. The true quantum analog of the H. Poincare receipt—the "jump-start"—was obtained as a *one-dimensional perturbation* of the unperturbed quantum Hamiltonian [**8, 27**] in the form of a solvable model, similar to the model for the junction, constructed in the previous section. Poincaré's proposal concerning the elimination of dangerous resonances was actually interpreted in [**8, 27**], in terms of scattering theory, the following way.

The basic difficulty of analysis of quantum resonance systems is defined by the typical structure of the corresponding scattering matrix containing a Blaschke product of resonance factors:

$$\mathcal{B}(\beta,p)\equiv\prod_{s=1}^{n}\left[\Theta_s\,\frac{\bar{p}_s(\beta)-p}{p_s(\beta)-p}P_s+P_s^{\perp}\right]\equiv\prod_{s=1}^{n}B_s(p,\beta),$$

with numerators and denominators containing resonances $p_s(\beta)$ depending analytically on the small parameter β and their adjoints $\bar{p}_s(\beta)$ approaching (as $\beta\to 0$) the embedded eigenvalues $p_s(0)$ of the unperturbed selfadjoint generator of the evolution group, respectively from the upper and lower half-planes of the spectral parameter: $p_s(\beta) \to p_s(0)$, $\bar{p}_s(\beta) \to \bar{p}_s(0) = p_s(0)$. The order of the factors in the above product may be chosen in various ways, but the projections in the factors depend on the order chosen; see [**4**]. If the factors B_s are ordered as

$$\mathcal{B}(p,\beta) = \prod_{s\neq r}^{n} B_s(p,\beta) \times B_r(p,\beta) = \mathcal{B}_r(p,\beta) \times B_r(p,\beta), \tag{5.1}$$

in a small neighborhood Ω_r of an isolated embedded eigenvalue $p_r(0)$ of the unperturbed generator, the first factor $\mathcal{B}_r(p,\beta)$ depends analytically on β on a direct product of a complement Ω_r' of the neighborhood Ω_r and the small neighborhood $\{|\beta| << \delta\}$ of the origin in the β -space. But the second factor $B_r(p,\beta)$ is not analytic with respect to β, since both zeros of the numerator and denominator converge as $\beta \to 0$ to the same real point $p_r(0) = \bar{p}_r(0) \in \Omega_r$. For a typical Lax-Phillips resonance system, modelling the acoustic junction, the above factors are often interpreted as scattering matrices $S(\mathcal{L}_2,\mathcal{L}_1)$, $S(\mathcal{L}_1,\mathcal{L}_0)$ between the corresponding components of the perturbed generator of the evolution and the unperturbed generator. The first factor depends analytically on β in Ω_r' and can be obtained via the standard analytic perturbation procedure, but the second factor plays a role as the jump-start scattering matrix of the operator $\mathcal{L}_0$, which is constructed as a zero- range solvable model of the resonance scattering system, *fitted in the selected interval* Δ *of the spectral parameter*. See an extended discussion in [**5**], where the non-analytic factor of the scattering matrix of the junction was selected in the same way as the approximate scattering matrix constructed in the previous section. The two-step analytic perturbation procedure for the Friedrichs model was discussed in [**27**]. The term "jump-start" was introduced in [**27**], following the recommendation of L. Faddeev, for the factor of the scattering matrix involving the "dangerous" resonances. This factor can be constructed in the course of the solution of the corresponding **local inverse scattering problem** for an acoustic junction based on the Lax-Phillips machinery. We discuss this in the next section. The case of quantum junction was considered in [**21**].

In fact the special role of the jump-start factor was observed much earlier. In [**10, 11**] Birman suggested a local version of the stationary scattering theory and introduced the corresponding wave-operators and the local scattering matrix. In particular he succeeded in calculating the local scattering matrix for a few pairs of differential operators which seemed, at that time, to be just an exotic phenomenon. Much later we discovered that the the local scattering matrix can be calculated for the admitted pair formed from the Hamiltonian of a junction and the Hamiltonian of a properly fitted solvable model of the junction, with a finite-dimensional inner structure; see [**4**]. The scattering matrix of the solvable model coincides exactly with the local scattering matrix. Here we accomplish this by noticing that the scattering matrix for the acoustic junction admits a decomposition into two factors where the right one—the jump-start—contains the dangerous resonances and can be interpreted as a local scattering matrix, and the complementary factor, which, after appropriate change of spectral variable, is interpreted as the scattering matrix between the jump-start and the full junction Hamiltonian. Thus we bridge Poincaré's original idea with dangerous resonances in acoustics.

6. Chain-rule for the scattering matrices and the local inverse problem

Consider the wave equation $u_{tt} = \Delta u$ on an acoustic junction $\Omega = \omega_1 \cup \omega_2 \cup \dots \omega_M \cup \partial\Omega_{int}$ as in section 1. The role of open channel of the corresponding Lax-Phillips evolution plays the subspace $L_2(0,\infty) \times \left[E_+^1 \oplus E_+^2 \oplus E_+^3 \oplus \dots E_+^M\right]$ spanned by the constant cross-section elements. We introduce the space $\mathcal{E}$ of all sufficiently smooth Cauchy data $(u, u_t) \equiv (u_0, u_1) \equiv \vec{(u)}$ for the wave equation with the energy norm

$$\| \vec{u} \|^2 = \frac{1}{2} \int_\Omega \left[|\nabla u_0^2 + |u_1|^2\right] d\Omega, \tag{6.1}$$

and re-write the wave equation on the junction in Lax-Phillips form (see [**20**]):

$$\frac{1}{i}\frac{\partial \vec{u}}{\partial t} = i \begin{pmatrix} 0 & -1 \\ -\Delta & 0 \end{pmatrix} \vec{u} \equiv \mathcal{L}\,\vec{u}, \tag{6.2}$$

with the selfadjoint in $\mathcal{E}$ generator $\mathcal{L}$. Consider the relevant unitary evolution group with mutually orthogonal in $\mathcal{E}$ invariant incoming and outgoing subspaces $\mathcal{D}_{in}$, $\mathcal{D}_{out}$ constituted by the data supported by the reservoir

$$\mathcal{D}_{in} \equiv \{(u_0(x), u_0'(x))\}\,, \quad \mathcal{D}_{out} \equiv \{(u_0(x), -u_0'(x))\}\,. \tag{6.3}$$

This evolution group $U(t) = e^{i\mathcal{L}t}$ is a nontrivial colligation of the invariant subspaces that also involves the orthogonal complement—the co-invariant subspace $\mathcal{E} \ominus [\mathcal{D}_{in} \oplus \mathcal{D}_{out}]$. The corresponding trivial colligation is just an evolution $U_0(t) = e^{i\mathcal{L}_0\, t}$ in the energy-normed space $\mathcal{E}$ on the reservoir ω, which has the same incoming and outgoing subspaces and no co-invariant subspace. Following [**2**], we introduce the wave operators for the unitary groups U, U_0 with projections P_{out}, P_{in} onto the incoming and outgoing subspaces:

$$\begin{aligned} W_{out}(U_0, U) &= s - lim_{t\to\infty} U_0(-t)\; P_{out}\; U(t),\; W_{in}(U_0, U) \\ &= s - lim_{t\to-\infty} U_0(-t)\; P_{in}\; U(t),\; W_{out}(U, U_0) = W^+ out(U_0, U). \end{aligned}$$

Then we construct the Lax-Phillips scattering operator as

$$S(U_0, U) = W_{in}(U_0, U)\; W_{out}(U, U_0) \equiv S_{LP}.$$

It commutes with the unperturbed group U_0, and coincides, in the *incoming* spectral representation of U_0 (see [**2**]) with multiplication by the adjoint transmission coefficient S^+ on the open spectral band Δ_0. It is important that the Lax-Phillips scattering matrix for the acoustic junction is unitary only on the first spectral band Δ_0 and is a contracting function on the higher spectral bands.

Hereafter we consider the chain rule for the Lax-Phillips scattering matrix. This is different from the Schrödinger scattering matrix of the junction on the higher spectral bands. The "chain rule" for scattering matrices can't be derived based on the above definition of scattering matrices because they act in different spaces. We will obtain the chain rule for similar objects after an appropriate change of the spectral variable.

Consider three Lax-Phillips groups $U_s \equiv e^{i\mathcal{L}_s t}$, $s = 0, 1, 2$, with equivalent incoming and outgoing subspaces $\mathcal{D}_{in,out}$, assuming that U_0 is a trivial colligation, and $U_{1,2}$ are non-trivial ones. Denote by $S(U_0, U_s) \equiv S_{0,s}$ the corresponding scattering operators, and by $s_{0,s}$ -the relevant scattering matrices obtained as spectral representations of the scattering operators with respect to eigenfunctions of the trivial colligation. Similarly we introduce the scattering operators $S_{r,l} \equiv S(U_r, U_l)$

and the scattering matrices s_{rl} by representing the scattering operators in some (for instance, incoming) spectral representation of the group U_r.

Notice that the functions $\Phi(\mathcal{L}_0), \Phi(\mathcal{L}_1)$ of two selfadjount generators $\mathcal{L}_0, \mathcal{L}_1$ of the unitary groups with equivalent $\mathcal{D}_{in,out}$ in the Hilbert spaces $\mathcal{E}_0, \mathcal{E}_1$ are connected by the wave operators $W_{in}(\mathcal{L}_0, \mathcal{L}_1)$, $W_{out}(\mathcal{L}_0, \mathcal{L}_1)$ and adjoint ones as:
(6.4)
$$\Phi(\mathcal{L}_0)W_{in}(\mathcal{L}_0, \mathcal{L}_1) = W_{in}(\mathcal{L}_0, \mathcal{L}_1)\Phi(\mathcal{L}_1), \Phi(\mathcal{L}_1)W_{out}(\mathcal{L}_1, \mathcal{L}_0) = W_{out}(\mathcal{L}_1, \mathcal{L}_0)\Phi(\mathcal{L}_0).$$

Denote by J_0, J_1 the spectral transforms from the Hilbert spaces $\mathcal{E}_0, \mathcal{E}_1$ into the corresponding spaces $L_2(R, E)$ of the square-integrable functions depending on the spectral parameter:

$$\Phi(p_0) \equiv J_0\Phi_0(\mathcal{L}_0)\, J_0^+, \quad \Phi(p_1) \equiv J_1\Phi_0(\mathcal{L}_1)\, J_1^+$$

and introduce the spectral images of the wave operators

$$\mathcal{W}_{in,out} = J_0\, W_{in,out}(\mathcal{L}_0, \mathcal{L}_1)\, J_0^+$$

and ones of the adjoints.

To derive the chain rule, consider three colligations $\mathcal{L}_0, \mathcal{L}_1, \mathcal{L}_2$ with equivalent incoming and outgoing subspaces $\mathcal{D}_{in,out}$, and construct the relative scattering operators $S_0 \equiv I, S_1 \equiv S_{0,1}, S_2 \equiv S_{0,2}$. Let $W_{in,out}(i,k) = \lim_{t\to\pm\infty} e^{-i\mathcal{L}_i t}\, P_{in,out}\, e^{i\mathcal{L}_k t}$ be the wave operators. Let $S_{i,k} = s(p_i)$ be the scattering operators written down in terms of the spectral representations of $\mathcal{L}_i$, (standard scattering matrices). The commutants $\mathbf{C}(\mathcal{L}_0)$, $\mathbf{C}(\mathcal{L}_1$, $\mathbf{C}(\mathcal{L}_2))$ of the generators are connected via corresponding wave operators. If, for instance, $\Phi(\mathcal{L}_0)$ belongs to the commutant of $\mathcal{L}_0$, and $\Phi(\mathcal{L}_1)$ belongs to the commutant of $\mathcal{L}_1$, then

$$\Phi(\mathcal{L}_0)\, W_{in,out}(0,1) = W_{in,out}(0,1)\, \Phi(\mathcal{L}_1),$$

$$\Phi(\mathcal{L}_1)W_{in,out}(1,0) = W_{in,out}(1,0)\, \Phi(\mathcal{L}_0).$$

Using the spectral representations J_0, J_1 of the operators $\mathcal{L}_0$, $\mathcal{L}_1$ for $\Phi(\mathcal{L}_0), \Phi(\mathcal{L}_i)$ we obtain, with use of the wave operators, a connection between the spectral images $J_0\, \Phi(\mathcal{L}_0)\, J_0^+ \equiv \Phi(p_0)$, $J_1\, \Phi(\mathcal{L}_1)\, J_1^+ \equiv \Phi(p_1)$:

$$\Phi(p_0)\mathcal{W}_{in}(\mathcal{L}_0, \mathcal{L}_1) = \mathcal{W}_{in}(\mathcal{L}_0, \mathcal{L}_1)\Phi(p_1),\ \mathcal{W}_{out}(\mathcal{L}_1, \mathcal{L}_0)\, \Phi(p_0) = \Phi(p_1)\, \mathcal{W}_{out}(\mathcal{L}_1, \mathcal{L}_0).$$

This establishes the relations between the scattering matrices depending initially on different spectral variables, just by the reduction all of them to the same spectral parameter, usually p_0. This corresponds to the unperturbed operator $\mathcal{L}_0$. Indeed, due to the chain rule for the wave operators and above equations (6.4), we have, with abridged notations for wave operators:

$$\begin{aligned} S(0,2) &= W_{in}(0,2)W_{out}(2,0) = W_{in}(0,1)W_{in}(1,2)W_{out}(2,1)W_{out}(1.0) \\ &= W_{in}(0,1)S(1,2)W_{out}(1,0). \end{aligned}$$

Transferring $S(1,2)$ on the right and passing to the variable p_0 we obtain
(6.5)
$$\begin{aligned} s_{0,2}(p_0) &= J_0W_{in}(0,1)S(1,2)W_{out}(1,0)J_0 = s_{1,2}(p_0)J_0W_{in}(0,1)W_{out}(1,0)J_0 \\ &= s_{1,2}(p_0)s_{0,1}(p_0). \end{aligned}$$

Thus the following theorem (see [**5**]) is proven:

THEOREM 6.1. *If $\mathcal{L}_0, \mathcal{L}_1, \mathcal{L}_3$ are three generators of the Lax-Phillips groups with equivalent orthogonal incoming and outgoing subspaces $\mathcal{D}_{out,in}$, where $\mathcal{L}_0$ is a trivial colligation in $\mathcal{D}_{in} \oplus \mathcal{D}_{out} \equiv \mathcal{E}_0$, the scattering matrix $s_{0,2}(p_0)$ is factorized as*

$$s_{0,2}(p_0) = s_{1,2}(p_0)s_{0,1}(p_0). \tag{6.6}$$

Here $s_{0,1}(p_0), s_{1,2}(p_0)$ are the corresponding s-matrices with properly changed variables $s_{1,2}(p_0)\mathcal{W}_{in}(0,1) = \mathcal{W}_{in}(0,1)s_{1,2}(p_1)$.

7. Local inverse scattering problem and elimination of dangerous resonances

Assume that the right multiplier $s_{0,1}(p)$ of the Lax-Phillips scattering matrix (6.6) of the acoustic junction at the resonance p_1 is equal to the matrix Blaschke factor

$$S_{LP}(p) = -\frac{(p_1 - p)(\bar{p}_1 + p)}{(p_1 + p)(\bar{p}_1 - p)} P_1 - P_1^{\perp} = -\frac{1 + \frac{2i\,\Im p_1\, p}{|p_1|^2 - p^2} P_1}{1 - \frac{2i\,\Im p_1\, p}{|p_1|^2 - p^2} P_1}, \tag{7.1}$$

with an orthogonal one-dimensional projection P_1 and the resonances $p_1 \equiv p_1(\beta)$, $-\bar{p}_1 = -\bar{p}_1(\beta)$ depending on a small parameter, for instance the width of the leads of the junction. Then the dual multiplier of the corresponding stationary transmission coefficient is just the adjoint:

$$S_{Sch}(p) = -\frac{1 - \frac{2i\,\Im p_1\, p}{|p_1|^2 - p^2} P_1}{1 + \frac{2i\,\Im p_1\, p}{|p_1|^2 - p^2} P_1}. \tag{7.2}$$

To implement Poincaré's idea of elimination of dangerous resonances $p_1, \bar{p_1}$ we need to construct a jump-start Hamiltonian $\mathcal{L}_1$ such that the Lax-Phillips scattering matrix $S_{\mathcal{L}_0,\mathcal{L}_1}$ coincides with the factor S_{LP} or the partial transmission coefficient (7.2) coincides with (7.2). We will construct C as a 1D perturbation of trivial colligation on the 1D laplacian on the open channel space $L_2(0,\infty) \times E_+$. We achieve that by choosing an appropriate modification of the energy metric.

We consider an orthogonal sum $\mathcal{E} \oplus E_1 \equiv \mathcal{E}_1$ of the energy-normed space on the reservoir and the energy-normed space associated with Cauchy data $(u, u_t) \in E_1$ of an oscillator $u_{tt} + \alpha^2 u = 0$ in 1D Hilbert space. We introduce a linear operator $B : E_+ \to E_1$ and define the interaction between the wave dynamics in the reservoir and the oscillator, with use of the projection P_1 onto the zero space of the Blaschke factor $S_{LP}(p_1)$

$$\begin{aligned} \Psi_{tt} - \Psi_{xx} &= i\delta(x - 0)P_1 B^+ \, u_t, \\ u_{tt} + \alpha^2 u &= iB \; P_1 \Psi_t(0, t). \end{aligned} \tag{7.3}$$

The corresponding energy norm taking into account the interaction constitutes the sum of the energy norms of the wave equation $\| \Psi \|_{\mathcal{E}}$ on the reservoir, with the Neumann boundary condition on the bottom section of the leads, and one of the oscillators $\| u \|_{E_1}$ and the additional boundary terms

$$\| \Psi \|^2_{\mathcal{E}} + \| u \|^2_{E_1} + \Re\langle \Psi(0), \; Bu\rangle \equiv \| (\vec{\Psi}, u) \|^2_{\mathcal{E}_1}, \text{ with } \vec{\Psi} = (\{\Psi, \Psi_t\}, \{u, u_t\})) \in \mathcal{E}_1.$$

Equation (7.3) corresponds to a Lax-Phillips scattering system with orthogonal incoming and outgoing subspaces in the open channel. The corresponding Lax-Phillips scattering matrix coincides with the adjoint of the stationary Schrödinger transmission coefficient.

THEOREM 7.1. *The transmission coefficient of the spectral problem (7.3) can be fitted via selection of the parameters α, β, P_1 based on the data of the Schrödinger right factor S_{Sch}.*

PROOF. The separation of harmonic time-dependence $\Psi(x,t) \to e^{ipt}\Psi^p(x)$, $u(t) \to e^{ipt}u^p$ reduces (7.3) to the corresponding spectral problem

$$-[\Psi^p]\,xx = p^2\Psi^p - p\delta(x-0)P_1\,B^+\,u^p, \alpha^2 u^p = p^2 u^p - p\,B\,P_1\Psi^p(0). \tag{7.4}$$

The substitution of u^p from the second equation

$$-\Psi^p_{xx} = p^2\Psi^p - \frac{p^2}{\alpha^2 - p^2}\delta(x-0)P_1\,B^+\,B\,P_1\,\Psi^p$$

and the subsequent integration on $(-0,+0)$ with regard to the Neumann boundary condition at the bottom sections Γ yields the transmission coefficient S in the scattering Ansatz

$$S_{Sch}(p) = -\frac{I - \frac{ip}{\alpha^2-p^2}P_1\,B^+\,B\,P_1}{I + \frac{ip}{\alpha^2-p^2}P_1\,B^+\,B\,P_1}. \tag{7.5}$$

Selecting $\alpha = |p_1|$, $B^+\,B = \Im p_1\,P_1$ and choosing P_1 as the projection onto the zero space of the Blaschke-factor S_{LP}, we fit the constructed solvable model to the data of the factor. □

This theorem then implements the elimination of dangerous resonances $p_1, -\bar{p}_1$ as a solution of the local inverse problem. The complementary factor depends analytically on the small parameter and can be obtained via standard analytic perturbation procedure.

Acknowledgement

We are grateful to Ioannis Antoniou, one of the closest collaborators with 1977 Nobel Prize winner Iliya Prigogine. From Antoniou we learned that the Solvay Institute, led by Prigogine (1959-2002), was actively involved in the search for a quantum version of the two-step analytic perturbation procedure, proposed by H. Poincaré for the elimination of dangerous resonances in the perturbative analysis of problems of celestial mechanics with small denominators. We consider our paper a continuation of this study, by means of harmonic analysis and the spectral theory of functions.

References

[1] V. Adamyan, *Scattering matrices for microschemes*, Operator Theory: Adv. and Appl. **59**, 1 (1992). MR1246807 (94m:47016)

[2] V. M. Adamyan and D. Z. Arov, *On scattering operators and contraction semigroups in Hilbert space* (Russian) Dokl. Akad. Nauk SSSR **165** (1965), 9–12. MR0188808 (32:6240)

[3] V. Adamyan, B. Pavlov and A. Yafyasov, *Modified Krein formula and analytic perturbation procedure for scattering on arbitrary junction*, International Newton Institute report series NI07016, Cambridge, 18 April 2007, 30 pp.

[4] V. Adamyan and B. Pavlov, *Local Scattering problem and a Solvable model of Quantum Network*, Operator Theory: Advances and Applications, Birkhäuser Verlag, Basel/Switzerland **198** (2009), 1–10. MR2648563 (2011i:47015)

[5] V. Adamyan and B. Pavlov, *Chain-rule for scattering matrices and two-steps analytic perturbation procedure*, Recent Trends in Analysis, Proceedings of the conference in honor of Nikolai Nikolski, Theta Foundation Bucharest, ISBN 978-606-8443-02-7. Edited by Alexander Borichev, Kenneth R. Davidson, Stanislas Kupin, Gilles Pisier, Florian-Horia Vasilescu, Vasiliy Vasyunin, 2013, pp. 1–16.

[6] S. Albeverio and P. Kurasov, *Singular Perturbations of Differential Operators*, London Math. Society Lecture Note Series, 271, Cambridge University Press, 2000. MR1752110 (2001g:47084)

[7] N. T. Bagraev, A. D. Bouravlev, V. K. Ivanov, L. E. Kljachkin, A. M. Maljarenko, S. A. Rykov, and I. A. Shelyh, *Interference of charge carriers in one–dimensional semiconductor rings*, Semiconductors **34** (2000), 7, 817–824.

[8] N. Bagraev, A. Mikhailova, B. Pavlov, L. Prokhorov, and A. Yafyasov, *Parameter regime of a resonance quantum switch*, Phys. Rev. B (2005), 165308.

[9] F. A. Berezin and L. D. Faddeev, *Remark on the Schrödinger equation with a singular potential*, Dokl. AN SSSR **137** (1961), 1011–1014. MR0129309 (23:B2345)

[10] M. Birman, *On existence conditions for wave operators*, Russian Academy Doklady **143** (1962), 506–509. Translated into English in "Soviet Math. Doklady". MR0149333 (26:6823)

[11] M. Birman, *A local criterion for the existence of wave operators*, Math. URSS Izvestia **2** (1968), 879–906.

[12] S. Datta and B. Das Sarma, *Electronic analog of the electro-optic modulator*, Appl. Phys. Lett. **56** (1990), No. 7, 665–667.

[13] E. Fermi, *Sul motto dei neutroni nelle sostance idrogenate* (in Italian) Richerka Scientifica **7** (2) (1936), 13–52.

[14] F. Gesztesy, Y. Latushkin, and M. Mitrea, M.Zinchenko, *Non–selfadjoint operators, infinite determinants and some applications*, Russian Journal of Mathematical Physics **12** (2005), 443–471. MR2201310 (2008f:47030)

[15] F. Gesztesy, M. Mitrea, and M. Zinchenko, *On Dirichlet–to–Neumann maps and some applications to modified Fredholm determinants*, Oper. Theory Adv. Appl. **186** (2006), 191–215. MR2732078 (2012b:47053)

[16] I. S. Gohberg and E. I. Sigal, *An operator generalization of the logarithmic residue theorem and Rouché's theorem*, Mat. Sbornik **84** (1971), 607. MR0313856 (47:2409)

[17] M. Harmer, B. Pavlov, A. Yafyasov, *Boundary condition at the junction*, Journal of Computational Electronics, DOI 10. 1007/s10825-006-0085-7

[18] V. Kruglov, K. Makarov, B. Pavlov and A. Yafyasov, *Exponential decay in Quantum Mechanics*, Computations, Physics and Beyond (M. Dinneen, B. Khousainov, and A. Nies, eds.), Lecture Notes in Computer Science, 7160, pp. 268–288. MR2965530

[19] P. Kuchment and H. Zeng, *Convergence of spectra of mesoscopic systems collapsing onto a graph*, Journ. Math. Anal. Appl. **258** (2001), 671–700. MR1835566 (2002d:35158)

[20] P. Lax and R. Phillips, *Scattering theory*, Academic Press, New York, 1967. MR0217440 (36:530)

[21] G. Martin, A. Yafyasov, and B. Pavlov, *Resonance one-body scattering on a junction*, Nanosystems: Physics, Chemistry, Mathematics. Vol. 1, No. 1 (2010), 108–147.

[22] R. Mennicken and A. Shkalikov, *Spectral decomposition of symmetric operator matrices*, Mathematische Nachrichten **179** (1996), 259–273. MR1389460 (97d:47012)

[23] A. Mikhailova, B. Pavlov, and L. Prokhorov, *Intermediate Hamiltonian via Glazman's splitting and analytic perturbation for meromorphic matrix-functions*, Mathematishe Nachrichten **280** (2007), 12, 1376–1416. MR2344872 (2008g:47147)

[24] R. Mittra and S. Lee, *Analytical techniques in the theory of guided waves*, The Macmillan Company, NY, Collier–Macmillan Limited, London, 1971.

[25] B. Pavlov, *S–Matrix and Dirichlet–to–Neumann Operators.* Encyclopedia of Scattering, edited by R. Pike and P. Sabatier, Academic Press, Harcourt Science and Tech. Company, 1678–1688, 2001.

[26] B. Pavlov, *A star–graph model via operator extension*, Mathematical Proceedings of the Cambridge Philosophical Society, Vol. 142, Issue 02, March 2007, 365–384. MR2314606 (2008d:81258)

[27] B. Pavlov and I. Antoniou, *Jump-start in analytic perturbation procedure for the Friedrichs model*, J. Phys. A: Math. Gen. **38** (2005), 4811–4823. MR2148624 (2005m:47029)

[28] B. Pavlov, *A solvable model for scattering on a junction and a modified analytic perturbation procedure*, Characteristic functions, scattering functions and transfer functions, Oper. Theory Adv. Appl., 197, Birkhauser Verlag, Basel, 2010, pp. 281–335. MR2647542 (2011f:47013)
[29] H. Poincaré, *Methodes nouvelles de la mećanique celeste*, Vol. 1 (1892), Dover, New York, 1957.
[30] C. Presilla and J. Sjöstrand *Transport properties in resonance tunnelling heterostructures*, J. Math. Phys. **37** (1996), no. 10, 4816–4844. MR1411610 (98g:81041)
[31] J. Sylvester and G. Uhlmann, *The Dirichlet to Neumann map and applications*, Proceedings of the Conference on Inverse problems in partial differential equations, Arcata, SIAM, Philadelphia, 1990, pp. 101–139. MR1046433 (91d:35063)
[32] E. P. Wigner, *On a class of analytic functions from the quantum theory of collisions*, Annals of Mathematics (2) **53** (1951), 36–67. MR0039059 (12:490d)

Department of Theoretical Physics, Odessa National I.I. Mechnikov University 65082 Odessa, Ukraine
E-mail address: `vadamyan@onu.edu.ua`

New Zealand Institute of Advanced Study, Massey University, Auckland, New Zealand
E-mail address: `G.J.Martin@massey.ac.nz`

V. A. Fock Physics Institute, Department of Physics, St. Petersburg University, Ulyanovskaja 2, Petrodvorets, 198504, St. Petersburg, Russia and New Zealand Institute of Advanced Study, Massey University, Auckland, New Zealand
E-mail address: `pavlov@math.auckland.ac.nz`

Amer. Math. Soc. Transl.
Volume **233**, 2014

The Continuous Theory of Dislocations for a Material Containing Dislocations to One Burgers Vector Only

Hans-Dieter Alber

Abstract. We review the continuous theory of dislocations from a mathematical point of view using mathematical tools, which were only partly available when the theory was developed several decades ago. We define a space of dislocation measures, which includes Hausdorff measures representing the dislocation measures of single dislocation curves. The evolution equation for dislocation measures is defined on this space. It is derived from four basic conditions, which must be satisfied by the model.

1. Introduction

The plastic deformation of metallic bodies is caused by the creation and movement of dislocations in the crystal lattice of the metallic material. Therefore the plastic deformation depends on the number of dislocations and on the restrictions of the dislocation movement by the crystal structure and by the geometry of the body. Standard phenomenological models for viscoplastic material behavior do not reflect these material properties depending on the microstructure of the material. A modeling approach that takes this microstructure into account is the continuous theory of dislocations, which was developed several decades ago and which is well understood in continuum mechanics. Of the many articles in this field we mention only the classical expositions [**8, 7, 4, 5**] and the articles that contain new developments [**1, 2**]. This theory has interesting and difficult mathematical aspects. We hope to make the mathematical aspects more accessible by reviewing the theory from a mathematical point of view using mathematical tools, which were only partly available when the theory was developed. We are convinced that the mathematical aspects deserve much more investigation and that the theory can be advanced and simulations based on the theory can be improved by such investigation.

We begin by stating the standard model for the deformation of a viscoplastic body which consists of a metallic material with dislocations moving only in one slip plane of the crystal structure. It consists of the equations

$$-\mathrm{div}_x\, T(x,t) = 0, \tag{1.1}$$

$$T(x,t) = D\Big(\varepsilon\big(\nabla_x u(x,t)\big) - m\varepsilon_p(x,t)\Big), \tag{1.2}$$

$$\partial_t\varepsilon_p = f\big(m : T(x,t)\big), \tag{1.3}$$

2010 *Mathematics Subject Classification.* Primary 35Q74, 74A20, 74E15.

which must hold for time $t \geq 0$ and for x varying in the open set $\Omega \subseteq \mathbb{R}^3$ representing the material points of the body. The unknowns are the displacement $u(x,t) \in \mathbb{R}^3$ of the material point x at time t, the Cauchy stress tensor $T(x,t) \in \mathcal{S}^3$, where $\mathcal{S}^3$ denotes the set of symmetric 3×3-matrices, and the plastic strain $\varepsilon_p(x,t) \in \mathbb{R}$ along the slip plane. We use the standard notation

$$\operatorname{div}_x T = \Big(\sum_{j=1}^{3} \partial_{x_j} T_{ij} \Big)_{i=1,2,3}.$$

$\nabla_x u$ denotes the 3×3-matrix of first order partial derivations of u, and

$$\varepsilon(\nabla_x u) = \frac{1}{2}\big(\nabla_x u + (\nabla_x u)^T\big) \in \mathcal{S}^3$$

is the linear strain tensor. We write A^T for the transpose of a matrix A. The elasticity tensor $D : \mathcal{S}^3 \to \mathcal{S}^3$ is a linear, symmetric, positive definite mapping and the constant matrix m is given by

$$m = \varepsilon(\hat{b} \otimes g) = \frac{1}{2}(\hat{b} \otimes g + g \otimes \hat{b}) \in \mathcal{S}^3, \tag{1.4}$$

where $g \in \mathbb{R}^3$ is the unit vector normal to the slip plane, and $\hat{b} \in \mathbb{R}^3$ is a unit vector in the direction of plastic slip; it is therefore a vector in the slip plane. For vectors $a, b \in \mathbb{R}^3$ we write $a \otimes b$ to denote the matrix $(a_i b_j)_{i,j=1,2,3}$. The scalar product of two 3×3-matrices A, B is denoted by $A : B = \sum_{i,j=1}^{3} a_{ij} b_{ij}$. Finally, $f : \mathcal{D}(f) \subseteq \mathbb{R} \to \mathbb{R}$ is a given function satisfying

$$s \cdot f(s) \geq 0,$$

for all $s \in \mathbb{R}$. We call f a constitutive function. A typical choice for f is $f(s) = C|s|^{\gamma-1}s$, with constants $C > 0$ and $\gamma > 1$. It is well known that if boundary conditions for u or T are imposed, if $0 \in \mathcal{D}(f)$ and $f(0) = 0$ hold, and if the function f satisfies suitable growth conditions and is maximal monotone, which means that it cannot be extended to a monotone function on a larger domain, then the initial-boundary value problem to (1.1)–(1.3) has a unique solution. This is proved for example in [**3**].

The constitutive equation (1.3) is an ordinary differential equation in time and does not reflect material behavior caused by the dislocation microstructure. The field equations of the continuous theory of dislocations do reflect this microstructure. To derive these field equations we start in Section 2 by discussing the Volterra model for dislocation curves in linear elasticity and by defining the space of dislocation measures. The evolution equation for dislocation measures is derived in Section 3 from four basic principles. In Section 3.3 we discuss the restrictions on the model equations following from the incompressibility constraint for the plastic part of the strain tensor and we obtain the final field equations. The simplification following from the assumption that dislocations move in slip planes is discussed in Section 4. At the end of that section we compare the field equations thus derived to the standard equations (1.1)–(1.3).

As usual in the derivation of model equations, we cannot define the function spaces, to which the solutions of the model equations belong, with the same level of rigor as in investigations of existence. In particular this concerns the space of dislocation measures. We must assume that the solutions have certain properties; the exact properties can be determined only after the model equations are known.

To keep the length acceptable, several technical proofs are not included in this article. These proofs will be published elsewhere.

2. The boundary value problem for the stress field of a dislocation curve

In this section we review the Volterra model for dislocation curves within the linear theory of elasticity (cf. [**5**]). This model suggests the definition of the space $\mathcal{M}_d(\Omega)$ of dislocation measures, which is given at the end of the section.

Let ℓ be a closed curve in the elastic body $\overline{\Omega}$, which represents a dislocation curve. It is allowed that a part of the curve belongs to the boundary $\partial\Omega$. We assume that an arc length parametrization $s \mapsto y(s)$ is given, which is continuously differentiable with the exception of at most finitely many points. This parametrization defines a unit tangent vector field $\tau = \frac{\mathrm{d}}{\mathrm{d}s} y$ along ℓ. Let Σ be a surface in Ω with $\partial\Sigma = \ell$ and let $n : \Sigma \to \mathbb{R}^3$ be a continuous normal vector field on Σ. We choose n such that at $x \in \partial\Sigma$ the vector $n(x) \times \tau(x)$ points into the surface Σ. For functions v defined on $\Omega \setminus (\Sigma \cup \ell)$ and for $x \in \Sigma$ we use the notations

$$v^{\pm}(x) = \lim_{\eta \searrow 0} v\big(x \pm \eta\, n(x)\big), \quad [v]_\Sigma(x) = v^+(x) - v^-(x). \tag{2.1}$$

With the dislocation curve ℓ we associate a fixed vector $b \in \mathbb{R}^3$, $b \neq 0$, the Burgers vector of the dislocation curve. We use the notation

$$\hat{b} = \frac{b}{|b|}$$

for the unit vector in direction of b.

To compute the stress field generated by the dislocation curve in the body Ω, consider the boundary and transmission problem for the displacement field $u : \overline{\Omega} \setminus (\Sigma \cup \ell) \to \mathbb{R}^3$ and the Cauchy stress tensor field $T : \overline{\Omega} \setminus (\Sigma \cup \ell) \to \mathcal{S}^3$:

$$-\operatorname{div} T = 0, \tag{2.2}$$

$$T = D\varepsilon(\nabla u), \tag{2.3}$$

$$[u]_\Sigma = -b, \tag{2.4}$$

$$[T]_\Sigma n = 0, \tag{2.5}$$

$$T_{|_{\partial\Omega}} n_B = \gamma. \tag{2.6}$$

The elasticity equations (2.2) and (2.3) must hold on $\Omega \setminus (\Sigma \cup \ell)$, equations (2.4) and (2.5) are jump relations on Σ, and (2.6) is the boundary condition on $\partial\Omega$, where $n_B(x) \in \mathbb{R}^3$ denotes the unit normal to $\partial\Omega$ at $x \in \partial\Omega$ pointing to the exterior of Ω and $\gamma : \partial\Omega \to \mathbb{R}^3$ are the given boundary data. This problem describes the displacement and stress fields in an elastic body, which is cut along the surface Σ. After cutting, the two boundary parts created by the cutting are displaced elastically against one another by the vector b and glued together again. Since the length of b is approximately equal to the lattice constant of the crystal lattice, after this procedure the atoms on both sides of Σ are again in the right positions to form an elastically stressed but otherwise perfect crystal at Σ. The crystal is only disturbed along the boundary ℓ of Σ, making (2.2)–(2.6) a model for the dislocation curve ℓ. The stress field generated by this dislocation curve is given by T. This stress field will have a singularity along ℓ.

Our goal is to start from this model and to generalize it to a model for bodies containing an array of dislocations described by a dislocation density. To do this rigorously, it would be necessary to solve the problem (2.2)–(2.6) in a suitable function space and to study the singularity of T along ℓ. This difficult task is out of the scope of this article. Instead, we consider the dislocation problem in a different situation with a simpler geometry, where an explicit solution is known.

Namely, we assume that Ω is equal to $\mathbb{R}^3$, that the dislocation curve ℓ is equal to the x_3–axis, and that the material is isotropically elastic, which means that the elasticity tensor is given by

$$D_I\varepsilon = \lambda\,\mathrm{trace}(\varepsilon)I + 2\mu\varepsilon, \tag{2.7}$$

for all $\varepsilon \in \mathcal{S}^3$, where I is the identity matrix and where λ, μ are material constants satisfying $\mu > 0$ and $3\lambda + 2\mu > 0$.

To formulate the boundary value problem for the stress field in this situation we assume that the Burgers vector is of the form $b = (b_1, 0, b_3)$. This can always be achieved by rotation of the coordinate system around the x_3–axis. We define the tangential vector field τ along the line ℓ by $\tau(x) = e_3 = (0, 0, 1)$. We are free to choose for Σ any half plane with boundary ℓ. Therefore we take

$$\Sigma = \{(x_1, 0, x_3) \mid x_1 > 0,\ x_3 \in \mathbb{R}\}$$

and define the unit normal vector field n on Σ by $n(x) = e_2$. With this definition we obtain for $x \in \ell$ that the vector $n(x) \times \tau(x) = e_2 \times e_3 = e_1$ points into Σ; hence our requirement for the orientation of n is satisfied. Moreover, from (2.1) we obtain for functions v defined on $\mathbb{R}^3 \setminus (\Sigma \cup \ell)$ and $x \in \Sigma$ that

$$v^{\pm}(x) = \lim_{\eta \searrow 0} v(x_1, \pm\eta, x_3), \quad [v]_\Sigma(x) = \lim_{\eta \searrow 0} \big(v(x_1, \eta, x_3) - v(x_1, -\eta, x_3)\big).$$

The problem corresponding to (2.2) – (2.6) for the stress field generated by the dislocation line ℓ consists of the equations

$$-\mathrm{div}\, T(x) = 0, \tag{2.8}$$
$$T(x) = D_I\varepsilon\big(\nabla u(x)\big), \tag{2.9}$$
$$[u]_\Sigma(x) = -(b_1, 0, b_3), \tag{2.10}$$
$$\big([T]_\Sigma(x)\big)e_2 = 0, \tag{2.11}$$

where the first two equations must hold for $x \in \mathbb{R}^3 \setminus (\Sigma \cup \ell)$ and the last two for $x \in \Sigma$. A solution of this problem is given in [**5**, pp. 49 – 53]. In the following theorem we state this solution and give some additional properties of it. We use the notation $x = (x', x_3) \in \mathbb{R}^3$ with $x' = (x_1, x_2)$.

THEOREM 2.1. *Let* $u = (u_1, u_2, u_3) : \mathbb{R}^3 \setminus \overline{\Sigma} \to \mathbb{R}^3$ *be defined by*

$$u_1(x) = \frac{b_1}{2\pi}\Big(\arctan\frac{x_2}{x_1} + \frac{1}{2(1-\nu)}\frac{x_1 x_2}{r^2}\Big), \tag{2.12}$$
$$u_2(x) = -\frac{b_1}{4\pi(1-\nu)}\Big((1-2\nu)\log r + \frac{x_1^2}{r^2}\Big), \tag{2.13}$$
$$u_3(x) = \frac{b_3}{2\pi}\arctan\frac{x_2}{x_1}, \tag{2.14}$$

where $\nu = \frac{\lambda}{2(\lambda+\mu)}$ *is Poisson's ratio and* $r^2 = |x'|^2 = x_1^2 + x_2^2$. *Also, let the symmetric tensor function* $T = (T_{ij})_{i,j=1,\dots,3} : \mathbb{R}^3 \setminus \overline{\Sigma} \to \mathcal{S}^3$ *be given by*

$$T_{11} = -D_1 \frac{x_2(3x_1^2 + x_2^2)}{r^4}, \quad T_{12} = D_1 \frac{x_1(x_1^2 - x_2^2)}{r^4}, \quad T_{13} = -D_2 \frac{x_2}{r^2}, \tag{2.15}$$

$$T_{22} = D_1 \frac{x_2(x_1^2 - x_2^2)}{r^4}, \qquad T_{23} = D_2 \frac{x_1}{r^2}, \qquad T_{33} = -2\nu D_1 \frac{x_2}{r^2}, \tag{2.16}$$

with $D_1 = \frac{\mu b_1}{2\pi(1-\nu)}$, $D_2 = \frac{\mu b_3}{2\pi}$. *Then the function* (u, T) *solves* (2.8) – (2.11), *has the asymptotic behavior*

$$\nabla u(x) = O\left(\frac{1}{|x'|}\right), \quad T(x) = O\left(\frac{1}{|x'|}\right), \quad \textit{for } |x'| \to 0, \tag{2.17}$$

and satisfies the condition

$$\lim_{r \searrow 0} \int_{\mathcal{C}_r} \big(T(x)n(x)\big) \cdot \varphi(x)\, dS = 0, \tag{2.18}$$

for every $\varphi \in C_0^\infty(\mathbb{R}^3, \mathbb{R}^3)$, *where* $\mathcal{C}_r = \{(x', x_3) \in \mathbb{R}^3 \mid |x'| = r\}$.

The proof of this theorem is omitted. (2.15) and (2.16) show that the stress tensor T is infinitely differentiable on $\mathbb{R}^3 \setminus \ell = \{(x', x_3) \in \mathbb{R}^3 \mid x' \neq 0\}$. In particular, it is infinitely differentiable at every point of Σ. A simple computation shows that this is also true for ∇u. Of course, this must be the case, since Σ is an artificially introduced surface: The crystal lattice is undisturbed at this surface.

Let $\Omega \subseteq \mathbb{R}^3$ be an open set. To every C^1–curve ℓ in Ω representing a dislocation curve with unit tangent vector field τ and Burgers vector b we define vector and tensor valued Radon measures ρ_ℓ, $\hat{b} \otimes \rho_\ell$ respectively, by setting for all $\varphi \in C_0(\Omega, \mathbb{R}^3)$, $\tilde{\varphi} \in C_0(\Omega, \mathbb{R}^{3\times 3})$,

$$\langle \rho_\ell, \varphi \rangle = |b| \int_\ell \tau(x) \cdot \varphi(x)\, ds_x, \tag{2.19}$$

$$\langle \hat{b} \otimes \rho_\ell, \tilde{\varphi} \rangle = \int_\ell \big(b \otimes \tau(x)\big) : \tilde{\varphi}(x)\, ds_x, \tag{2.20}$$

with the Nye dislocation tensor $b \otimes \tau(x)$. As usual, for a tensor valued distribution w and for $\varphi \in C_0^\infty(\Omega, \mathbb{R}^3)$, $\tilde{\varphi} \in C_0^\infty(\Omega, \mathbb{R}^{3\times 3})$ we define

$$\langle \operatorname{div} w, \varphi \rangle = -\langle w, \nabla \varphi \rangle, \quad \langle \varepsilon(w), \varphi \rangle = \langle w, \varepsilon(\varphi) \rangle, \quad \langle \operatorname{rot} w, \tilde{\varphi} \rangle = \langle w, \operatorname{rot} \tilde{\varphi} \rangle,$$

where $\operatorname{rot} \tilde{\varphi} \in C_0^\infty(\Omega, \mathbb{R}^{3\times 3})$ is obtained by applying the standard rot–operator to the rows of $\tilde{\varphi}$.

LEMMA 2.2. *Let* ℓ *be the* x_3*–axis with* $\tau(x) = e_3$, *and let* u, T *be the functions given in* (2.12)–(2.16) *to the Burgers vector* $b = (b_1, 0, b_3)$. *Then the tensor valued distributions* $\tilde{h}_e$ *and* T *defined by*

$$\langle \tilde{h}_e, \tilde{\varphi} \rangle = \int_{\mathbb{R}^3} \nabla u(x) : \tilde{\varphi}(x)\, dx, \tag{2.21}$$

$$\langle T, \tilde{\varphi} \rangle = \int_{\mathbb{R}^3} T(x) : \tilde{\varphi}(x)\, dx \tag{2.22}$$

satisfy the equations

$$-\operatorname{div} T = 0, \tag{2.23}$$

$$T = D_I\,\varepsilon(\tilde{h}_e), \tag{2.24}$$

$$\operatorname{rot} \tilde{h}_e = -\hat{b} \otimes \rho_\ell. \tag{2.25}$$

Note that the integrals in (2.21) and (2.22) exist because of (2.17). D_I is defined in (2.7). We must also omit the proof of this lemma.

We call the two Radon measures ρ_ℓ and $\hat{b} \otimes \rho_\ell$ defined in (2.19) and (2.20) vector valued and tensor valued dislocation measures of the dislocation curve ℓ. The form of these measures and the equations (2.21) and (2.25) provide the idea for the definition of general dislocation measures given now.

Let $\Omega \subseteq \mathbb{R}^3$ be an open set and let μ be a scalar Radon measure on Ω. As usual we say that μ vanishes in a neighborhood $\mathcal{U}$ of a point x, if $\langle \mu, \varphi \rangle = 0$ for all $\varphi \in C_0(\mathcal{U}, \mathbb{R})$. The support $\operatorname{supp} \mu$ of μ is defined as the set of all points in Ω which have no neighborhood where μ vanishes. Let $\beta \in C(\operatorname{supp} \mu, \mathbb{R}^3)$ be a given function. Since $\operatorname{supp} \mu$ is a relatively closed set in Ω, the theorem of Tietze-Urysohn implies that we can extend β to a function $\hat{\beta} \in C(\Omega, \mathbb{R}^3)$ with $\|\hat{\beta}\|_{L^\infty(\Omega)} \le \|\beta\|_{L^\infty(\operatorname{supp} \mu)}$. The linear mapping

$$\big(x \mapsto \varphi(x)\big) \mapsto \big(x \mapsto \hat{\beta}(x) \cdot \varphi(x)\big) : C_0(\Omega, \mathbb{R}^3) \to C_0(\Omega, \mathbb{R})$$

satisfies $\|\hat{\beta} \cdot \varphi\|_{L^\infty(\Omega)} \le \|\hat{\beta}\|_{L^\infty(\Omega)} \|\varphi\|_{L^\infty(\Omega)}$; hence it is continuous. Therefore a vector valued Radon measure $\beta\mu$ on Ω is defined by

$$\langle \beta\mu, \varphi \rangle = \langle \mu, \hat{\beta} \cdot \varphi \rangle, \qquad \varphi \in C_0(\Omega, \mathbb{R}^3).$$

It is not difficult to see that this definition does not depend on the special continuation $\hat{\beta}$ that we have chosen. Therefore the notation $\beta\mu$ is justified.

DEFINITION 2.3. The set of all measures $\beta\mu$ on Ω with a scalar Radon measure μ and $\beta \in C(\operatorname{supp} \mu, \mathbb{R}^3)$ is denoted by $\mathcal{M}(\Omega)$. We call $\rho = \tau\mu \in \mathcal{M}(\Omega)$ a vector valued dislocation measure, if μ is a nonnegative measure, if $\tau \in C(\operatorname{supp} \mu, \mathbb{R}^3)$ satisfies $|\tau(x)| = 1$ for all $x \in \operatorname{supp} \mu$ and if a function $h \in L^{1,\mathrm{loc}}(\Omega, \mathbb{R}^3)$ exists such that $\rho = \operatorname{rot} h$. The set of vector valued dislocation measures is denoted by $\mathcal{M}_d(\Omega)$. For $\rho = \tau\mu \in \mathcal{M}_d(\Omega)$ and $\tilde{\varphi} \in C_0(\Omega, \mathbb{R}^{3\times 3})$ we also use the notation and definition

$$|\rho| = \mu, \quad \frac{\rho}{|\rho|} = \tau, \quad \langle \hat{b} \otimes \rho, \tilde{\varphi} \rangle = \langle |\rho|, (\hat{b} \otimes \tau) : \tilde{\varphi} \rangle.$$

$\hat{b} \otimes \rho$ is called a tensor valued dislocation measure. If a dislocation measure has a density in $L^1(\Omega)$, we call it dislocation density and denote the measure and the density by the same symbol.

REMARK 2.4. The dislocation measure of a dislocation curve ℓ defined in (2.19) belongs to $\mathcal{M}_d(\Omega)$ and has the form $\rho_\ell = \tau|\rho|$ with $|\rho| = |b|\,\mathcal{H}_\ell$, where $\mathcal{H}_\ell = \mathcal{H}^1 \lfloor \ell$ is the one-dimensional Hausdorff measure restricted to the curve ℓ.

This definition suggests that we generalize the problem (2.23)–(2.25) to a boundary value problem in a domain $\Omega \subseteq \mathbb{R}^3$ with an arbitrarily given tensor valued dislocation measure of the form $-\hat{b} \otimes \rho$ on the right-hand side of (2.25). Before we formulate this general problem we discuss the meaning of $\tilde{h}_e$ in the context of the theory of viscoplasticity at small strains. In this theory one uses the additive decomposition

$$\nabla u = (\nabla u - \tilde{h}_p) + \tilde{h}_p$$

of the deformation gradient ∇u into a plastic part $\tilde{h}_p$ and an elastic part $\nabla u - \tilde{h}_p$, where only the elastic part generates the stress field:

$$T = D\varepsilon(\nabla u - \tilde{h}_p).$$

Comparing this equation with (2.24) we see that $\tilde{h}_e$ is the elastic part of the deformation gradient:

$$\tilde{h}_e = \nabla u - \tilde{h}_p \,. \tag{2.26}$$

In the following we work with $\tilde{h}_p$ instead of $\tilde{h}_e$. We eliminate $\tilde{h}_e$ in (2.25) by using (2.26). If we also replace ρ_ℓ on the right-hand side of (2.25) by an arbitrary dislocation measure $\rho \in \mathcal{M}_d(\Omega)$ we obtain

$$\operatorname{rot} \tilde{h}_p = -\operatorname{rot} \tilde{h}_e = \hat{b} \otimes \rho. \tag{2.27}$$

This equation can be simplified slightly by noting that if the function $h_p \in L^{1,\mathrm{loc}}(\Omega, \mathbb{R}^3)$ satisfies

$$\operatorname{rot} h_p = \rho, \tag{2.28}$$

then (2.27) is fulfilled with $\tilde{h}_p = \hat{b} \otimes h_p$. Taking this expression for $\tilde{h}_p$, inserting (2.26) into (2.24), and replacing (2.25) by (2.28), we obtain the boundary value problem for the displacement field u and the stress field T in an open set $\Omega \subseteq \mathbb{R}^3$ representing the material points of a viscoplastic body:

$$-\operatorname{div} T = 0, \tag{2.29}$$

$$T = D\big(\varepsilon(\nabla u) - \varepsilon(\hat{b} \otimes h_p)\big), \tag{2.30}$$

$$\operatorname{rot} h_p = \rho, \tag{2.31}$$

$$T\big|_{\partial\Omega} n_B = \gamma. \tag{2.32}$$

The elasticity tensor $D : \mathcal{S}^3 \to \mathcal{S}^3$ can be any linear, symmetric, positive definite mapping. The dislocation measure $\rho \in \mathcal{M}_d(\Omega)$ and the boundary data γ are given, and $\hat{b}$ is the unit vector in the direction of the Burgers vector.

We note that the splitting (2.26) of $\tilde{h}_e$ into a gradient field and a field $\tilde{h}_p$ satisfying $\operatorname{rot} \tilde{h}_p = -\operatorname{rot} \tilde{h}_e$ is not unique, since we can add the same gradient field to ∇u and $\tilde{h}_p$ and obtain a new splitting. This means, in particular, that if (u, T, h_p) is a solution of (2.29)–(2.32) and if $\Gamma \in L^{1,\mathrm{loc}}(\Omega, \mathbb{R})$, then we obtain another solution (u', T', h_p') by setting

$$u' = u + \hat{b}\Gamma, \quad T' = T, \quad h_p' = h_p + \nabla\Gamma,$$

since these equations imply $\operatorname{rot} h_p' = \operatorname{rot} h_p = \rho$ and

$$\varepsilon(\nabla u' - \hat{b} \otimes h_p') = \varepsilon(\nabla u + \hat{b} \otimes \nabla\Gamma - \hat{b} \otimes h_p - \hat{b} \otimes \nabla\Gamma) = \varepsilon(\nabla u - \hat{b} \otimes h_p).$$

3. The evolution equation for dislocation measures

Let $[0, T_e)$ be a time interval with $T_e > 0$. In the following we formulate an evolution equation for dislocation measures

$$\rho : [0, T_e) \to \mathcal{M}_d(\Omega)$$

depending on the time. We base this formulation on four principles:

(P1) By Definition 2.3, the dislocation measure $\rho(t)$ and the time derivative $\partial_t \rho(t)$ must be a rotation field. Therefore the evolution equation must be of the form

$$\partial_t \rho = \mathrm{rot}_x \alpha[T, \rho, b], \tag{3.1}$$

with a function $\alpha : C(\overline{\Omega}, \mathcal{S}^3) \times \mathcal{M}_d(\Omega) \times \mathbb{R}^3 \to \mathcal{M}(\Omega)$ to be determined.

(P2) There must exist a free energy $\psi(\varepsilon, h_p)$ and a flux $q(\varepsilon, u_t, h_p)$ of the free energy such that the Clausius–Duhem inequality holds:

$$\partial_t \psi + \mathrm{div}_x q \leq 0. \tag{3.2}$$

(P3) The evolution equation (3.1) must allow for solutions $t \to \rho_{\ell(t)}$, which are dislocation measures of dislocation curves, which move with driving force given by the Peach–Koehler force

$$F = \tau \times Tb. \tag{3.3}$$

(P4) Plastic deformation is volume conserving. This means that we must have

$$\mathrm{trace}\, \tilde{h}_p = \mathrm{trace}(\hat{b} \otimes h_p) = 0. \tag{3.4}$$

Condition (P2) is a consequence of the second law of thermodynamics, which requires that the Clausius–Duhem inequality be satisfied, which under the assumption of constant temperature takes the form (3.2); we refer to standard texts in thermodynamics, for example [**6**]. The derivation of the Peach–Koehler force can be found in [**5**, pp. 62–66]. It is in fact also related to the second law of thermodynamics. Condition (P4) is a standard requirement in plasticity theory. This condition is motivated by the fact that plastic deformation is caused by the movement of dislocations, which are faults in the crystal lattice. The movement of such faults does not change the volume of the body.

3.1. Conditions (P1) and (P2). We first discuss the consequences of (P1) and (P2). The combination of (2.29)–(2.32) with (3.1) yields the closed system of partial differential equations governing the evolution of the dislocation measure and the stress field in a viscoplastic body:

$$-\mathrm{div}_x T(x,t) = 0, \tag{3.5}$$

$$T(x,t) = D\Big(\varepsilon\big(\nabla_x u(x,t)\big) - \varepsilon\big(\hat{b} \otimes h_p(x,t)\big)\Big), \tag{3.6}$$

$$\mathrm{rot}_x h_p(x,t) = \rho(x,t), \tag{3.7}$$

$$\partial_t \rho(x,t) = \mathrm{rot}_x \alpha[T, \rho, b](x,t), \tag{3.8}$$

$$T(x,t) n_B(x,t) = \gamma(x,t), \qquad (x,t) \in \partial\Omega \times [0, T_e), \tag{3.9}$$

where the first four equations must hold for $(x,t) \in \Omega \times [0, T_e)$.

LEMMA 3.1. *(i) Let $\Omega \subseteq \mathbb{R}^3$ be an open, bounded, simply connected set. Then (u, T, h_p, ρ) satisfies the equations* (3.5)–(3.8) *if and only if there is a function $\Gamma : \Omega \times [0, T_e) \to \mathbb{R}$ such that (u, T, h_p, Γ) solves the equations*

$$-\mathrm{div}_x T = 0, \tag{3.10}$$

$$T = D\big(\varepsilon(\nabla_x u) - \varepsilon(\hat{b} \otimes h_p)\big), \tag{3.11}$$

$$\partial_t h_p = \alpha[T, \mathrm{rot}_x h_p, b] + \nabla_x \Gamma. \tag{3.12}$$

On the other hand, if (u, T, h_p, Γ) solves (3.10)–(3.12) *we define $\rho = \mathrm{rot}_x h_p$. Then (u, T, h_p, ρ) satisfies* (3.5)–(3.8).

(ii) Let the free energy and the flux be given by

$$\psi\big(\varepsilon(\nabla_x u), h_p\big) = \frac{1}{2}\big(D\varepsilon(\nabla_x u - \hat{b} \otimes h_p)\big) : \varepsilon(\nabla_x u - \hat{b} \otimes h_p), \tag{3.13}$$

$$q\big(\varepsilon, u_t.h_p, \Gamma\big) = -T(u_t - \hat{b}\Gamma). \tag{3.14}$$

Then the Clausius–Duhem inequality (3.2) *holds for every solution (u, T, h_p, Γ) of* (3.10) – (3.12) *if α satisfies for all points $(T, \rho) \in \mathcal{S}^3 \times \mathbb{R}^3$ the inequality*

$$(T\hat{b}) \cdot \alpha[T, \rho, b] \geq 0. \tag{3.15}$$

PROOF. Let (u, T, h_p, ρ) be a solution of (3.5)–(3.8). The combination of (3.7) and (3.8) yields

$$\mathrm{rot}_x\big(\partial_t h_p - \alpha\big) = \partial_t \rho - \mathrm{rot}_x \alpha = 0.$$

Since Ω is simply connected, this equation implies that $\partial_t h_p - \alpha$ is a gradient field. Consequently there is a function $\Gamma : \Omega \times [0, T_e) \to \mathbb{R}$ such that $\partial_t h_p - \alpha[T, \rho, b] = \nabla_x \Gamma$ holds. From this equation we obtain (3.12) if we use (3.7) to eliminate ρ in the argument of α. On the other hand, if (u, T, h_p, Γ) solves (3.10)–(3.12), then we obtain from (3.12) for $\rho = \mathrm{rot}_x h_p$ that

$$\partial_t \rho = \mathrm{rot}_x \partial_t h_p = \mathrm{rot}_x \alpha[T, \rho, b],$$

since $\mathrm{rot}_x \nabla_x \Gamma = 0$. This proves (i). To prove (ii) we infer from (3.13) and (3.11) that

$$\begin{aligned}\partial_t \psi\big(\varepsilon(\nabla_x u), h_p\big) &= \nabla_\varepsilon \psi : \varepsilon(\nabla_x u_t - \hat{b} \otimes \partial_t h_p) = T : \varepsilon(\nabla_x u_t - \hat{b} \otimes \partial_t h_p) \\ &= T : (\nabla_x u_t - \hat{b} \otimes \partial_t h_p) = T : \nabla_x u_t - (T\hat{b}) \cdot (\partial_t h_p),\end{aligned}$$

where we used several times that $T(x, t)$ is a symmetric matrix. (3.14) yields

$$\begin{aligned}\mathrm{div}_x\, q &= -\mathrm{div}_x\big(T(u_t - \hat{b}\Gamma)\big) \\ &= -(\mathrm{div}_x\, T^T) \cdot (u_t - \hat{b}\Gamma) - T^T : \nabla_x u_t + T^T : (\hat{b} \otimes \nabla_x \Gamma) \\ &= -T : \nabla_x u_t + (T\hat{b}) \cdot \nabla_x \Gamma,\end{aligned}$$

where we employed that $\mathrm{div}_x T^T = \mathrm{div}_x T = 0$, by (3.10). The combination of the last two equations with (3.12) and (3.15) results in

$$\frac{\partial}{\partial t}\psi + \mathrm{div}_x q = -(T\hat{b}) \cdot (\partial_t h_p - \nabla_x \Gamma) = -(T\hat{b}) \cdot \alpha \leq 0.$$

□

REMARK 3.2. This lemma shows that we are free to choose any function Γ in the evolution equation (3.12). This freedom is used in [**1**] to introduce an additional field variable to include dislocation nucleation. In our investigation we choose for simplicity $\Gamma = 0$, which avoids the unusual term $(T\hat{b})\Gamma$ in the free energy flux (3.14).

3.2. Condition (P3). Next we construct a function α, for which condition (P3) is satisfied. Let $\mathbb{S}^2$ denote the unit sphere in $\mathbb{R}^3$ and let $\tilde{\alpha} \in C(\mathbb{S}^2 \times \mathbb{R}^3, \mathbb{R}^3)$ be a given function. For $T \in C(\overline{\Omega}, \mathcal{S}^3)$, $\rho = \tau|\rho| \in \mathcal{M}_d(\Omega)$ and $b \in \mathbb{R}^3$ we set

$$\alpha[T, \rho, , b] = \tilde{\alpha}(\tau, \tau \times Tb) \times \tau\, |\rho|. \tag{3.16}$$

This defines a function $\alpha : C(\overline{\Omega}, \mathcal{S}^3) \times \mathcal{M}_d(\Omega) \times \mathbb{R}^3 \to \mathcal{M}(\Omega)$, since by Definition 2.3 we have $\tau = \frac{\rho}{|\rho|} \in C(\text{supp}\,|\rho|, \mathbb{R}^3)$, whence $\tilde{\alpha}(\tau, \tau \times Tb) \times \tau \in C(\text{supp}\,|\rho|, \mathbb{R}^3)$, which implies that the right-hand side of (3.16) is in $\mathcal{M}(\Omega)$, again by Definition 2.3. With the notation introduced in Definition 2.3 we write the right-hand side of (3.16) as $\tilde{\alpha}\big(\frac{\rho}{|\rho|}, \frac{\rho}{|\rho|} \times Tb\big) \times \rho$.

DEFINITION 3.3. Let α be defined by (3.16), let $b \in \mathbb{R}^3$, and let $T : \Omega \times [0, T_e) \to \mathcal{S}^3$ with $T(t) \in C(\overline{\Omega}, \mathcal{S}^3)$ be given. The time-dependent dislocation measure $\rho : [0, T_e) \to \mathcal{M}_d(\Omega)$ is a solution of

$$\partial_t \rho = \text{rot}_x \Big(\tilde{\alpha}\big(\frac{\rho}{|\rho|}, \frac{\rho}{|\rho|} \times Tb\big) \times \rho \Big), \tag{3.17}$$

if for all $\varphi \in C_0^\infty\big((0, T_e) \times \Omega, \mathbb{R}^3\big)$ the integrals in the following equation exist and satisfy

$$-\int_0^{T_e} \langle \rho(t), \partial_t \varphi(t) \rangle\, dt = \int_0^{T_e} \Big\langle \tilde{\alpha}\big(\frac{\rho}{|\rho|}, \frac{\rho}{|\rho|} \times Tb\big) \times \rho, \text{rot}_x \varphi(t) \Big\rangle\, dt.$$

As a consequence of the next theorem we see that condition (P3) is satisfied for the evolution equation (3.17). We need two definitions to state this theorem. Let the family $t \mapsto \ell(t)$ represent a moving dislocation curve with tangent vector $\tau(x, t)$ at the point $x \in \ell(t)$. By $\text{proj}_{\tau(x,t)^\perp}$ we denote the orthogonal projection

$$\text{proj}_{\tau(x,t)^\perp} : \mathbb{R}^3 \to H(x, t) = \{\xi \in \mathbb{R}^3 \mid \xi \cdot \tau(x, t) = 0\} \subseteq \mathbb{R}^3 \tag{3.18}$$

to the orthogonal space of $\tau(x, t)$. To define the normal velocity $v(x_0, t_0)$ of the dislocation curve at time t_0 at $x_0 \in \ell(t_0)$, let $x(t)$ be the intersection point of $\ell(t)$ with $H(x_0, t_0) + x_0$. Set

$$v(x_0, t_0) = \frac{d}{dt} x(t)\big|_{t=t_0}.$$

THEOREM 3.4. *Let α be defined by* (3.16) *with a given function $\tilde{\alpha} \in C(\mathbb{S}^2 \times \mathbb{R}^3, \mathbb{R}^3)$.*
(i) α satisfies the inequality (3.15) *if*

$$\xi \cdot \tilde{\alpha}(\tau, \xi) \geq 0, \quad \textit{for all } (\tau, \xi) \in \mathbb{S}^2 \times \mathbb{R}^3. \tag{3.19}$$

(ii) Assume that $T \in C\big([0, T_e] \times \overline{\Omega}, \mathcal{S}^3\big)$ is a given stress field and that $\rho_{\ell(t)} = \tau\, |\rho_{\ell(t)}| \in \mathcal{M}_d(\Omega)$ is the dislocation measure of a dislocation curve $\ell(t)$ to the Burgers vector $b \in \mathbb{R}^3$. For $x \in \ell(t)$ let $F(x, t) = \tau(x, t) \times T(x, t)b$ be the Peach–Koehler force. Then $\rho_{\ell(t)}$ solves the evolution equation (3.17), *if and only if for every $x \in \ell(t)$ the normal speed is*

$$v(x, t) = \text{proj}_{\tau(x,t)^\perp} \tilde{\alpha}\Big(\tau(x, t), F(x, t)\Big). \tag{3.20}$$

We must omit the proof of this theorem. (3.20) shows that $\tilde{\alpha}$ can be considered to be a constitutive function determining the relation between the normal velocity and the Peach–Koehler force F. Since the continuity of $\tilde{\alpha}$ and the inequality (3.19) imply $\tilde{\alpha}(\tau, 0) = 0$, the normal velocity is equal to zero if $F = 0$. Therefore we can consider F to be the driving force for the movement of $\ell(t)$. In this sense, condition (P3) is satisfied by the evolution equation (3.17). The evolution equation (3.17) was in principle derived in [**7**], following ideas in [**8**]; cf. also [**4**].

With the function α determined in (3.16), with $\Gamma = 0$ and with $\operatorname{rot}_x h_p = \tau\,|\operatorname{rot}_x h_p| \in \mathcal{M}_d(\Omega)$, the system (3.10)–(3.12), combined with the boundary condition (3.9), takes the form

$$-\operatorname{div}_x T = 0, \tag{3.21}$$

$$T = D\big(\varepsilon(\nabla_x u) - \varepsilon(\hat{b} \otimes h_p)\big), \tag{3.22}$$

$$\partial_t h_p = \tilde{\alpha}\big(\tau, \tau \times Tb\big) \times \operatorname{rot}_x h_p\,, \tag{3.23}$$

$$T_{|_{\partial\Omega}} n_B = \gamma, \qquad \text{on } \partial\Omega \times [0, T_e). \tag{3.24}$$

3.3. Condition (P4). To determine the consequences of condition (P4) we differentiate (3.4) with respect to t and use (3.23) to compute

$$0 = \operatorname{trace}(\hat{b} \otimes \partial_t h_p) = \hat{b} \cdot \partial_t h_p = \hat{b} \cdot (\tilde{\alpha} \times \tau)|\operatorname{rot} h_p| = 0.$$

Here we used that $\operatorname{rot}_x h_p = \tau|\operatorname{rot} h_p|$. From this equation we infer that

$$b \cdot \Big(\tilde{\alpha}\big(\tau(x,t), \tau(x,t) \times T(x,t)b\big) \times \tau(x,t)\Big) = 0 \tag{3.25}$$

must hold for all $(x,t) \in \Omega \times [0, T_e)$ if condition (P4) is satisfied. This equation has the following consequence for the movement of dislocation curves:

COROLLARY 3.5. *Let the stress field $T : \Omega \times [0, T_e) \to \mathcal{S}^3$ with $T(t) \in C(\overline{\Omega}, \mathcal{S}^3)$ be given, and let $\rho_{\ell(t)} = \tau\,|\rho_{\ell(t)}|$ be the dislocation measure of a moving dislocation curve $\ell(t)$. If $\rho_{\ell(t)}$ solves the evolution equation* (3.17) *and if the condition* (3.25) *holds for a point $x \in \ell(t)$ at which $\tau(x,t)$ is not parallel to b, then the normal speed $v(x,t)$ of the dislocation curve is parallel to the vector* $\operatorname{proj}_{\tau(x,t)^\perp} b$.

PROOF. τ and b are parallel if and only if $b \times \tau = 0$. If this product is not zero, then (3.25) implies that the vector $\tilde{\alpha}(\tau, \tau \times Tb)$ must lie in the plane spanned by τ and b. In this case the vectors $\operatorname{proj}_{\tau(x,t)^\perp} b$ and $\operatorname{proj}_{\tau(x,t)^\perp} \tilde{\alpha}(\tau, \tau \times Tb)$ lie on the same line, so they are parallel. Since by Theorem 3.4(ii) we have $v = \operatorname{proj}_{\tau(x,t)^\perp} \tilde{\alpha}(\tau, F)$, the corollary follows. □

REMARK 3.6. If the tangent vector $\tau(x,t)$ is parallel to the Burgers vector b, then $\ell(t)$ is called a screw dislocation at $x \in \ell(t)$. The condition that the volume is not changed by plastic deformation thus requires that if $\ell(t)$ is not a screw dislocation at $x \in \ell(t)$, then the normal speed $v(x,t)$ must be a linear combination of b and $\tau(x,t)$. Only for screw dislocations the direction of the normal speed is not determined by the Burgers vector. However, a dislocation curve can move freely in any direction only if it is a screw dislocation at every point, which means that it must be a straight dislocation line in direction of the Burgers vector, and if the movement is a parallel shift. As soon as the movement ceases to be a parallel shift, points appear on $\ell(t)$, at which τ and b are not parallel, restricting the direction of the movement of $\ell(t)$ at these points.

From (3.25) we see that in order to guarantee that condition (P4) is satisfied, the function $\tilde{\alpha}$ must be such that

$$b \cdot \big(\tilde{\alpha}(\tau,\xi) \times \tau\big) = 0,$$

for all $(\tau,\xi) \in \mathbb{S}^2 \times \mathbb{R}^3$. If b and τ are linearly independent, this holds if and only if $\tilde{\alpha}(\tau,\xi)$ belongs to the linear span of b and τ, hence there are real valued functions f_1 and f_2 such that $\tilde{\alpha}(\tau,\xi) = bf_1(\tau,\xi) + \tau f_2(\tau,\xi)$. This implies $\tilde{\alpha}(\tau,\xi) \times \tau = (b \times \tau) f_1(\tau,\xi)$, hence τf_2 does not contribute to the evolution equation and we can omit this term and construct $\tilde{\alpha}$ in the form

$$\tilde{\alpha}(\tau,\xi) = bf_1(\tau,\xi),$$

where the function f_1 must be chosen such that the inequality (3.19) holds. Under many possibilities a simple choice is

$$\tilde{\alpha}(\tau,\xi) = \frac{b}{|b \times \tau|} f\Big(\frac{b}{|b \times \tau|} \cdot \xi\Big), \tag{3.26}$$

with a function $f \in C(\mathbb{R},\mathbb{R})$ satisfying $r \cdot f(r) \geq 0$ for all $r \in \mathbb{R}$. We insert (3.26) into (3.17) and obtain the evolution equation

$$\partial_t \rho = \mathrm{rot}_x \left(f\Big(\frac{b \times \tau}{|b \times \tau|} \cdot Tb\Big) \frac{b \times \tau}{|b \times \tau|} |\rho| \right), \tag{3.27}$$

for $\rho = \tau|\rho| \in \mathcal{M}_d(\Omega)$. Here we used that $b \cdot (\tau \times Tb) = (b \times \tau) \cdot Tb$ to rewrite the argument of f. The choice of the $\tilde{\alpha}$ in (3.26) is justified by the following result:

THEOREM 3.7. *(i) For the function $\tilde{\alpha}$ defined in* (3.26) *the inequality* (3.19) *holds.*
(ii) Assume that $T \in C\big([0,T_e] \times \overline{\Omega}, \mathcal{S}^3\big)$ is a given stress field and that $\rho_{\ell(t)} = \tau\,|\rho_{\ell(t)}| \in \mathcal{M}_d(\Omega)$ is the dislocation measure of a dislocation curve $\ell(t)$ to the Burgers vector $b \in \mathbb{R}^3$. Let $F = \tau \times Tb$ be the Peach–Koehler force, and for $x \in \ell(t)$ denote the unit vector in the direction of $\mathrm{proj}_{\tau(x,t)^\perp} b$ *by*

$$b^{\tau^\perp}(x,t) = \frac{\mathrm{proj}_{\tau(x,t)^\perp} b}{|\mathrm{proj}_{\tau(x,t)^\perp} b|}.$$

Then $\rho_{\ell(t)}$ solves the evolution equation (3.27)*, if and only if $\ell(t)$ moves with the normal speed*

$$v(x,t) = f\big(b^{\tau^\perp}(x,t) \cdot F(x,t)\big)\, b^{\tau^\perp}(x,t). \tag{3.28}$$

(iii) Let (u,T,h_p) be a solution of the system

$$-\mathrm{div}_x T = 0, \tag{3.29}$$

$$T = D\big(\varepsilon(\nabla_x u) - \varepsilon(\hat{b} \otimes h_p)\big), \tag{3.30}$$

$$\partial_t h_p = f\Big(\frac{b \times \tau}{|b \times \tau|} \cdot Tb\Big) \frac{b \times \tau}{|b \times \tau|} |\mathrm{rot}_x h_p| \tag{3.31}$$

in $\Omega \times [0,T_e)$, where $\mathrm{rot}_x h_p = \tau|\mathrm{rot}_x h_p|$. If at time $t = 0$ we have $b \cdot h_p(x,0) = 0$ for all $x \in \Omega$, then condition (3.4) *holds for all $(x,t) \in \Omega \times [0,T_e)$.*

REMARK 3.8. (3.28) shows that under the choice of $\tilde{\alpha}$ given in (3.26) the driving force for the movement of dislocation curves is the component $b^{\tau^\perp} \cdot F$ of the Peach–Koehler force in the direction of the vector $\mathrm{proj}_{\tau^\perp} b$. This justifies (3.26). The equation (3.31) is obtained by inserting the function $\tilde{\alpha}$ from (3.26) into (3.23), using again that $b \cdot (\tau \times Tb) = (b \times \tau) \cdot Tb$.

PROOF. For the proof of (i) note that the assumption $r \cdot f(r) \geq 0$ implies

$$\xi \cdot \tilde{\alpha}(\tau, \xi) = \frac{\xi \cdot b}{|b \times \tau|} f\Big(\frac{b \cdot \xi}{|b \times \tau|}\Big) \geq 0, \quad (\tau, \xi) \in \mathbb{S}^2 \times \mathbb{R}^3.$$

To verify (ii) note that since the evolution equation (3.27) is obtained from (3.17) by insertion of the function $\tilde{\alpha}$ defined in (3.26), we obtain from Theorem 3.4(ii) that $\rho_{\ell(t)}$ is a solution of (3.27) if and only if the dislocation curve moves with normal velocity

$$v = \mathrm{proj}_{\tau^\perp}\Big(\frac{b}{|b \times \tau|} f\Big(\frac{b}{|b \times \tau|} \cdot F\Big)\Big) = f\Big(\frac{b}{|b \times \tau|} \cdot F\Big) \frac{\mathrm{proj}_{\tau^\perp} b}{|b \times \tau|}. \tag{3.32}$$

Since $|b \times \tau| = |\mathrm{proj}_{\tau^\perp} b|$, we have $\frac{\mathrm{proj}_{\tau^\perp} b}{|b \times \tau|} = b^{\tau^\perp}$. Noting that $F = \tau \times Tb$ is orthogonal to τ, we thus conclude

$$\frac{1}{|b \times \tau|} b \cdot F = \frac{1}{|b \times \tau|} (\mathrm{proj}_{\tau^\perp} b) \cdot F = b^{\tau^\perp} \cdot F\,.$$

(3.28) is obtained by insertion of this equation into (3.32).

To prove (iii) observe that the right-hand side of (3.31) is orthogonal to the vector b, which implies that $\partial_t(\hat{b} \cdot h_p) = \hat{b} \cdot \partial_t h_p = 0$. This equation and the assumption $b \cdot h_p(x, 0) = 0$ together imply $\hat{b} \cdot h_p(x,t) = 0$ for all $(x,t) \in \Omega \times [0, T_e)$, from which we conclude that $\mathrm{trace}(\hat{b} \otimes h_p) = \hat{b} \cdot h_p = 0$. This is (3.4). □

It is possible to simplify (3.31) slightly by working with a suitable coordinate system. To show this we assume that T is a given stress field and that h_p is a solution of (3.31) satisfying $\hat{b} \cdot h_p = \mathrm{trace}(\hat{b} \otimes h_p) = 0$. Choose the Cartesian coordinates (x_1, x_2, x_3) such that the x_2–axis points into the direction of the Burgers vector b and let $h_p = (h_{p1}, h_{p2}, h_{p3})$ be the components of h_p in this coordinate system. Since $\hat{b} = (0,1,0)$, the equation $\hat{b} \cdot h_p = 0$ is equivalent to $h_{p2} = 0$. Using this property of h_p and noting that $\tau = \frac{\mathrm{rot}_x h_p}{|\mathrm{rot}_x h_p|}$, we obtain by a computation that in these coordinates

$$\mathrm{rot}_x h_p = \big(\partial_{x_2} h_{p3}, \partial_{x_3} h_{p1} - \partial_{x_1} h_{p3}, -\partial_{x_2} h_{p1}\big), \qquad \hat{b} \times \mathrm{rot}_x h_p = -\partial_{x_2} h_p.$$

$$\frac{b \times \tau}{|b \times \tau|} = \frac{b \times \mathrm{rot}_x h_p}{|b \times \mathrm{rot}_x h_p|} = -\frac{\partial_{x_2} h_p}{|\partial_{x_2} h_p|}, \qquad |\mathrm{rot}_x h_p| = \sqrt{|\partial_{x_2} h_p|^2 + (\mathrm{rot}_2 h_p)^2},$$

where $\mathrm{rot}_2 h_p = \partial_{x_3} h_{p1} - \partial_{x_1} h_{p3}$. With these expressions (3.31) takes the form

$$\partial_t h_p = -f\Big(-\frac{\partial_{x_2} h_p}{|\partial_{x_2} h_p|} \cdot Tb\Big) \frac{\partial_{x_2} h_p}{|\partial_{x_2} h_p|} \sqrt{|\partial_{x_2} h_p|^2 + (\mathrm{rot}_2 h_p)^2}. \tag{3.33}$$

This is a system of two equations, since h_{p2} vanishes identically.

4. Dislocations moving in slip planes

We finally consider the situation, where every dislocation curve is contained in a plane and moves within this plane. Different dislocation curves can be contained in the same plane or in parallel planes. The planes are called slip planes.

Let $g \in \mathbb{R}^3$ be a unit vector, which is normal to all slip planes. Every tangent vector τ to a dislocation curve in a slip plane is orthogonal to the vector g. Since the dislocation curve moves in the slip plane, the normal velocity v of the dislocation curve must also be orthogonal to g. By Theorem 3.7(ii), the vector v is parallel to the vector $\mathrm{proj}_{\tau^\perp} b$, which is the orthogonal projection of the Burgers vector b to the orthogonal space of τ. This implies that b itself is orthogonal to g. Therefore the vectors τ and b span the two-dimensional subspace parallel to the slip planes. The vector $b \times \tau$ is normal to this subspace; hence

$$\frac{b \times \tau}{|b \times \tau|} = \pm g.$$

We insert this equation into (3.27) and obtain the evolution equation

$$\partial_t \rho = \operatorname{rot}_x\Big(\pm g\, f\big(\pm g \cdot Tb\big)|\rho|\Big). \tag{4.1}$$

The change of sign can occur at points, where the dislocation curve is a screw dislocation, that is, at points where τ and b are parallel. We have not accounted for this situation in the definition (3.26) of the constitutive function $\tilde{\alpha}$, which is used to obtain (3.31). From Corollary 3.5 we know that at these points the direction of the normal velocity is not determined by τ and b, and a separate definition of $\tilde{\alpha}$ should be given for such points. However, under our present assumption that dislocations move in slip planes, we can obtain an evolution equation valid everywhere by simply taking the $+$–signs in (4.1). We do not have to discuss the justification of this choice in general, but can simply assume in the following that f is chosen as an odd function, in which case the $\pm$–signs in (4.1) can obviously be replaced by $+$–signs.

COROLLARY 4.1. *Assume that* $T \in C\big([0, T_e] \times \overline{\Omega}, \mathcal{S}^3\big)$ *is a given stress field and that* $\rho_{\ell(t)} = \tau\, |\rho_{\ell(t)}| \in \mathcal{M}_d(\Omega)$ *is the dislocation measure of a dislocation curve* $\ell(t)$ *to the Burgers vector* $b \in \mathbb{R}^3$, *which for all* $t \in [0, T_e)$ *is contained in a plane normal to the vector* g. *Then* $\rho_{\ell(t)}$ *is a solution of the evolution equation*

$$\partial_t \rho = \operatorname{rot}_x\Big(g\, f\big(g \cdot Tb\big)|\rho|\Big), \tag{4.2}$$

if and only if it moves with the normal speed given in (3.28).

This corollary follows immediately from Theorem 3.7(ii) and the construction of (4.2) given above.

We can simplify (4.2) by introducing a Cartesian coordinate system with the x_2–axis pointing in the direction of b and the x_3–axis pointing in the direction of g. For a scalar function w we then have $\operatorname{rot}_x(wg) = -(g \times \nabla_x)w = \nabla_g^{\perp} w$, with $\nabla_g^{\perp} = (\partial_{x_2}, -\partial_{x_1}, 0)^T$. Equation (4.2) thus becomes

$$\partial_t \rho = \nabla_g^{\perp}\Big(f(g \cdot Tb)|\rho|\Big). \tag{4.3}$$

The equation for h_p corresponding to the evolution equation (4.2) is obtained by inserting the vector g for $\frac{b\times\tau}{|b\times\tau|}$ in (3.31). We obtain

$$\partial_t h_p = g\, f\big(g \cdot Tb\big)|\operatorname{rot}_x h_p|. \tag{4.4}$$

From this equation we see that $\partial_t h_p$ is a scalar multiple of the vector g. We can assume that this is also true for the initial data $h_p(x, 0)$, from which we conclude that there is a function $\varepsilon_p : \Omega \times [0, T_e) \to \mathbb{R}$ such that for all $(x, t) \in \Omega \times [0, T_e)$

$$h_p(x, t) = \varepsilon_p(x, t)\, g. \tag{4.5}$$

With the coordinate system chosen as above and with the tangential gradient $\nabla_g = (\partial_{x_1}, \partial_{x_2}, 0)^T$ of the slip plane we obtain from (4.5) that

$$\operatorname{rot}_x h_p = \operatorname{rot}_x(\varepsilon_p g) = \nabla_g^{\perp} \varepsilon_p, \qquad |\operatorname{rot}_x(\varepsilon_p g)| = |\nabla_g^{\perp} \varepsilon_p| = |\nabla_g \varepsilon_p|. \tag{4.6}$$

We insert the second expression and (4.5) into (4.4) and obtain the evolution equation for ε_p:

$$\partial_t \varepsilon_p = f(g \cdot Tb)\, |\nabla_g \varepsilon_p|.$$

For dislocations moving in slip planes this equation replaces the third equation in the system (3.29)–(3.31). The model equations for plastic deformation of materials with dislocations moving in slip planes thus take the form

$$
\begin{aligned}
&(4.7) && -\mathrm{div}_x T &&= 0,\\
&(4.8) && T &&= D\big(\varepsilon(\nabla_x u) - m\varepsilon_p\big),\\
&(4.9) && \partial_t \varepsilon_p &&= f(|b|\, m : T)\, |\nabla_g \varepsilon_p|,\\
&(4.10) && T_{|_{\partial\Omega}} n_B &&= \gamma, \qquad \text{on } \partial\Omega \times [0, T_e),
\end{aligned}
$$

where m is defined in (1.4) and where we used that $\varepsilon(\hat{b} \otimes h_p) = \varepsilon(\hat{b} \otimes \varepsilon_p g) = m\varepsilon_p$ and $g \cdot Tb = |b|\, m : T$. Of course, this system has to be supplemented by a boundary condition and an initial condition for the function ε_p. We do not discuss this here.

Remark 4.2. Let (u, T, ε_p) be a solution of (4.7)–(4.9). By (3.7) the dislocation measure to this solution is given by $\rho = \mathrm{rot}_x h_p \in \mathcal{M}_d(\Omega)$, from which we obtain by (4.6) that $\rho = \tau|\rho| = \nabla_g^{\perp} \varepsilon_p$. Since $\nabla_g^{\perp} \varepsilon_p$ is orthogonal to $g = (0, 0, 1)^T$, we conclude that also the tangential vector field τ of the dislocation measure ρ is orthogonal to g, and this implies that the dislocation curves corresponding to solutions of the system (4.7) – (4.9) lie in slip planes $x_3 = c$ and that the function $(x_1, x_2) \mapsto \varepsilon_p(x_1, x_2, c, t) : \mathbb{R}^2 \to \mathbb{R}$ plays the role of a stream function of the dislocation measure ρ on the slip plane.

If $\rho = \rho_\ell = \tau|\rho_\ell|$ is the dislocation measure of a dislocation curve, then ε_p has a jump along the dislocation curve. By definition of ρ_ℓ in (2.19), the height of the jump is equal to the absolute value $|b|$ of the Burgers vector. If ρ is a dislocation density, then ε_p is smooth and the level curves of $(x_1, x_2) \mapsto \varepsilon_p(x_1, x_2, c, t)$ are the averaged dislocation curves.

Summary

The systems (3.29), (3.30), (3.33), and (4.7)–(4.9), together with the corresponding evolution equations (3.27) and (4.3), respectively, satisfy the conditions (P1) – (P4). The simpler system (4.7)–(4.9) models dislocation curves which lie in slip planes. If this restrictive assumption is not made, then the more complicated equations (3.29), (3.30), and (3.33) must be used.

The system (4.7)–(4.9) differs from the standard plasticity model (1.1)–(1.3) only by the term $|\nabla_g \varepsilon_p|$. From (4.6) we see that $|\rho| = |\nabla_g \varepsilon_p|$, hence $|\nabla_g \varepsilon_p|$ is the total variation measure to the dislocation measure and represents the "absolute value" of the dislocation density. If this density is high throughout the material, then $|\nabla_g \varepsilon_p|$ can be replaced approximately by a constant. In this case the system (4.7)–(4.9) reduces to the standard plasticity model (1.1)–(1.3). However, if this assumption is not valid, then (4.7)–(4.9) should be used as model for plastic deformation.

Acknowledgment

The author thanks the reviewer for helpful remarks.

References

[1] A. Acharya, A. Beaudoin, and R. Miller. *New perspectives in plasticity theory. Dislocation nucleation, waves, and partial continuity of plastic strain rate.* Math. Mech. Solids **13** (2008), 292–315. MR2412008 (2009f:74018)

[2] A. Acharya. *Microcanonical entropy and mesoscale dislocation mechanics and plasicity.* J. Elast. **104** (2011), 23–44. MR2806903 (2012g:74007)

[3] H.-D. Alber and K. Chełmiński. *Quasi-static problems in viscoplasticity theory II: Models with nonlinear hardening.* Math. Models Meth. Appl. Sci. **17** (2007), no. 2, 189–213. MR2292355 (2007k:74050)

[4] N. Fox. *A continuum theory of dislocations for single crystals.* IMA J. Appl. Math. **2** (1966), 285–298.

[5] A. M. Kosevich. *Crystal dislocations and the theory of elasticity.* Dislocations in Solids (F.R.N. Nabarro, ed.), North-Holland 1979, 33–141.

[6] I. Müller. *Thermodynamics.* Boston, Pitman 1985.

[7] T. Mura. *Continuous distribution of moving dislocations.* Philosophical Magazine **89** (1963), 843–857.

[8] E. Kröner. *Kontinuumstheorie der Versetzungen und Eigenspannungen.* Ergebnisse der Angewandten Mathematik **5**, Springer-Verlag, Berlin-Göttingen-Heidelberg, 1958. MR0095615 (20:2117)

Fachbereich Mathematik, Technische Universität Darmstadt, Schlossgartenstr. 7, 64289 Darmstadt, Germany

E-mail address: alber@mathematik.tu-darmstadt.de

Amer. Math. Soc. Transl.
Volume **233**, 2014

Spectral Problems of High-Order Recurrences

A. I. Aptekarev

ABSTRACT. We discuss the spectral properties of a simple nonsymmetric difference operator of higher order associated with the three-term, (p+1)-order recurrence

$$Q_{n+1} = zQ_n - a_{n-p+1}Q_{n-p},\ p \in \mathbb{N},\ n \geq p, \quad Q_j(z) = z^j,\ j = 0, 1, \dots, p.$$

For the positive coefficients a_k there are p positive measures ("spectral" measures) supported on a starlike set S of $p+1$ rays emanating from the origin on which the Q_n satisfies a multiple orthogonality condition involving these p measures. We consider compact and l_1 perturbations of the constant coefficients case, $a_n := a > 0$, $n = 1, 2, \dots$. In the l_1 class we relate analytic properties of the vector of the "resolvent" functions with the spectral curve of the constant coefficients case. The connection between spectral and scattering data using of BVPs on the Riemann surface will be observed.

Finally we survey examples and applications to various problems in special functions theory, combinatorics, discrete integrable systems, and normal ensembles of random matrices.

1. Introduction

The following nonsymmetric $(p+2)$-banded (lower Hessenberg) infinite matrix

$$(1.1)\qquad A := \begin{pmatrix} a_{0,0} & a_{0,1} & 0 & 0 & 0 & \dots & \dots \\ a_{1,0} & a_{1,1} & a_{1,2} & 0 & 0 & \dots & \dots \\ \dots & \dots & \dots & \dots & \dots & \dots & \dots \\ a_{p,0} & a_{p,1} & a_{p,2} & \dots & a_{p,p+1} & 0 & \dots \\ 0 & a_{p+1,1} & a_{p+1,2} & \dots & a_{p+1,p+1} & a_{p+1,p+2} & \dots \\ \dots & \dots & \dots & \dots & \dots & \dots & \dots \\ & & & & & & \end{pmatrix}$$

defines in the space l^2 an operator that we will denote by the same symbol A. Let $\{e_n\}_0^\infty$ be the standard basis of l^2. At first the operator A is defined on the linear

---

2010 *Mathematics Subject Classification.* Primary 41A20, 41A21, 41A10; Secondary 47B99, 30B70.

Key words and phrases. Higher-order recurrences, polynomials, multiple orthogonality, Hermite-Padé approximants, vector continued fractions, spectral measures, difference operators, Weyl's theorem, Nikishin systems, Szegő function, matrix models, and normal matrices ensembles.

The author was supported by grants RFBR-13-01-12430 OFIm, RFBR-14-01-00604, and Program 1 DMS RAS.

space span(e_j) by

$$\begin{aligned} Ae_0 &= a_{0,0}e_0 + a_{1,0}e_1 + \cdots + a_{p,0}e_p, \\ Ae_k &= a_{k-1,k}e_{k-1} + a_{k,k}e_k + \cdots + a_{k+p,k}e_{k+p}, \ k \geq 1 \end{aligned}$$

and then we take its closure, which always exists. The following functions are called *resolvent* or *Weyl functions* of the operator A :

$$f_j(z) := (R_z e_{j-1}, e_0), \quad j = 1, 2, ..., p, \tag{1.2}$$

where $R_z := (zI - A)^{-1}$ is the resolvent operator of A, and $(\cdot, \cdot)$ denotes the inner product in the space l^2 of complex sequences. We remark that the operator A is a bounded operator if and only if $\sup_{i,j} |a_{i,j}| < \infty$. In this case, the Weyl functions (1.2) admit the following power (Neumann) series expansion at infinity:

$$f_j(z) = \sum_{k=0}^{\infty} \frac{(A^k e_{j-1}, e_0)}{z^{k+1}}, \quad |z| > ||A||. \tag{1.3}$$

For a fixed $j = 1, 2, ..., p$ the quantities $f_k^{(j)} := (A^k e_{j-1}, e_0)$ are called the *moments of operator* A associated with the Weyl function $f_j(z)$. The data f, $(f_k)_{k \geq 0}$ are spectral characteristics of the operator L. The coefficients (1.1) of the matrix A are operator data.

During the past thirty years there has been a stable interest in the study of the spectral problems for the high order difference operator. These problems are intimately related with the theory of Hermite-Padé rational approximants and the multiple orthogonal polynomials. Recently a connection was discovered between these notions and the theory of random matrices.

In the present survey paper we mainly deal with a special case of (1.1) associated with the following nonsymmetric difference operator with only 2 nonzero diagonals :

$$L := \begin{pmatrix} 0 & 1 & 0 & 0 & 0 & \dots & \dots \\ 0 & 0 & 1 & 0 & 0 & \dots & \dots \\ 0 & 0 & 0 & 1 & 0 & \dots & \dots \\ \dots & \dots & \dots & \dots & \dots & \dots & \dots \\ a_1 & 0 & 0 & 0 & \dots & \dots & \dots \\ 0 & a_2 & 0 & 0 & 0 & \dots & \dots \\ 0 & 0 & a_3 & 0 & 0 & 0 & \dots \\ \dots & \dots & \dots & \dots & \dots & \dots & \dots \end{pmatrix}. \tag{1.4}$$

We choose this model case for two reasons. The first reason is a relation with the Stieltjes approach to the continued fractions theory and its very natural generalization to the vector case. The second reason is an application of the spectral properties of the operator (1.4) to the important models of the ensembles of normal random matrices.

The special structure of the matrix (1.4) implies that for the moments of this operator we have for fixed $j = 1, 2, \dots, p$ and $k \geq 0$

$$f_n^{(j)} = (L^n e_{j-1}, e_0) = 0, \quad n \neq (p+1)k + j - 1, \qquad f^{(j)}_{(p+1)k+j-1} =: S_{k,j} \neq 0, \tag{1.5}$$

and the Weyl functions (1.2) are invariant under rotation on the angle $2\pi i/(p+1)$. Thus, we can define a new system of functions (or formal power series) $S_1, S_2, \dots, S_p$

by

$$f_j(z) = z^{p-j+1} S_j(z^{p+1}), \qquad S_j(z) = \sum_{n=0}^{\infty} \frac{S_{n,j}}{z^{n+1}}, \qquad j = 1, 2, ..., p. \tag{1.6}$$

In the next two sections we consider this new set of the spectral data $\{S_j(z)\}_{j=1}^p$ from the point of view of the remarkable memoir of Stieltjes [**1**] on continued fractions.

2. The Stieltjes continued fraction ($p = 1$) and formula for moments

We start with the case of a second order difference operator when $p = 1$ in (1.1) and (1.4):

$$L := \begin{pmatrix} 0 & 1 & 0 & \cdots & & & & \\ a_1 & 0 & 1 & 0 & \cdots & & & \\ 0 & a_2 & 0 & 1 & 0 & \cdots & & \\ 0 & 0 & a_3 & 0 & 1 & 0 & \cdots & \\ 0 & 0 & 0 & \ddots & & \ddots & & \cdots \end{pmatrix}. \tag{2.1}$$

For this case one can associate with the sequence of the coefficients $(a_n)_{n\geq 1}$ (the spectral data) a Stieltjes continued fraction as in [**1**] and [**2**] (see also[**44**] and [**20**]):

$$S(z) = \cfrac{1}{z - \cfrac{a_1}{1 - \cfrac{a_2}{z - \cfrac{a_3}{\ddots}}}} = \frac{1|}{|z} + \frac{-a_1|}{|1} + \frac{-a_2|}{|z} + \cdots. \tag{2.2}$$

The classical theorem of Stieltjes states that the positivity of the coefficients a_n is equivalent to the existence of the positive measure $d\bar{S}(x)$ on $\mathbb{R}_+$ such that

$$a_k > 0 \iff S(z) = \int_0^{\infty} \frac{d\bar{S}(x)}{z - x} = \frac{1}{z} + \frac{S_1}{z^2} + \frac{S_2}{z^3} + \cdots, \qquad z \notin [0, +\infty). \tag{2.3}$$

One can easily check the relations (1.6) between the Stieltjes function $S(z)$ in (2.2) and the Weyl function f of the operator (1.5)

$$\begin{cases} f(z) = zS(z^2), & z \notin (-\infty, +\infty), \\ f_{2k} = S_k, \quad f_{2k+1} = 0, & k \geq 0, \\ d\mu(x) = d\bar{S}(x^2), & x \in (-\infty, +\infty), \end{cases} \tag{2.4}$$

where the positive measure μ

$$a_k > 0 \iff f(z) = \int_{-\infty}^{+\infty} \frac{d\mu(x)}{z - x}, \quad f_k = \int_{-\infty}^{+\infty} x^k d\mu(x), \tag{2.5}$$

is the spectral measure of the operator L in (2.1). (For the definitions and connections between these notions, see for example [**3**], [**4**], [**5**], [**31**], and [**7**].) The data f, $(f_k)_{k\geq 0}$ and μ (when it exists) are spectral characteristics of the operator L. Thus for L, the direct-inverse problems

$$L \longleftrightarrow \{f, \{f_k\}, d\mu\}$$

reduce to the connections between the classical notions from the Stieltjes theory

$$(a_i)_{i\geq 1} \longleftrightarrow \{S(z),\ \{S_k\}_{k\geq 0},\ d\bar{S}(x)\}.$$

In the theories of continued fractions, orthogonal polynomials and rational approximants, whose modern state has been developed from the classical works of P. L. Chebychev, T. Stieltjes and A. A. Markov, there is a variety of methods for solving the direct-inverse problems. We use as the basis of our approach the representation of the moments S_k in terms of multisums involving the coefficients of the continued fraction (2.2):

$$S_n = a_1 \sum_{i_2=1}^{2} a_{i_2} \sum_{i_3=1}^{i_2+1} a_{i_3} \cdots \sum_{i_n=1}^{i_{n-1}+1} a_{i_n}, \qquad n \geq 1, \quad S_0 = 1\,. \tag{2.6}$$

Stieltjes in [**1**, Chapitre II, §7]) and [**2**] suggested recurrence relations to determine moments S_k in terms of the coefficients $(a_i)_{i\geq 1}$ of the continued fraction (2.2). In [**8**] we noted that the Stieltjes recurrences take the explicit form (2.6). We did not see this simple formula in the monographs [**44**], [**20**] and other papers on the subject of Stieltjes continued fractions. However, more complicated multisums formulae (such as the Euler-Minding formula) occur. See [**36**] and [**23**].

Equation (2.6) gives a straightforward procedure for the solution of the direct spectral problem. Conversely, the procedure of expanding a function in its Stieltjes continued fraction solves, from the algorithmic point of view, the inverse problem.

To demonstrate the capabilities of the calculus of the multisums we present a proof of (2.6).

PROPOSITION 2.1. *The formal power series expansion* (2.3) *of the Stieltjes continued fraction* (2.2) *has representation* (2.6) *for its coefficients.*

PROOF. We check the implication (2.6) $\Rightarrow$ (2.2). We define an associated multisums for (2.6):

$$S_0^{(1)} = 1, \quad S_n^{(1)} = a_2 \sum_{i_2=2}^{3} a_{i_2} \sum_{i_3=2}^{i_2+1} a_{i_3} \cdots \sum_{i_n=2}^{i_{n-1}+1} a_{i_n}, \quad n \geq 1\,, \quad S^{(1)}(z) = \sum_{n=0}^{\infty} \frac{S_n^{(1)}}{z^{n+1}}\,.$$

Expanding the last sum in the representation (2.6) for S_{n+1}

$$S_{n+1} = a_1 [S_n + a_2 a_1 \sum_{i_3=1}^{2} a_{i_2} \cdots \sum_{i_n=1}^{i_{n-1}+1} a_{i_n} + a_2(a_2+a_3)a_1 \sum_{i_4=1}^{2} a_{i_2} \cdots \sum_{i_n=1}^{i_{n-1}+1} a_{i_n}$$

$$+ \cdots + \left(a_2 \sum_{i_2=2}^{3} a_{i_2} \sum_{i_3=2}^{i_2+1} a_{i_3} \cdots \sum_{i_p=2}^{i_{p-1}+1} a_{i_p} \right) \left(a_1 \sum_{i_{p+2}=1}^{2} a_{i_{p+2}} \cdots \sum_{i_n=1}^{i_{n-1}+1} a_{i_n} \right) + \ldots],$$

we obtain

$$S_{n+1} = a_1 \sum_{\nu=0}^{n} S_\nu^{(1)} S_{n-\nu}\,,$$

and for the corresponding power series

$$S(z) = a_1\, S(z) S^{(1)}(z) + \frac{1}{z}\,.$$

Eventually the subsequent application of this transformation

$$S(z) = \frac{1}{z - z\, a_1\, S^{(1)}(z)}\,,$$

gives us the expansion of $S(z)$ in the Stieltjes continued fraction (2.2). We remark also that the uniqueness of the regular expansion (2.2) implies formula (2.6) for the moments of the Stieltjes functions. □

3. Vector ($p > 1$) spectral measure – direct and inverse problem

An extension of the Stieltjes approach for the high order discrete operator (1.4) has been developed in [**10**]. The following generalization of Proposition 2.1 holds true.

THEOREM 3.1 (cf. [**10**]). *The nonzero moments (see* (1.5)–(1.6)*) of the operator L* (1.4) *with the coefficients $(a_i)_{i\geq 1}$ have the following multisums representation:*

$$S_{0,j} = 1, \qquad S_{k,j} = \sum_{i_1=1}^{j} a_{i_1} \sum_{i_2=1}^{i_1+p} a_{i_2} \sum_{i_3=1}^{i_2+p} a_{i_3} \cdots \sum_{i_k=1}^{i_{k-1}+p} a_{i_k}, \quad j = 1, 2, \dots, p. \tag{3.1}$$

There is a natural sufficient (but not necessary for $p > 1$) condition for the existence of the vector of the Stieltjes measures that generates the vector of moments (1.6).

THEOREM 3.2 (cf. [**10**]). *If in the multisums representation* (3.1) *of $S_{n,j}$ the condition*

$$a_n > 0, \qquad n \geq 1, \tag{3.2}$$

is fulfilled, then there exists a system of positive measures $\bar{S}_j$, $j = 1, \cdots, p$, with supports contained in $[0, +\infty)$, such that the $S_{n,j}$ are the moments of these measures

$$S_{n,j} = \int_0^{+\infty} x^n \, d\bar{S}_j(x) \ . \tag{3.3}$$

Thus for the vector of Weyl functions (see (1.3) and (1.6)) we have an integral representation

$$f_j(z) = \int_S \frac{d\mu_j(t)}{z-t}, \qquad d\mu_j(t) = d\bar{S}_j(t^{(p+1)}), \qquad j = 1, 2, \dots, p\,. \tag{3.4}$$

The vector $(\mu_1, \mu_2, \dots, \mu_p)$ with common support of the components on the starlike set

$$S := \bigcup_{k=0}^{p} \exp(2\pi i k/(p+1)) \times [0, \infty), \tag{3.5}$$

can be called a spectral vector measure of the operator L – (1.4). Compare with (2.3) and (2.5).

Vector continued fractions were introduced by Jacobi and Perron to solve the problem of characterizing irrationalities of order higher than two [**36**]. See also [**15**]. The Stieltjes continued fraction (2.2) can be extended for the vector case.

Let a, b be two vectors of $\mathbb{C}^p$. Then the vector product and quotient are defined by the following formulas

$$a \cdot b = \operatorname{diag}(a) \cdot b = (a_1 b_1, a_2 b_2, \cdots, a_p b_p)^t, \qquad \frac{1}{a} = (\frac{1}{a_p}, \frac{a_1}{a_p}, \cdots, \frac{a_{p-1}}{a_p}).$$

THEOREM 3.3 (cf. [**10**]). *Let $(a_n)_{n\in\mathbb{N}}$ be a sequence of nonzero terms. If the vector of formal power series $\mathcal{S}(z) = (S_1, \cdots, S_p)$ is defined by (1.6) and (3.1), then*

(3.6)
$$\mathcal{S}(z) = \frac{(1,\cdots,1)}{(0,\cdots 0,z)+}\ \frac{(1,\cdots,1,-a_1)}{(0,\cdots 0,1)+\cdots}\cdots\frac{(1,\cdots,1,-a_p)}{(0,\cdots 0,1)+}\ \frac{(1,\cdots,1,-a_{p+1})}{(0,\cdots 0,z)+\cdots}\ .$$

Equations (3.1) and (3.6) are the solutions to the direct and inverse spectral problems. We also present a continued fraction for the vector of the Weyl functions (1.3) and (1.6):

(3.7)
$$\mathcal{F}(z) := (f_1,\cdots,f_p)(z) = \frac{(1,\cdots,1)}{(0,\cdots,0,z)+\cdots}\cdots\frac{(1,\cdots,1,-a_1)}{(0,\cdots,0,z)+}\ \frac{(1,\cdots,1,-a_2)}{(0,\cdots,0,z)+\cdots}\ .$$

4. Hermite-Padé approximants and multiple orthogonal polynomials

Again we recall the classical case ($p = 1$) of a second order difference operator (2.1). The Weyl functions (see (1.3) and (1.6)) have the following Jacobi continued fraction expansion

$$f(z) = \frac{1}{z-}\ \frac{a_1}{z-}\ \frac{a_2}{z-\cdots}\ . \tag{4.1}$$

It is known (see [**4**], [**44**]) that the n-th convergent of this continued fraction is a rational function, diagonal Padé approximant $\pi_n := (P_n/Q_n)$ with numerator P_n and denominator Q_n which are defined by the recurrences

$$Y_{n+1} = zY_n - a_{n-p+1}Y_{n-p}, \quad n \ge p\ , \tag{4.2}$$

for $p = 1$ and with initial conditions $Q_0 = 1,\ Q_1 = z,\quad P_0 = 0,\ P_1 = 1$.

For the case under consideration ($p \ge 1$) the n-th convergent of the vector continued fraction (3.7) is a vector of rational functions with common denominator $\vec{\pi}_n := \left(\frac{P_n^{(1)}}{Q_n}, \frac{P_n^{(2)}}{Q_n}, \ldots, \frac{P_n^{(p)}}{Q_n}\right)$ which are defined by the recurrences (4.2) with initial conditions

$$Q_0 \;=\; 1,\ \ Q_1 \;=\; z,\ \ Q_2 \;=\; z^2,\ \ \ldots\ \ Q_p \;=\; z^p\ , \tag{4.3}$$

and

$$P_k^{(j)} = \frac{1}{(k-j+1)_{p-k+j}}\frac{d^{p-k+j}}{z^{p-k+j}}z^p\ , \quad k = 0,\ldots,p,\ j = 1,\ldots,p\,.$$

For this case (see [**29**], [**30**], [**9**], and [**10**]) the above vector $\vec{\pi}_n$ is the Hermite-Padé approximant of the system $\mathcal{F}$ of the Weyl functions (3.7). We recall that by definition a Hermite-Padé approximant of multi-index $\vec{n} = (n_1, n_2, \ldots, n_p)$, $n_j \in \mathbb{N}$, $j = 1, 2, \ldots, p$, the numerators $P^{(1)}$, $P^{(2)}$, ..., $P^{(p)}$ and denominator Q of the (simultaneous) Hermite-Padé approximant to $\vec{f} = (f_1, \ldots, f_p)$ associated with $\vec{n}$ satisfy to the following relations:

$$Q(z)f_j(z) - P^{(j)}(z) = \frac{c_j}{z^{n_j+1}} + \cdots\ , \qquad j = 1, 2, \ldots, p\ ,$$

where $Q \not\equiv 0$, $\deg Q \le n$, and $n = n_1 + n_2 + \cdots + n_p$. Thus the n-th ($n = kp + d$) convergent of the vector continued fraction (3.7) is the Hermite-Padé approximant of $\mathcal{F}$ corresponding to the multi-index $\vec{n} = (k+1, k+1, \ldots, k+1, k, \ldots, k)$.

If our operator (1.4) has the vector spectral measure (3.4), then common denominator Q_n satisfies multiple orthogonality relations and the following *Favard-type* theorem holds true.

THEOREM 4.1 (cf. [**10**]). *If* $a_n > 0$ *for* $n \geq 1$ *in* (4.2), *then there exists a system of positive measures* $\{\mu_j\}_{j=1}^p$ *such that the polynomials* Q_n *defined by* (4.2)–(4.3) *satisfy the following multiple-orthogonal (MO) relations:*

$$\begin{aligned} &\textstyle\int_S Q_n(t)t^r d\mu_j(t) = 0, \quad r = 0,1,\dots,k, \quad j = 1,2,\dots,d\ , \\ &\textstyle\int_S Q_n(t)t^r d\mu_j(t) = 0, \quad r = 0,1,\dots,k-1, \quad j = d+1, d+2,\dots,p\ , \end{aligned} \tag{4.4}$$

where $n = kp + d$, $0 \leq d \leq (p-1)$, *and* S *is the starlike set* (3.5). *Moreover, the measures* μ_j, $j = 1,\dots,p$, *have common support which is a subset of* S *and they are invariant under rotations in the angles* $2\pi k/(p+1)$, $k = 1,2,\dots,p$.

The asymptotics of the MO polynomials can be treated as a *scattering data*. Establishing a relation between the spectral and scattering data is an important problem in spectral analysis.

5. Recurrences with constant coefficients and compact perturbation

The polynomials U_n defined by the recurrence (see (4.2)–(4.3)) with the constant coefficients

$$U_n(z) := Q_n(z)\ : \qquad a_n := a > 0\,, \qquad n = 1,2,\dots, \tag{5.1}$$

were studied in several papers (see [**24**], [**28**], and [**11**]). These polynomials generalize the classical Tchebyshev polynomials of the second kind (for $p = 1$). The vector of the rational Hermite-Padé approximants for the system $\mathcal{F}$ of the the Weyl functions (see (1.3) and (1.6)) has the form

$$\overrightarrow{\pi_n} = \left(\frac{U_{n-1}}{U_n}, \frac{U_{n-2}}{U_n}, \dots, \frac{U_{n-p}}{U_n}\right)$$

and the vector of the Weyl functions has the explicit representation

$$\mathcal{F} = \left(\frac{1}{w_0}, \frac{1}{w_0^2}, \dots, \frac{1}{w_0^p}\right), \tag{5.2}$$

where w_0 is the main ("physical") branch of the algebraic function $w(z) := \{w_j(z)\}_{j=0}^p$:

$$w^{p+1} - zw^p + a = 0, \qquad |w_0(z)| \geq |w_1(z)| \geq \dots \geq |w_p(z)|, \quad \forall z \in \mathbb{C}. \tag{5.3}$$

The genus of f is equal to zero. The finite branch points of this function are

$$\mathcal{A} = \{\alpha_0,\ \alpha_1 := \varepsilon_{(p+1)}^{(1)}\alpha_0,\ \alpha_2 := \varepsilon_{(p+1)}^{(2)}\alpha_0, \quad \dots,\ \alpha_p := \varepsilon_{(p+1)}^{(p)}\alpha_0\}\,, \tag{5.4}$$

with $\alpha_0 = [(p+1)/p^{p/(p+1)}]a^{1/(p+1)}$, $\varepsilon_{(p+1)}^{(k)} := \exp(k2\pi i/(p+1))$, $k = 1,\dots,p$ and at infinity

$$w_0(z) = z + \cdots\,, \qquad w_j(z) = \mathcal{O}(1/z^{1/p})\,, \qquad j = 1,\dots,p, \quad z \to \infty. \tag{5.5}$$

The structure of the Riemann surface $\mathfrak{R}$ of w is $w_j \in \mathrm{H}(\Omega_j)$, $j = 1,\dots,p$, where

$$\begin{aligned} &\Omega_0 := \mathbb{C}\backslash S_0, && \Omega_1 := \begin{cases} \mathbb{C}\backslash S_0, & p = 1 \\ \mathbb{C}\backslash\{S_0 \cup s\}, & p > 1 \end{cases}, \\ &\Omega_k := \mathbb{C}\backslash\{s \cup S\}, \quad (k < p), && \Omega_p := \begin{cases} \mathbb{C}\backslash s, & p = 2l \\ \mathbb{C}\backslash\ S, & p = 2l+1 \end{cases}, \end{aligned} \tag{5.6}$$

and

(5.7)
$$S_0 := [0,\alpha]\cup[0,\alpha_1]\cup[0,\alpha_2]\ldots\cup[0,\alpha_p]\,, \qquad s := \{z : z = \xi\exp(\frac{i\pi}{p+1}),\ \xi\in S\}.$$

Using the notation π for the projection of $\mathfrak{R}$ on $\mathbb{C}$ and π_k^{-1} for its inverse branches, we have

$$\mathfrak{R} = \overline{\cup_{j=0}^{p}\mathfrak{R}_j},\quad \mathfrak{R}_0 := \pi_0^{-1}(\Omega_0\cup\{\infty\}),\ \mathfrak{R}_j := \pi_j^{-1}(\Omega_j),\ j = 1,\ldots,p\,. \tag{5.8}$$

The case of the compact perturbations

$$\lim_{n\to\infty} a_n = a(>0)\,. \tag{5.9}$$

was studied in [**11**], where a *Weyl-type* theorem characterizing the support of the measures of the multiple orthogonality $\{\mu_j\}_{j=1}^{p}$ has been proven.

THEOREM 5.1 (cf. [**11**]). *Suppose that condition* (5.9) *is fulfilled. Then the measures* μ_j, $j = 1,2,\ldots,p$, *of multiple orthogonality* (4.4) *for the polynomials* ((4.2)–(4.3)) *have a common support consisting of the set* S_0 *in* (5.7) *plus a countable set of mass points that lie on the set* S *in* (3.5) *with only accumulation points at the ends of the intervals of* S_0.

6. Direct problems in l_1 perturbation class

The next step is to impose an extra condition on the speed of convergence in (5.9); namely,

$$\sum_{n=1}^{\infty}|\,a_n - a\,| < \infty, \qquad a_n > 0\,, \tag{6.1}$$

in order to obtain the strong asymptotics for the multiple orthogonal polynomials Q_n and further investigate the analytic properties of the measures $\{\mu_j\}_{j=1}^{p}$. In other words, we deal with *direct scattering and spectral problems* in the l^1 *perturbation* class (6.1). In the classical case $p = 1$ these problems were quite well understood (see for example [**27**] and [**42**]). A development for $p \geq 1$ was perfomed in [**11**].

The starting point in [**11**] is a *comparison equation* (Gelfand-Levitan-Marchenko equation) which connects the perturbed polynomials Q_n defined by (4.2)-(4.3) with the polynomials U_n defined by the recurrence with constant coefficients $a_n = a$ in (5.1). Namely,

$$Q_n = U_n + \sum_{k=0}^{n-1}(a - a_{k+1})U_{n-p-1-k}Q_k\,, \qquad U_{-1} = U_{-2} = \ldots = U_{-p} = 0. \tag{6.2}$$

Using bounds for the polynomials U_n, this equation allows us to obtain corresponding bounds for the polynomials Q_n and to introduce the series

$$\Phi_0(z) \;:=\; 1 + \sum_{k=0}^{\infty}\frac{a - a_{k+1}}{w_0^{p+1}}\cdot\frac{Q_k(z)}{w_0^k(z)}$$
$$\Phi_j(z) \;:=\; 1 + \sum_{k=0}^{\infty}\frac{a - a_{k+1}}{w_0^{p+1}}\cdot\frac{P_k^{(j)}(z)}{w_0^{k-j}(z)},\quad j = 1,2,\ldots,p\,,$$

which converges on compact subsets of $\Omega = \mathbb{C}\backslash S_0$ as well as on "both sides" of S_0, except for the set of end points $\mathcal{A}$. The following strong asymptotics hold true.

THEOREM 6.1 (cf. [**11**]). *For the polynomials $Q_n(z)$ defined by (4.2)–(4.3) with coefficients satisfying (6.1), we have*

$$\lim_{n\to\infty}\frac{Q_n(z)}{w_0^n(z)} = F_0(z)\ ,\quad z\in K\subset\Omega_0\,,\qquad F_0(z):=\frac{w_0(z)^p\ \Phi_0(z)}{\prod_{k=1}^p(w_0-w_k)} \tag{6.3}$$

uniformly on compact sets in Ω_0, and we have uniformly on compact subsets $\widetilde{K}$ of $S_0\setminus\mathcal{A}$

$$\frac{Q_n(t)}{|w_0(t)|^n} = \left(\frac{w_{0,+}(t)}{|w_0(t)|}\right)^n F_{0,+}(t)+\left(\frac{w_{0,-}(t)}{|w_0(t)|}\right)^n F_{0,-}(t)+o(1)\,,\qquad t\in\widetilde{K}\subset S_0\backslash\mathcal{A}. \tag{6.4}$$

We remark that the function F_0 introduced in (6.3) plays the same role as the Szegő function for the classical case $p=1$.

The asymptotics obtained in [**11**] (including strong asymptotics of $P_n^{(j)}$) imply the convergence of the Hermite-Padé approximants to the vector of resolvent (Weyl) functions $\{f_j\}_{j=1}^p$, and the jumps of these functions give us the measures $\{\mu_j\}_{j=1}^p$ in (4.4), which are absolutely continuous on S_0 and have a form for $j=1,2,\ldots,p$:

$$d\mu_j(t)=\rho_j(t)|dt|,\qquad \rho_j(t)=\left[\frac{1}{w_0(t)^j}\frac{\Phi_j(t)}{\Phi_0(t)}\right]^+-\left[\frac{1}{w_0(t)^j}\frac{\Phi_j(t)}{\Phi_0(t)}\right]^-\,,\qquad t\in S_0\,. \tag{6.5}$$

Thus we can say that in the l^1 perturbation class formulae (6.3)–(6.4) solve the direct scattering problem and (6.5) solves the direct spectral problem. In the next section we proceed with an analysis of the scattering and spectral data.

7. BVP connection of the spectral and scattering data

To prepare to state and solve the inverse problems, we have to first analyze the solution of the direct problems in the l_1 perturbation class in order to find their appropriate characterization. Then these classes of the spectral and scattering data will be used as input for the inverse problems and our goal will be to prove that the obtained operator data are in the l_1 perturbation class.

For this characterization we are going to use some boundary values problems (BVPs) for vector-analytic functions. These BVPs were developed in the asymptotical theory of the Hermite-Padé approximants and multiple orthogonal polynomials. In the following subsections we introduce two such BVPs. The BVP for spectral data (vector of the Weyl functions (1.2)–(1.6) and the spectral measure (6.5)) is related to the so-called *Nikishin systems*. For the Szegő function in (6.3) we get BVP on the Riemann surface of the algebraic function (5.3). Comparing these two BVPs we obtain an immediate connection between spectral and scattering data.

7.1. Nikishin systems on the stars. Let $\{S_{k-1}\}_{k=1}^p$ be a set of contours (3.5), (5.7)

$$S_{2l-1}=s\,,\quad S_{2l}=S\,,\quad l=1,\ldots,[\tfrac{p}{2}]\,. \tag{7.1}$$

Let $\{\sigma_j\}_{j=1}^p$ be a system of locally integrable functions on $\{S_{j-1}\}_{j=1}^p$, respectively, which are non-vanishing and have constant argument, so that

$$\operatorname{supp}\sigma_j=S_{j-1}\ ,\quad j=1,\ldots,p.$$

We say that the system of functions

$$(7.2) \qquad \{f_j\}\,, \quad f_j \in H(\overline{\mathbb{C}}\backslash S_0)\,, \quad j = 1, \ldots, p\,,$$

forms a *Nikishin system with respect to the system of weights* $\{\sigma_k\}_{k=1}^p$ *on* $\{S_{k-1}\}_{k=1}^p$, if $\{f_j\}$ have locally integrable boundary values on S_0, such that

$$f_{j+} - f_{j-}\Big|_{S_0} = \sigma_1 \cdot f_j^{(2)}\,, \quad j = 1, \ldots, p\,,$$

where

$$f_1^{(2)} \equiv 1\,, \quad f_j^{(2)} \in H(\overline{\mathbb{C}}\backslash S_1)\,, \quad j = 2, \ldots, p\,,$$

and the functions $\{f_j^{(2)}\}_{j=2}^p$ in their turn form the Nikishin system with respect to $\{\sigma_k\}_{k=2}^p$ on $\{S_{k-1}\}_{k=2}^p$.

We see that the Nikishin system starting from $f_j^{(1)} := f_j$, $j = 1, \ldots, p$, defines inductively a hierarchy of analytic functions

$$(7.3) \qquad \{f_j^{(k)}\}\,, \quad k = 1, \ldots, p\,, \quad j = k, \ldots, p\,,$$

by means of analytic continuation of their jumps

$$(7.4) \qquad f_{j+}^{(k)} - f_{j-}^{(k)}\Big|_{S_{k-1}} = \sigma_k \cdot f_j^{(k+1)}\,,$$

from S_{k-1} to the whole domain of their holomorphicity:

$$(7.5) \qquad f_k^{(k+1)} \equiv 1\,, \quad f_j^{(k+1)} \in H(\overline{\mathbb{C}}\backslash S_k)\,, \quad k = 1, \ldots, p\,, \quad j = k, \ldots, p\,.$$

We remark that we use the notion of Nikishin system with respect to the system of sets $\{S_{k-1}\}_{k=1}^p$, which includes the unbounded sets, and weight functions $\{\sigma_k\}_{k=2}^p$ that are not necessarily globally integrable on $\{S_{k-1}\}_{k=2}^p$. This differs from the usual definition of the Nikishin system by means of Cauchy integrals on a system of intervals of the real axis (see [**32**], [**33**], and [**14**]). We have

THEOREM 7.1 (cf. [**11**]). *The system of the Weyl functions ((1.3) and (1.6)) for the operator (1.4) with coefficients from the l_1 perturbation class (6.1) forms the Nikishin system (see (7.3)–(7.5)) with respect to a system of weights $\{\sigma_k\}_{k=1}^p$.*

For this case in [**11**] some expressions for the weights $\{\sigma_k\}_{k=1}^p$ are presented. For the case of the constant coefficients (5.1) the expressions have a simple form:

$$(7.6) \qquad \sigma_k := \frac{1}{w_{k-1,+}} - \frac{1}{w_{k-1,-}} \text{ on } \quad S_{k-1}\,, \quad k = 1, \ldots, p$$

and in (7.3) for the Nikishin functions hierarchy $f_j^{(k+1)} \in H(\overline{\mathbb{C}}\backslash S_k)$, $j = k + 1, \ldots, p$,

$$(7.7) \quad f_j^{(k+1)} = \sum_{\nu_0=k-1}^{j-1} \frac{1}{w_0^{j-1-\nu_0}} \sum_{\nu_1=k-2}^{\nu_0-1} \frac{1}{w_1^{\nu_0-1-\nu_1}} \cdots \sum_{\nu_{k-1}=0}^{\nu_{k-2}-1} \frac{1}{w_{k-1}^{\nu_{k-2}-1-\nu_{k-1}} w_k^{\nu_{k-1}}}\,.$$

7.2. BVP on $\mathfrak{R}$ for the Szego functions. Given a set of locally integrable weight functions

$$\sigma_l \quad \text{on} \quad S_{l-1}\,, \quad l = 1, 2, \ldots, p\,,$$

we set

$$\omega_l := (w_{l-1,+} - w_{l-1,-})\Big|_{S_{l-1}}\,, \quad l = 1, 2, \ldots, p\,. \tag{7.8}$$

Consider the Riemann surface $\mathfrak{R}$ (see (5.8)) and a closed contour $\partial\mathfrak{R}_{l-1,l}$ in $\mathbb{C}^2$ separating sheets $\mathfrak{R}_j$ and $\mathfrak{R}_k$. We assume that the contour $\partial\mathfrak{R}_{j,k}$ is orientated such that sheet $\mathfrak{R}_j$ lies to the left $(+)$ side. We have

$$\pi(\partial\mathfrak{R}_{j-1,j}) \;=\; S_{(j-1)+} \cup S_{(j-1)-}\,, \quad j = 1, \ldots, p. \tag{7.9}$$

We define on each such contour the function

$$\overset{\circ}{\sigma}_l := \begin{cases} \sigma_l\omega_l & \text{on} \quad S_{l-1,-} \\ \sigma_l\omega_l & \text{on} \quad S_{l-1,+} \end{cases}\,, \tag{7.10}$$

lift the values of $\overset{\circ}{\sigma}_l$ to $\partial\mathfrak{R}_{l-1,l}$ in accordance with (7.9) and (5.8), and formulate the following BVP on $\mathfrak{R}$:

We find a piecewise holomorphic function $\mathcal{F}$ on $\mathfrak{R}$ with locally integrable boundary values such that

$$\begin{aligned} &1)\ \mathcal{F} \in H\left(\mathfrak{R}\setminus\bigcup_{l=1}^{p}\partial\mathfrak{R}_{l-1,l}\right)\,,\ \exists\,\mathcal{F}_\pm \in L^1_{loc}\left(\bigcup_{l=1}^{p}\partial\mathfrak{R}_{l-1,l}\right)\,,\\ &2)\ \mathcal{F}_+ = \mathcal{F}_-\,\frac{1}{\overset{\circ}{\sigma}_l} \quad \text{on} \quad \partial\mathfrak{R}_{l-1,l}\,,\ l = 1, \ldots, p\,,\\ &3)\ \mathcal{F}(\infty^{(0)})\,\mathcal{F}(\infty^{(1)})\cdots\mathcal{F}(\infty^{(p)}) = 1\,. \end{aligned} \tag{7.11}$$

THEOREM 7.2. *Let $\{\sigma_k\}_{k=1}^p$ be a system of weights generating the Nikishin system of the Weyl functions in Theorem* 7.1. *Then the Szegő function F_0, which stands in the asymptotics* (6.3), *is represented by the solution of the BVP* (7.11) *on $\mathfrak{R}$ with jumps* (7.10), *forming by $\{\sigma_k\}_{k=1}^p$; namely*

$$F_0 = \mathcal{F}\Big|_{\mathfrak{R}_0}\,. \tag{7.12}$$

7.3. Algebraic curve and vector equilibrium for the limiting zero distribution. The weak limit $(n \to \infty)$ of the zero counting measure ν_{Q_n} of the MO polynomials $Q_n(z)$:

$$\nu_{Q_n}(z) = \frac{1}{n}\sum_{k=1}^{n}\delta(z - z_{k,n}) \overset{*}{\longrightarrow} \lambda(z)\,, \qquad Q_n(z) = \prod_{k=1}^{n}(z - z_{k,n})\,, \tag{7.13}$$

plays an important role in various applications. Taking from Theorem 6.1 the main term of the asymptotics $(n \to \infty)$ of the polynomials $Q_n(z)$ defined in (6.3) we have

$$|Q_n(z)|^{1/n} \to |w_0(z)|\,, \qquad \frac{w_0'(z)}{w_0(z)} =: h_0(z)\,, \qquad h_0(z) = \int_{S_0}\frac{d\lambda(t)}{z - t}. \tag{7.14}$$

Here for the Cauchy transform h_0 of the limiting zero distribution measure λ we have from (5.3)–(5.4) an algebraic equation

$$(p\,zh\,+\,1)^p\,(zh\,-\,1) \;=\; a\,(p+1)^{1-p}\,h^{p+1}\,, \tag{7.15}$$

and at infinity as $z \to \infty$

(7.16)
$$h := \{h_j\}_{j=0}^p : \quad h_0(z) = -1/z + \cdots , \qquad h_j(z) = 1/(pz) + \cdots , \quad j = 1, \dots, p.$$

The function h is a rational function on the Riemann surface (see (5.6)–(5.8)) of order $(p+1)$ with poles at the finite branch points $\mathcal{A}$ and zeros at infinity as in (7.16). Other branches $\{h_j\}_{j=1}^p$ can be presented as a Cauchy transform of charges supported by the starlike sets from (7.1):

$$\text{(7.17)} \qquad h_j(z) = \int_{S_j} \frac{d\lambda_{j+1}(t)}{z-t} - \int_{S_{j-1}} \frac{d\lambda_j(t)}{z-t}, \qquad h_p(z) = -\int_{S_{p-1}} \frac{d\lambda_p(t)}{z-t},$$

where measures $\lambda_j > 0$, $|\lambda_j| = \frac{1}{2}$, $j = 2, \dots, p$, together with the measure $\lambda_1 := \lambda$ from (7.14), provide a unique solution of a *vector potential equilibrium* problem for a vector of measures

(7.18)
$$\vec{\mu} = (\mu_1, \dots, \mu_p) : \quad \operatorname{supp} \mu_1 \subset S_0, |\mu_1| = 1, \quad \operatorname{supp} \mu_j \subset S_{j-1}, |\mu_j| = \frac{1}{2}, j = 2, \dots, p,$$

and the energy functional $I(\vec{\mu})$

(7.19)
$$I(\vec{\mu}) = \sum_{i=1}^p \sum_{j=1}^p d_{i,j} I(\mu_i, \mu_j), \qquad I(\mu_i, \mu_j) = \int_{\Delta_i} \int_{\Delta_j} \log \frac{1}{|x-t|} \, d\mu_i(x) \, d\mu_j(t),$$

with the interaction matrix $D = (d_{i,j})_{i,j=1}^p$ of the form

(7.20)
$$d_{j,j} = 1, \; j = 1, \dots, p, \quad d_{j,j-1} = d_{j-1,j} = \frac{1}{2}, \; j = 2, \dots, p, \quad d_{j,i} = 0, \; |i-j| > 1.$$

The extremal vector measure $\vec{\lambda} := (\lambda_1, \dots, \lambda_p)$ minimizing the energy functional (7.19) possesses the equilibrium for the *vector potential* $\vec{U}^{\vec{\lambda}} = (U_1^{\vec{\lambda}}, \dots, U_p^{\vec{\lambda}})$

(7.21)
$$U_j^{\vec{\lambda}}(x) = \sum_{i=1}^p d_{i,j} V^{\lambda_i}(x) \begin{cases} = \kappa_j, & x \in \operatorname{supp} \lambda_j =: S_j^* \\ \geq \kappa_j, & x \in S_j \setminus S_j^* \end{cases}, \quad V^{\lambda}(z) = \int \log \frac{1}{|z-t|} \, d\lambda(t),$$

where uniquely defined by (7.21) constants κ_j are called equilibrium constants. We note that, for the vector potential problem with matrix of interaction (7.20), with normalization of the measures as in (7.18) and satisfying the condition $S_j \cap S_{j-1} = \{0\}$, $j = 2, \dots, p$ (such problems are named after Nikishin), we have $S_j = S_j^*$ (see [**33**]). Moreover, the occasion of $\infty \in S_j$, $j = 1, \dots, p-1$ in (7.18) implies $\kappa_j = 0$, $j = 2, \dots, p$ in (7.21).

8. Examples and applications

8.1. Explicit solutions and special functions. In [**10**] for a set of the Stieltjes spectral measures (see (3.3)-(3.4)) on $[0, 1]$

$$\text{(8.1)} \qquad d\bar{S}_j(x) = x^{\alpha_j}(1-x)^{\alpha_0} dx, \qquad \alpha_j > -1, \; \alpha_i - \alpha_j \notin Z, \quad j = 1, 2, \dots, p,$$

explicit expressions for the coefficients of the operator (1.4) have been obtained for p=2. Here we present them for the values of the parameters $\alpha_0 = \alpha_1 = 0$ and

$\alpha_2 = \alpha$:

$$a_{6k+1} = \frac{(2k+1)(2k+1+\alpha)(k+1)}{(3k+1)(3k+2)(3k+1+\alpha)}, \; a_{6k+2} = \frac{(2k+1+\alpha)(2k+1)(k+\alpha)}{(3k+1+\alpha)(3k+2+\alpha)(3k+2)},$$

$$a_{6k+3} = \frac{(2k+2)(2k+1)(k+1-\alpha)}{(3k+2)(3k+3)(3k+2+\alpha)}, \; a_{6k+4} = \frac{(2k+2+\alpha)(2k+2)(k+1+\alpha)}{(3k+2+\alpha)(3k+3+\alpha)(3k+3)},$$

$$a_{6k+5} = \frac{(2k+2+\alpha_2)(2k+2)(k+1)}{(3k+3+\alpha_2)(3k+3)(3k+4)}, \; a_{6k} = \frac{(2k+1)(2k)k}{(3k+\alpha_2)(3k+1+\alpha_2)(3k+1)}.$$

Multiple orthogonal polynomials (MOPs) associated with the system (8.1) were investigated in [**37**].

(8.2)

$$\int \bar{Q}_{\vec{n}}(z) z^{\nu} w_j(z)\, dz = 0\,, \quad \nu = 0, \ldots, n_j - 1\,, \quad j = 1, \ldots, p\,, \quad \deg \bar{Q}_{\vec{n}} = \sum_{j=1}^{p} n_j\,.$$

In [**12**] MOPs with respect to classical weight functions, which are solutions of Pearson's differential equation with polynomial coefficients, were studied in detail. Here we present an example from [**12**] (see also [**41**]) of MOPs with respect to the system of Laguerre weights

(8.3)

$$d\bar{S}_j(x) = w_j(x)\, dx = x^{\alpha_j} \exp(\beta x)\, dx\,, \qquad \beta < 0\,, \quad \alpha_1, \alpha_2 > -1\,, \quad \alpha_1 - \alpha_2 \notin \mathbb{Z}\,.$$

The recurrences for the monic polynomials (see (8.2)–(8.3)) when $p = 2$, $\alpha_1 = 0\,, \beta = -1$ are:

$$\begin{aligned} x\bar{Q}_{(n_1,n_2)}(x) =& \bar{Q}_{(n_1+1,n_2)}(x) + u_{\vec{n}}\, \bar{Q}_{(n_1,n_2)}(x) + v_{\vec{n}}\, \bar{Q}_{(n_1,n_2-1)}(x) \\ &+ w_{\vec{n}}\, \bar{Q}_{(n_1-1,n_2-1)}(x)\,, \end{aligned}$$

$$u_{\vec{n}} \;=\; (2n_1 + n_2 + 1)\,, \quad v_{\vec{n}} \;=\; (n_1^2 + n_2^2 + n_1 n_2 + n_2\alpha_2)\,, \quad w_{\vec{n}} \;=\; n_1^2(n_1 - \alpha_2)\,.$$

8.2. Connection with combinatorics. If we put $a_n := 1$ in equation (2.6) for the moments of the classical (p=1) Stieltjes continued fraction, then we get a sequence of numbers:

$$S_0 = 1, \quad S_1 = 1, \quad S_2 = 2, \quad S_3 = 5, \quad S_4 = 14, \quad S_5 = 42, \quad \ldots\,.$$

These numbers are called Catalan numbers, and they are probably the most famous ones in combinatorics. They appeared in the works of Euler and Segner in the second half of the eighteenth century, and now there are more than fifty different types of combinatorial structures that are enumerated by the Catalan numbers (see [**26**]). One of the models is arrangements of parentheses. For each moment of the discreet time $t = 1, 2 \ldots k$, you may open one parenthesis $\sharp[op|_k] \le 1$ or close any number of the parentheses not exceeding the number of the already opened ones $\sharp[cl|_k] \le \sum_{j=1}^{k} \sharp[op|_j]$. Thus the Catalan numbers S_k express how many parenthesis arrangements you will get at the moment $t = k$. Here is the example of all possible arrangements for $t = 3$:

$$((()));\quad (()());\quad (())();\quad ()(());\quad ()()()\,.$$

Let us change the model, allowing at each moment $t = k$ to open up to $p \ge 1$ parentheses: $\sharp[op|_k] \le p$, $\sharp[cl|_k] \le \sum_{j=1}^{k} \sharp[op|_j]$. Then total number of arrangements will be represented by the sum $\sum_{j=1}^{p} S_{k,j}$ of the vector moments (3.1) for

$a_n := 1$. We note that the expression of the generating function (see (5.2)–(5.3)) and many other explicit and asymptotical formulae make the analytic approach to this generalization of the Catalan numbers very promising.

8.3. Discrete KdV equation. The following semi-infinite system of nonlinear differential equations with unknown functions $a_n(t)$ was investigated in [**21**], [**22**], [**35**], and [**10**]:

(8.4)

$$\dot{a}_n = a_n \left(\sum_{j=1}^{p} a_{n+j} - \sum_{j=1}^{p} a_{n-j} \right), \qquad a_0 = \ldots = a_{-p+1} = 0, \quad a_n(0) = a_n^0, \quad n \geq 1.$$

A similar system was studied in [**38**]. We associate with the system (8.4) the operator L defined in (1.1)–(1.4). The system (8.4) can be written (formally) in the Lax pair form

$$\dot{L} = LA - AL, \qquad A = (L^{p+1})_- := L^{p+1} - (L^{p+1})_{diag} - (L^{p+1})_+ .$$

If $0 < a_n(0) < M$, $n \geq 1$, then the operator $L(0)$ has the Stieltjes vector spectral measure (3.3) supported on a compact $K \subset \mathbb{R}_+$, and this measure has a simple evolution (see [**10**]):

$$d\bar{S}_j(x;t) = \frac{e^{xt} d\bar{S}_j(x;0)}{\int e^{xt} d\bar{S}_j(x;0)} \qquad \Leftrightarrow \qquad \dot{S}_{n,j} = S_{n+1,j} - S_{1,j} S_{n,j}, \quad j = 1, \ldots, p,$$

where $d\bar{S}_j(x;t)$ remains to be a positive measures $supp\,(\bar{S}_j) \subseteq K$ at any time $t \in [0, +\infty)$. It leads to the existence of a global, analytic (in t) solution of the system (8.4) for any positive and bounded initial data.

8.4. Multiple Orthogonal Polynomial ensembles. An MOP ensemble is a random vector variable with probability density function on $\mathbb{R}^n$:

$$\mathcal{P}(x_1, \ldots, x_n) = \frac{1}{Z_n} \prod_{j>i} (x_j - x_i) \det \left[\varphi_i(x_j) \right]_{i,j=1,\ldots,n}, \tag{8.5}$$

where $\text{span}\{\varphi_1, ..., \varphi_n\} = \text{span}\{x^k w_j(x) \mid k = 0, ..., n_j - 1,\ j = 1, ..., p,\ \sum_{j=1}^{p} n_j = n\}$, and Z_n is the normalization constant. The MOP ensemble is generated by the p weights $w_1, \ldots, w_p$ and $\vec{n} = (n_1, \ldots, n_p)$. The average characteristic polynomial

$$Q_{\vec{n}}(z) = \mathbb{E} \left[\prod_{j=1} (z - x_j) \right]$$

is an MOP with respect to the set $d\mu_j(x) = w_j(x)dx$. For more details, see the survey in [**13**].

During the last decade MOP ensembles have appeared in a natural way in certain models of random matrices and non-intersecting paths. For some ensembles of matrices the joint probability density for the eigenvalues of the random matrix can be presented, as in (8.5). This was first observed in [**16**] for the random matrix model with an external source matrix A

$$(H_n, \mu): \qquad \mu(dM) := \frac{1}{\tilde{Z}_n} e^{-N\,\mathrm{Tr}(V(M) - AM)} dM, \qquad M \in H_n, \tag{8.6}$$

on the space of $n \times n$ Hermitian matrix H_n. Here V is a function, N is the scaling parameter, and $\tilde{Z}_n$ is the normalization constant. Without loss of generality we consider

$$A = \operatorname{diag}(\underbrace{a_1, \ldots, a_1}_{n_1}, \cdots, \underbrace{a_p, \ldots, a_p}_{n_p}), \quad \sum_{j=1}^{p} n_j = n, \quad a_i \in \mathbb{R}, \quad i = 1, \ldots, p.$$

Then the joint probability density for the eigenvalues of (8.6) will be the probability density of a multiple orthogonal ensemble (8.5) generated by the p weight functions

$$w_j(x) = e^{-(V(x) - a_j x)}, \qquad j = 1, \ldots, p$$

and the multi-index $(n_1, \ldots, n_p)$. For interesting large n asymptotics in the random matrix model with external source one replaces V by nV and A by nA in (8.6).

The Gaussian case $V(M) = \frac{1}{2}M^2$ corresponds to the weights

$$\{w_j(x) = e^{-n(\frac{1}{2}x^2 - a_j x)}\}_{j=1}^{p}.$$

The MOPs for these weights are known as multiple Hermite polynomials see [**12**]. This case, with a very special external source $p = 2$, $a_1 = a$, $a_2 = -a$, was analyzed in [**17, 6, 18**]. The limiting ($n \to \infty$) mean density (7.13) of eigenvalues is given by

$$\rho(x) = \frac{1}{\pi} |\Im \xi(x)|, \quad x \in \mathbb{R}, \qquad \xi^3 - x\xi^2 - (a^2 - 1)\xi + xa^2 = 0,$$

where $\xi(x)$ is known as Pastur's curve [**34**].

8.5. Normal matrices ensembles and high order recurrences. A normal matrix model is defined by a probability density function on the space $\mathbf{N}_n \times \mathbf{N}_n$, $(M \in \mathbf{N}_n : [M, M^*] = 0)$:

$$\mu(dM, dM^*) := \frac{1}{Z_n} \exp\left(N \operatorname{Tr}(V(M) + \overline{V}(M^*) - MM^*)\right) dM dM^*,$$

where V is a function, N is the scaling parameter, and the normalization constant Z_n makes μ probabilistic measure. The joint probability density of eigenvalues for this model (see [**45**], [**25**], and [**19**]) is

$$\mathcal{P}(z_1, \ldots, z_n) = \frac{1}{\widetilde{Z}_n} \prod_{j>i} |z_j - z_i|\,|z_j - \overline{z}_i| \prod_{j=1}^{n} e^{N[2\,\mathrm{Re}\ V(z) - |z_j|^2]}, \tag{8.7}$$

where $\widetilde{Z}_n$ is the normalization constant.

In the space $\mathbf{P}$ of polynomials with respect to the variables z and $\bar{z}$ we introduce an inner product:

$$\langle f, g \rangle := \widetilde{\iint_{\mathbb{C}}} f\,\overline{g}\ W\, dz\, d\overline{z}, \quad W := e^{N\left(2\,\mathrm{Re}\ V(z) - |z|^2\right)}, \qquad f, g \in \mathbf{P}, \tag{8.8}$$

where by definition the integral $\widetilde{\iint_{\mathbb{C}}}$ requires

$$\langle \frac{\partial F}{\partial z}, 1 \rangle = \langle \frac{\partial F}{\partial \overline{z}}, 1 \rangle = 0, \qquad F := f\,W, \quad f \in \mathbf{P}. \tag{8.9}$$

This requirement is always fulfilled for the convergent integrals in (8.8) (it follows from the Fubini theorem and the Newton-Leibnitz formula). For general potentials V condition (8.9) serves to eliminate the divergent parts of the integral in the definition of functional (8.8). This regularization is similar to the "cut off" procedure (see details in [**25**]). The eigenvalue distribution (8.7) is a determinantal process with the reproducing kernel formed by an orthogonal polynomials system

$$\{Q_{n,N}(z)\}\ : \qquad \langle Q_{k,N}(z),\, Q_{l,N}(z)\rangle = m_k\,\delta_{k,\,l}\,. \tag{8.10}$$

It is known (see [**45**], [**25**]) that for polynomial potentials $V(z)$ the orthogonal polynomials (8.10) satisfy a high order recurrence relation. Let us demonstrate it for the special case $V(z) := \frac{t_3}{3}\,z^3$. Let A be the matrix (in the basis (8.10)) of the operator of multiplication by the variable z and B be the matrix of $\frac{\partial}{\partial z}$. For the elements of these matrices we have

$$A_{m+2,\,n} = 0\,,\ B_{m,\,n} = 0\,, \quad m \geq n\,, \qquad A_{k+1,\,k} = 1\,,\ B_{k,\,k+1} = k+1\,. \tag{8.11}$$

On the other hand the regularization requirement of the inner product (8.8) implies

$$D\,B\ +\ N\,t_3\,D\,A^2\ -\ N\,A^*\,D\ =\ 0\,, \qquad D := \operatorname{diag}(m_0,\, m_1,\, \cdots)\,. \tag{8.12}$$

Thus from (8.11) and (8.12) we see that orthogonal (with respect to the inner product (8.8)) polynomials (8.10) satisfy the high order recurrence relations

$$Q_{n+1,N} = z\,Q_{n,N} - a_{n-1}\,Q_{n-2,N}\,, \qquad a_{n-1} := \overline{t_3}\frac{m_n}{m_{n-2}}\,, \qquad n \geq 2, \quad z \in \mathbb{C}\,, \tag{8.13}$$

which are the same as the recurrences for the multiple orthogonal polynomials (see (4.2)–(4.4)).

The simple identity $BA - AB = E$ yields a nonlinear discrete equation for $\{a_n\}$:

$$a_k = \left[(k+1)\frac{\overline{t_3}}{N} + |t_3|^2(a_k + a_{k+1})\right]\left[k\frac{\overline{t_3}}{N} + |t_3|^2(a_{k-1} + a_k)\right].$$

This equation is equivalent to a discrete analog of the Painlevé 1 equation (see [**45**] and [**25**]):

$$r_k^2\,(1 - |t_3|^2(r_{k-1}^2 + r_{k+1}^2))\ =\ \frac{k}{N}\,, \qquad r_k^2\ =\ \frac{m_k}{m_{k-1}}\,. \tag{8.14}$$

It is a very interesting problem to find the vector measure for the MOPs defined by (8.13)–(8.14). The preliminary analysis shows that for fixed N the equation (8.14) implies the non-positivity of the a_k for the big k. Thus the sufficient condition for the existence of a system of positive measures $\{\mu_j\}_{j=1}^p$ (see Theorem 4.1) is not fulfilled. Although this sufficient condition is not necessary, it is nevertheless rather sharp and the chances of having $\{\mu_j\}_{j=1}^p$ produce an infinite system of MOPs for fixed N are negligible. Therefore one can expect a finite system of MOPs $\{Q_{k,N}(z)\}_{k=0}^N$ orthogonal with respect to a discrete vector measure (such as the Kravtchuk MOPs). Moreover, "freezing" solutions of (8.14) we can find a condition for parameters to guarantee the positivity of a_n:

$$r_{k-1}^2 \approx r_k^2 \approx r_{k+1}^2; \qquad 2\,|t_3|^2\,\frac{k}{N} < \frac{1}{4} \quad \Rightarrow \quad a_k > 0\,. \tag{8.15}$$

We note that the critical value of the parameter $t_3^2 < \frac{1}{8}$ already appeared in other approaches to the problem (see [**45**], [**25**], and [**19**]). Moreover, in this "freezing" regime it is not difficult to derive a well-known algebraic curve $\xi(z)$ for (8.13)–(8.14) and the density ρ of the limiting (as $n \to \infty$) zero distribution of $\{Q_{n,n}(z)\}$

$$\rho(z) = \frac{1}{\pi}\,|\Im\xi(z)|\,, \quad z \in S\,, \qquad \xi^3 - t_3 z^2 \xi^2 - (t_3 + \frac{1}{t_3})z\xi + z^3 + C(t_3) = 0\,.$$

An analysis of this algebraic curve leads to a Nikishin vector potential equilibrium problem with external field (as in (7.18)–(7.21)) for a vector of measures (the first coordinate is $\rho(z)dz$). See [**19**] for details.

References

[1] T. J. Stieltjes, *Recherches sur les fractions continues*, Ann. Fac. Sci. Toulouse **8** (1894), J1–122; **9** (1895), A1–47; Œuvres Complètes-Collected Papers, Vol. II (G. van Dijk, ed.), Springer Verlag, Berlin, 1993, pp. 406–570.

[2] T. J. Stieltjes, *Sur la réduction en fraction continue d'une série procédant suivant les puissances descendantes d'une variable*, Ann. Fac. Sci. Toulouse **3** (1889), H1–17; Œuvres Complètes-Collected Papers, Vol. II (G. van Dijk, ed.), Springer Verlag, Berlin, 1993, pp. 184–200.

[3] V. A. Marchenko, *Sturm-Liouville Operators and Applications* (revised edition), Amer. Math. Soc., Providence RI, 2011. MR2798059 (2012b:34002)

[4] N. I. Akhiezer, *The Classical Moment Problem and Some Related Topics in Analysis*, Fitzmatgiz, Moscow, 1961 (in Russian); English transl.: Oliver & Boyd, London, 1965. MR0184042 (32:1518)

[5] N. I. Akhiezer and I. M. Glazman, *Theory of Linear Operators in a Hilbert Space*, volumes I, II, Nauka, Moscow 1966 (in Russian); English transl.: Pitman, Boston 1981. MR615736 (83i:47001a)

[6] A. I. Aptekarev, P. M. Bleher and A. B. J. Kuijlaars, *Large n limit of Gaussian random matrices with external source, Part II*, Comm. Math. Phys., 259 (2), 2005, 367–389. MR2172687 (2006f:82036)

[7] A. Aptekarev, V. Kaliaguine and W. Van Assche, *Criterion for the resolvent set of nonsymmetric tridiagonal operators*, Proc. Amer. Math. Soc., **123** (1995), 2423–2430. MR1254830 (96c:47041)

[8] A. Aptekarev, *Forced motion of the semi-infinite Langmuir lattice*, in Function Theory and Approximation (P. L. Ulianov, ed.), volume 2, Saratov University Edition, 1986, pp. 17–21 (in Russian). MR990923

[9] A. I. Aptekarev and V.A. Kaliaguine, *Complex rational approximation and difference operators*, Rend. Circ. Mathem. Palermo, Ser. II, suppl. **52** (1998), 3–21. MR1644539 (99h:41019)

[10] A. I. Aptekarev, V. A. Kaliaguine and J. Van Iseghem, *The Genetic Sums' Representation for the Moments of a System of Stieltjes Functions and its Application,* Constr. Approx. **16** (2000), 487–524. MR1771693 (2001g:41021)

[11] A. I. Aptekarev, V. A. Kaliaguine and E. B. Saff, *Higher-Order Three-Term Recurrences and Asymptotics of Multiple Orthogonal Polynomials,* Constr. Approx. **30** (2009), 175–223. MR2519660 (2010h:42049)

[12] A. I. Aptekarev, A. Branquinho, W. Van Assche, *Multiple orthogonal polynomials for classical weights*, Trans. Amer. Math. Soc. 355 (2003), 3887–3914. MR1990569 (2004g:33014)

[13] A. I. Aptekarev and A. B. J. Kuijlaars, *Hermite-Padè approximations and multiple orthogonal polynomial ensembles,* Uspekhi Mat. Nauk **66**:6 (2011), 123-190, (in Russian); English transl. in Russian Math. Surveys **66**:6 (2011), 1133–1199. MR2963452

[14] A. I. Aptekarev, G. Lopes Lagomasino and I. A. Rocha, *Asymptotic behavior of the ratio of Hermite-Padè polynomials for Nikishin systems*, Mat. Sb. **196** (2005), no. 8, 3–20 (in Russian); English transl. in Sb. Math. **196** (2005), no. 7-8, 1089–1107. MR2188362 (2006h:41040)

[15] L. Bernstein, *The Jacobi-Perron Algorithm*, Lecture Notes in Mathematics, **207**, Springer-Verlag, Berlin, 1971. MR0285478 (44:2696)

[16] P. M. Bleher and A. B. J. Kuijlaars, *Random matrices with external source and multiple orthogonal polynomials*, Int. Math. Res. Not., 2004, no. 3, 109–129. MR2038771 (2004k:82034)

[17] P. M. Bleher and A. B. J. Kuijlaars, *Large n limit of Gaussian random matrices with external source, Part I*, Comm. Math. Phys. **252** (2004), 43–76. MR2103904 (2005i:82031)

[18] P. M. Bleher and A. B. J. Kuijlaars, *Large n limit of Gaussian random matrices with external source, Part III. Double scaling limit*, Comm. Math. Phys. **270** (2007), 481–517. MR2276453 (2008b:82032)

[19] P. M. Bleher and A. B. J. Kuijlaars, *Orthogonal polynomials in the normal matrix model with a cubic potential*, Adv. Math. **230** (2012), 1272–1321. MR2921180

[20] W. B. Jones and W. J. Thron, *Continued Fractions, Analytic Theory and Applications*, Addison-Wesley, Reading, MA, 1979. MR595864 (82c:30001)
[21] O. Bogoyavlenskii, *Some constructions of integrable dynamical systems*, Izv. Acad. Nauk SSSR, Ser. Mat. **51** (1987), n. 4; English transl.: Math. USSR Izvestiya **31** (1988), n. 1, 47–75. MR914858 (89g:58087)
[22] O. Bogoyavlenskii, *Integrable dynamical systems associated with the KdV equation*, Izv. Acad. Nauk SSSR, Ser. Mat. **51** (1987), n. 6.; English transl.: Math. USSR Izvestiya **31** (1988), n. 3, 435–454. MR933958 (89h:58075)
[23] W. B. Jones, W. J. Thron and N. J. Wyshinski, *A family of classical determinate Stieltjes moment problems with discrete solutions*, in Nonlinear Numerical Methods and Rational Approximation II (A. Cuyt, ed.), Kluwer, Dordrecht, 1994, pp. 409–422. MR1307214 (96e:33016)
[24] M. Eiermann and R. S. Varga, *Zeros and local extreme points of Faber polynomials associated with hypocycloidal domains,* ETNA **1** (1993), 49–71. MR1234356 (94i:30006)
[25] P. Elbau and G. Felder, *Density of eigenvalues of random normal matrices,* Comm. Math. Phys. **259**, 2, (2005), 433–450. MR2172690 (2006i:82034)
[26] P. Flajolet and R. Sedgewick, *Analytic Combinatorics*, Cambridge University Press, Cambridge, 2009. MR2483235 (2010h:05005)
[27] J.S. Geronimo and K. M. Case, *Scattering theory and polynomials orthogonal on the real line*, Trans. Amer. Math. Soc. **258** (1980), 467–494. MR558185 (82c:81138)
[28] M. He and E. B. Saff, *The zeros of Faber polynomials for an m-cusped hypocycloid*, J. Approx. Theory **78** (1994), 410–432. MR1292970 (95k:30009)
[29] V. A. Kaliaguine, *Hermite-Padé approximants and spectral analysis of nonsymmetric operators*, Math. Sb. Russian Acad. Sci. **185** (1994), 79–100; English transl.: Russian Acad. Sci. Sb. Math. **82** (1995), 199–216. MR1280397 (95d:47038)
[30] V. A. Kaliaguine, *The operator moment problem, vector continued fractions and an explicit form of the Favard theorem for vector orthogonal polynomials*, J. Comput. Appl. Math. **65** (1995), 181–193. MR1379130 (97c:44002)
[31] E. M. Nikishin, *The discrete Sturm-Liouville operator and some problems in function theory*, Trudy Semin. Petrovsk. **10** (1984), 3–77; English transl.: J. Soviet Math. **35** (1986), 2679–2744. MR778879 (86h:39007)
[32] E. M. Nikishin, *On simultaneous Padé approximants,* Math. Sb. **113** (155), (1980), 499-519. English transl. in Math. USSR Sb. **41** (1980), 409–421. MR602272 (82e:30053)
[33] E. M. Nikishin and V. N. Sorokin, Rational Approximations and Orthogonality, Nauka, Moscow, 1988; English transl.: Translations of Mathematical Monographs, vol. **92**, Amer. Math. Soc., Providence RI, 1991. MR1130396 (92i:30037)
[34] L. Pastur, *The spectrum of random matrices*, Theoret. Mat. Fiz. **10** (1972), 102–112. MR0475502 (57:15106)
[35] A. Osipov, *Discrete Korteweg-de Vries equation*, Math. Zametki **56** (1994), no. 6, 141–143; English transl.: Math. Notes **56** (1994), no. 5–6, 1312–1314. MR1330610 (96f:39016)
[36] O. Perron, Die Lehre von den Kettenbruchen, Teubner, Leipzig, 1929.
[37] L. Pineiro, *On simultaneous approximants of a system of Markov type functions*, Vestnik Mosk. Gos. Univer., Ser. Mat. **42** (1987), no. 2, 67–70; English transl.: Moscow Univ. Math. Bull. **42** (1987), n. 2, 52–55. MR884516 (88c:41033)
[38] V. N. Sorokin, *Integrable nonlinear dynamical systems of Langmuir lattice type*, Math. Zametki **62** (1997), no. 4, 588–601; English transl. Math. Notes **62** (1997), no. 4, 488–500. MR1620166 (99c:35236)
[39] V. N. Sorokin and J. Van Iseghem, *Algebraic aspects of matrix orthogonality for vector polynomials*, J. Approx. Theory **90** (1997), 97–116. MR1458824 (98m:42040)
[40] V. N. Sorokin and J. Van Iseghem, *Matrix continued fractions*, J. Approx. Theory **96** (1997), 237–257. MR1671197 (2000i:41027)
[41] W. Van Assche and E. Coussement, *Some classical multiple orthogonal polynomials,* J. Comput. Appl. Math. **127** (2001), 317–347. MR1808581 (2001i:33012)
[42] W. Van Assche, *Asymptotics for orthogonal polynomials and three-term recurrences*, in: Orthogonal Polynomials: Theory and Practice (P. Nevai, ed.), Kluwer Academic Publishers, 1990, pp. 435–462. MR1100305 (92k:42038)
[43] J. Van Iseghem, *Vector orthogonal relations, vector QD-algorithm*, J. Comput. Appl. Math. **19** (1987), 141–150. MR901223 (89c:65025)
[44] H. S. Wall, *Analytic Theory of Continued Fractions*, Chelsea, Bronx, NY, 1973.

[45] P. Wiegmann and A. Zabrodin, *Conformal maps and integrable hierarchies,* Comm. Math. Phys. **213** (2000), no. 3, 523–538. MR1785428 (2002g:37105)
[46] V. I. Yurko, *On higher-order difference operators*, J. Difference Equations Appl. **1** (1995), 347–352. MR1350449 (96k:39025)

Keldysh Institute of Applied Mathematics, Russian Academy of Sciences, Miusskaya Sq. 4, Moscow 125047, Russia
E-mail address: aptekaa@keldysh.ru

Amer. Math. Soc. Transl.
Volume **233**, 2014

On the Multi-Fractal Spectrum of the Whole-Plane Lévy–Loewner Evolution

Anne Boutet de Monvel, Igor Loutsenko, and Oksana Yermolayeva

Dedicated to Vladimir A. Marchenko, a mentor and a father

ABSTRACT. Lévy–Loewner evolution (LLE) is a generalization of Schramm–Loewner evolution (SLE) where the branching is possible in the course of growth process. We consider a class of radial Lévy–Loewner evolutions for which sets of points of the average means beta-spectrum can be found exactly. In this paper we show how to overcome difficulties that have arisen in previous work on multi-fractal spectrum of SLE/LLE.

1. Introduction

We start with a simple introduction to the radial Lévy–Loewner evolution (LLE). A good introduction to the chordal LLE can be found in [**15, 16**]. For an introduction to Lévy processes, see e.g. [**1**] and the references therein.

Let us consider iterative conformal mappings $z = F_n(w)$ from the exterior of the unit disc in the w-plane to the exterior of a bounded, simply connected domain in the z-plane: The nth mapping F_n is the composite of n elementary "spike" mappings $z = f_i(w, \delta t_i)$, $i = 1, \ldots, n$,

$$
\begin{aligned}
F_n(w) &= F_{n-1}(f_n(w, \delta t_n)), \\
F_0(w) &= w,
\end{aligned}
\tag{1.1}
$$

where

$$
\begin{aligned}
f_n(w,t) &= e^{i\varphi_n} h\left(e^{-i\varphi_n} w, t\right), \\
h(w,t) &= e^t (w+1)\frac{w+1+\sqrt{(w+1)^2 - 4e^{-t}w}}{2w} - 1.
\end{aligned}
\tag{1.2}
$$

The elementary mapping $z = f_n(w, \delta t_n)$ attaches to the unit disc a radial "spike" of length $\sqrt{\delta t_n}\big(1 + O(\delta t_n)\big)$ located at the angle φ_n. Here, the point $w = e^{i\varphi_n}$ on the unit circle in the w-plane is mapped to the tip of the spike in the z-plane (see Figure 1).

2010 *Mathematics Subject Classification.* Primary 60J67; Secondary 28A80.

Key words and phrases. Lévy–Loewner evolution (LLE), Schramm–Loewner evolution (SLE), multi-fractal spectrum.

The mapping $z = h(w,t)$ that attaches a spike to the unit disc at the angle $\varphi = 0$ satisfies the simplest Loewner equation

$$\partial_t h(w,t) = w\,\frac{w+1}{w-1}\,\partial_w h(w,t),$$

and as a consequence

$$\partial_t f_n(w,t) = w\,\frac{w+e^{i\varphi_n}}{w-e^{i\varphi_n}}\,\partial_w f_n(w,t).$$

This equation is invariant w.r.t. any conformal transformation $f_n(w,t) \to F\,(f_n(w,t))$ and therefore the iterative compound mapping F_n in (1.1) can be represented as the solution of the Loewner equation at time $t = \sum_{i=1}^{n} \delta t_i$:

$$\begin{aligned} F_n(w) &= F\Bigl(w, \sum_{i=1}^{n} \delta t_i\Bigr), \qquad F(w,0) = w, \\ \partial_t F(w,t) &= w\,\frac{w+e^{iL(t)}}{w-e^{iL(t)}}\,\partial_w F(w,t), \end{aligned} \tag{1.3}$$

where $L(t)$ is the piecewise constant function $(i = 1, \dots, n)$

$$L(t) = \varphi_i,$$
$$\sum_{j=1}^{i-1} \delta t_j < t < \sum_{j=1}^{i} \delta t_j.$$

One can consider continuous-time Loewner evolutions as limits of the above iterative processes when $n \to \infty$ and $\delta t_n \to 0$. This iterative picture will be useful for the simplest derivation of a linear integro-differential equation for the moments of derivatives of conformal mappings for the Lévy–Loewner evolution by the method proposed by M. Hastings [**9**].

We are interested in the case when $L(t)$ is a stochastic process without drift. Without loss of generality we set

$$\langle L(t) \rangle = 0, \tag{1.4}$$

where $\langle \cdot \rangle$ denotes expectation (ensemble average).

When a stochastic process $L(t)$ is continuous in time, the conformal mapping $z = F(w,t)$ describes the growth of a random continuous curve $\Gamma = \Gamma(t)$ starting from a point on the unit circle $|z| = 1$ at $t = 0$. On the other hand, when $L(t)$ is discontinuous in time, the growth branches in the z-plane.

If one requires that the Loewner evolution (1.3) is a conformally invariant Markovian process, in the sense that the time evolution is consistent with the composition of conformal maps, then a necessary condition for such an evolution is that $L(t)$ must have independent stationary increments, i.e. $L(t)$ must be a Markovian process with the probability distribution of $L(t+\tau) - L(t)$ depending only on τ. Among such processes there is the class of Lévy processes considered in the present paper.

The only continuous (modulo uniform drift) process of Lévy type is the Brownian motion

$$L(t) = B(t). \tag{1.5}$$

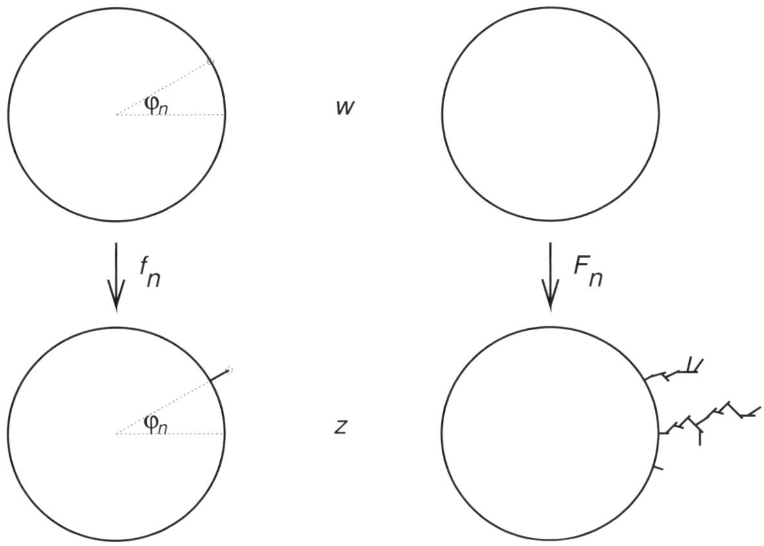

FIGURE 1. The elementary mapping $z = f_n(w, \delta t_n)$ (left) and the compound iterative mapping $z = F_n(w)$ (right) from the exterior of the unit circle in the w-plane to the exterior of a simply connected domain in the z-plane.

The Brownian motion is characterized by a single parameter — the "temperature" κ:

$$\langle (B(t+\tau) - B(t))^2 \rangle = \kappa |\tau|. \tag{1.6}$$

A stochastic Loewner evolution driven by Brownian motion is called a Schramm–Loewner Evolution (SLE or SLE_κ). Since it describes non-branching planar stochastic curves with conformally-invariant probability distribution, SLE is a useful tool for the description of boundaries of critical clusters in two-dimensional equilibrium statistical mechanics. In this picture, different κ's correspond to different classes of models of statistical mechanics (a good introduction to SLE for physicists can be found e.g. in [**3**, **7**] and mathematical reviews can be found e.g. in [**10**, **11**]).

By now, SLE is well studied and many exact results have been obtained. In particular, SLE is the only non-trivial example where multi-fractal spectrum as well as related β-spectrum of the curve is described exactly.

In this article we would like to present new results in the study of multi-fractal spectrum of SLE/LLE. Mainly, these results concern the "unbounded" version of the whole-plane LLE/SLE, where we determine exactly the points of integral means β-spectra for certain classes of Lévy processes. The studies of such processes were initiated by B. Duplantier et al. in [**5**]. In previous works [**12**, **13**] we proposed an approach to the problem based on the derivation of exact solutions of equations for the moments of derivatives of conformal mappings.

The β-spectrum is the Legendre transform of the dimension spectrum that quantitatively describes subsets of the domain boundary where certain scaling laws apply[1] to derivatives of conformal mappings $F'(w)$ along the boundary. The integral means $\beta(q)$-spectrum of the domain is defined through the qth moment of the derivative at the unit circle (i.e. as $|w| \to 1$) as follows:

$$\beta(q) = \limsup_{\epsilon\downarrow 0} \frac{\log \int_0^{2\pi} \langle |F'(e^{\epsilon+i\varphi})|^q \rangle \, d\varphi}{-\log \epsilon}. \tag{1.7}$$

This spectrum can be written explicitly for the whole-plane SLE, and several points of the spectrum can be found for certain types of the whole-plane LLE.

The "bounded" whole-plane LLE is a properly scaled infinite-time limit of the radial LLE,

$$\mathcal{F}(w,t) = \lim_{T\to\infty} e^{-T} F(w, T+t), \tag{1.8}$$

which describes the growth of a stochastic curve out of the point of origin in the plane. Obviously, $\mathcal{F}(w,t)$ also satisfies the radial Loewner equation (1.3) with time running from $t = -\infty$ to $t = \infty$.

The "unbounded" version of the whole-plane LLE is derived from the bounded version by inversion $\mathcal{F}(w,t) \mapsto 1/\mathcal{F}(1/w,t)$ of the bounded version. Obviously, this mapping from the unit disc in the w-plane to the complement of the stochastic curve which grows from infinity towards the origin in the z-plane satisfies the Loewner equation (1.3). It has been studied mainly due to its relationship with the problem of Bieberbach coefficients of conformal mappings for such processes [**5**, **12**]: Expectations of the squares of the Bieberbach coefficients as well as two non-trivial points of the β-spectrum have been conjectured for two special examples of unbounded whole-plane LLEs in [**5**]. The proof of this conjecture has first been proposed by one of the authors in [**12**] for the first example and partially for the second one. In this article we complete the proof of this conjecture.

Note that a partial (unsuccessful) proof of the second case has been attempted by B. Duplantier et al. in [**6**]. This proof is written for Lévy processes whose first N characteristic coefficients coincide with those of Brownian motion. The proof of non-trivial cases involves $N > 1$. But, Lévy processes on the circle with such a property for $N > 1$ don't exist except for Brownian motions.

In the sequel we also sketch the derivation of the full $\beta(q)$-spectrum

$$\beta(q) = \limsup_{\epsilon\downarrow 0} \frac{\log \int_0^{2\pi} \langle |\mathcal{F}'(e^{-\epsilon+i\varphi})|^q \rangle d\varphi}{-\log \epsilon} \tag{1.9}$$

of the "unbounded" whole-plane SLE.

It is important to note that the study of the unbounded version of the whole-plane SLE turns out to be useful not only in the coefficient problem, but also for a complete derivation of the version-independent multi-fractal spectrum of the bulk of SLE, since the attempt of rigorous derivation of the bulk spectrum by D. Belyaev and S. Smirnov [**2**] based on the bounded version happens to be incomplete (see e.g. [**14**, Appendix]).

2. An equation for the moments of the derivative

Let us consider the "exterior" radial LLE, i.e. a mapping from the exterior of the unit disc in the w-plane to the exterior of a simply connected domain in the z-plane (as shown in Figure 1). As has been mentioned above, the bounded version

[1] For an introduction to multi-fractal analysis see e.g. [**2**, **8**] and the references therein.

of the whole-plane LLE is a properly scaled infinite-time limit of this process. So, let us call the bounded version the "exterior problem".

The derivation for the unbounded version, which is referred to as the "interior problem", is similar.

To find the $\beta(q)$-spectrum (1.7) one needs to estimate the moments of the derivative $\langle |F'(w,t)|^q \rangle$.

It is now convenient to change the variable w to $\mathrm{e}^{\mathrm{i}L(t)}w$ and consider the mapping

$$\tilde{F}(w,t) = F\left(\mathrm{e}^{\mathrm{i}L(t)}w, t\right)$$

for which the point $w = 1$ is the pre-image of the growing tip of the curve. Obviously,

$$\beta(q) = \limsup_{\epsilon \downarrow 0} \frac{\log \int_0^{2\pi} \tilde{\rho}\left(\mathrm{e}^{\epsilon + \mathrm{i}\varphi}, \mathrm{e}^{\epsilon - \mathrm{i}\varphi}; t|q\right) \mathrm{d}\varphi}{-\log \epsilon},$$

where

$$\tilde{\rho}(w, \bar{w}; t|q) = \left\langle \left| \tilde{F}'(w,t) \right|^q \right\rangle. \tag{2.1}$$

It turns out that the moments $\tilde{\rho}(w, \bar{w}; t|q)$ satisfy an integral linear equation. The simplest way to obtain such an equation uses M. Hasting's iterative approach [**9**, **12**]: In the iterative picture (1.1) we consider the mappings

$$\tilde{F}_n(w) = \tilde{F}_{n-1}\left(\mathrm{e}^{\mathrm{i}(\varphi_n - \varphi_{n-1})} h(w, \delta t_n)\right).$$

By the chain rule

$$\left|\tilde{F}'_n(w)\right|^q = \left|\tilde{F}'_{n-1}\left(\mathrm{e}^{\mathrm{i}(\varphi_n - \varphi_{n-1})} h(w, \delta t_n)\right)\right|^q |h'(w, \delta t_n)|^q.$$

Taking expectations of both sides of the above equation, and taking into account (2.1), we get

$$\begin{aligned}
&\tilde{\rho}(w, \bar{w}, t_{n-1} + \delta t_n) \\
&= |h'(w, \delta t_n)|^q \int_0^{2\pi} P(\varphi, \delta t_n) \tilde{\rho}\left(\mathrm{e}^{\mathrm{i}\varphi} h(w, \delta t_n), \mathrm{e}^{-\mathrm{i}\varphi} \bar{h}(\bar{w}, \delta t_n), t_{n-1}\right) \mathrm{d}\varphi,
\end{aligned}$$

where $P(\varphi, t)$ is the probability density that $L(t) = \varphi$ under the condition that $L(0) = 0$.

Taking into account that for small δt

$$h(w, \delta t) = w + w\frac{w+1}{w-1}\delta t + \mathrm{o}(\delta t),$$

we obtain

$$\begin{aligned}
\tilde{\rho}(w, \bar{w}, t + \delta t) = &\left[1 + \delta t\Big(w\frac{w+1}{w-1}\partial_w + \bar{w}\frac{\bar{w}+1}{\bar{w}-1}\partial_{\bar{w}} - \frac{q}{(w-1)^2} - \frac{q}{(\bar{w}-1)^2} + q\Big)\right] \\
&\times \int_0^{2\pi} P(\varphi, \delta t) \tilde{\rho}\left(\mathrm{e}^{\mathrm{i}\varphi} w, \mathrm{e}^{-\mathrm{i}\varphi} \bar{w}, t\right) \mathrm{d}\varphi,
\end{aligned}$$

so, in the $\delta t \to 0$ limit we arrive at the equation for $\tilde{\rho}$,

$$\partial_t \tilde{\rho} = \left(-\hat{\eta} + w\frac{w+1}{w-1}\partial_w + \bar{w}\frac{\bar{w}+1}{\bar{w}-1}\partial_{\bar{w}} - \frac{q}{(w-1)^2} - \frac{q}{(\bar{w}-1)^2} + q\right)\tilde{\rho}. \tag{2.2}$$

Here the operator $\hat{\eta}$ acts on a function ρ of $w, \bar{w}$ as follows:

$$\hat{\eta}[\rho](w,\bar{w}) = \lim_{t\to 0}\frac{1}{t}\int_0^{2\pi}\left(\rho(w,\bar{w}) - \rho\left(e^{i\varphi}w, e^{-i\varphi}\bar{w}\right)\right)P(\varphi,t)d\varphi.$$

We consider only processes for which the above limit exists. On the unit circle, these Lévy processes are defined by their characteristic exponents η_m:

$$e^{-t\eta_m} = \langle e^{imL(t)}\rangle = \int_0^{2\pi} e^{im\varphi}P(\varphi,t)\,d\varphi,$$

i.e. $e^{-t\eta_m}$ is the Fourier transform of the probability distribution of $L(t)$.

According to the above definitions, the operator $\hat{\eta}$ acts diagonally on the basis of the two-dimensional Taylor/Laurent expansions $w^n\bar{w}^m$, $n, m \in \mathbb{Z}$:

$$\hat{\eta}[w^n\bar{w}^m] = \eta_{n-m}w^n\bar{w}^m. \tag{2.3}$$

Since we consider processes without drift, the characteristic exponents are real and symmetric:

$$\eta_m = \bar{\eta}_m, \quad \eta_m = \eta_{-m}, \quad \eta_0 = 0.$$

In the case of the Brownian motion (1.6), $P(\varphi,t)$ is the fundamental solution of the heat equation on the circle and

$$\eta_m = \frac{\kappa m^2}{2}, \quad \hat{\eta} = \frac{\kappa}{2}\left(w\partial_w - \bar{w}\partial_{\bar{w}}\right)^2, \tag{2.4}$$

i.e. (2.2) becomes a second order linear PDE.

3. The whole-plane LLE

Equation (2.2) simplifies in the case of the whole-plane LLE which is a properly scaled infinite-time limit (see (1.8)) of the radial LLE.

Now consider the Laurent expansion of $\tilde{\mathcal{F}}(w,t) = \mathcal{F}\left(e^{iL(t)}w, t\right)$ at $w = \infty$:

$$\tilde{\mathcal{F}}(w,t) = e^{t+iL(t)}\left(w + \sum_{i=0}^{\infty}\Phi_i(t)w^{-i}\right).$$

According to the above equation and (2.1), we have

$$\begin{aligned}\tilde{\rho} &= e^{qt}\left\langle \left|1 - \sum_{i=1}^{\infty} i\Phi_i w^{-i-1}\right|^q \right\rangle = e^{qt}\rho,\\ \rho &= \sum_{i,j=-1}^{\infty}\frac{\rho_{i,j}}{w^{i+1}\bar{w}^{j+1}}, \quad \rho_{-1,-1} = 1.\end{aligned} \tag{3.1}$$

By the existence of the limit (1.8), the probability distribution of

$$e^{-t-iL(t)}\tilde{\mathcal{F}}(w,t) = w + \sum_{i=0}^{\infty}\Phi_i(t)w^{-i}$$

is time-independent.[2] So, in the case of the whole-plane LLE, the function

$$\rho = e^{-qt}\tilde{\rho} = \left\langle \left|1 - \sum_{i=1}^{\infty} i\Phi_i w^{-i-1}\right|^q \right\rangle$$

[2]For details on the existence of this infinite-time limit, see [**12**, Section 4.3] or [**2**, Section 1.2].

does not depend on time, and

$$\partial_t \tilde{\rho} = q\tilde{\rho}.$$

Therefore, for the whole-plane LLE, the equation (2.2) becomes

$$L\rho = q\rho, \tag{3.2a}$$

$$L = -\hat{\eta} + w\frac{w+1}{w-1}\partial_w + \bar{w}\frac{\bar{w}+1}{\bar{w}-1}\partial_{\bar{w}} - \frac{q}{(w-1)^2} - \frac{q}{(\bar{w}-1)^2} + q. \tag{3.2b}$$

To find the moments of derivatives we have to look for a solution of the above equation which is analytic and non-vanishing at $w = \infty$. Note that such a solution is (up to a constant factor) unique.

Indeed, by analyticity of a conformal mapping at $w = \infty$, the function $\rho(w, \bar{w})$ is analytic at infinity and $\rho(\infty) = 1$. Then substituting the expansion (3.1) into (3.2a) we get a nine-term "two-dimensional" recurrence relation for the expansion coefficients $\rho_{i,j}$:

$$\sum_{l,m=0}^{2} C_{i,j}^{l,m}\, \rho_{i-l,j-m} = 0 \tag{3.3}$$

with the boundary conditions (see (3.1))

$$\rho_{-1,-1} = 1,$$
$$\rho_{i,j} = \rho_{j,i} = 0 \text{ for } j < -1,$$

and the recurrence coefficients (note the symmetry $C_{i,j}^{l,m} = C_{j,i}^{m,l}$)

$$\begin{aligned}
C_{i,j}^{0,0} &= -\eta_{i-j} - i - j - 2, & C_{i,j}^{1,1} &= -4\eta_{i-j}, \\
C_{i,j}^{2,2} &= -\eta_{i-j} + i + j - 2 - 2q, & C_{i,j}^{0,1} &= 2\left(\eta_{i-j+1} + i + 1\right), \\
C_{i,j}^{0,2} &= -\eta_{i-j+2} + j - i - 2 - q, & C_{i,j}^{1,2} &= 2\left(\eta_{i-j+1} - j + 1 + q\right).
\end{aligned}$$

Any element $\rho_{k,n}$ can be found by consecutively starting from $(i,j) = (-1,-1)$ and going along the "row" up to $j = n$ and repeating this procedure for consecutive rows up to $i = k$ by expressing $\rho_{i,j}$ as a linear combination of 8 elements: $\rho_{i-l,j-m}$, $l, m = 0, 1, 2$, $(l,m) \neq (0,0)$. This fixes $\rho(w, \bar{w})$ uniquely. Therefore, once a solution of (3.2) is found which is analytic and non-vanishing at $w = \infty$, it will correspond to the moments of derivatives.

Recall that in the exterior problem (i.e. the bounded whole-plane LLE) considered above, the curve grows from the origin towards infinity. In the interior problem (i.e. the unbounded whole-plane LLE), the curve grows from infinity towards the origin. Both versions are related by the inversion $\mathcal{F}(w,t) \mapsto 1/\mathcal{F}(1/w,t)$. Obviously, the unbounded mapping satisfies the same Loewner equation (1.3), but now $\mathcal{F}$ maps the interior of the unit disc to the complement of the unbounded curve in the plane:

$$\mathcal{F}(w,t) = \mathrm{e}^{-t}\left(w + \sum_{i=2}^{\infty} \mathcal{F}_i(t) w^i\right), \quad |w| < 1.$$

By analogy with the exterior problem, the moments of derivatives

$$\tilde{\rho} = \langle |\mathcal{F}'(\mathrm{e}^{\mathrm{i}L(t)} w, t)|^q \rangle$$

depend on time only through exponential scaling, i.e. (note the difference in the exponent sign in comparison with (3.1))

$$\tilde{\rho} = \mathrm{e}^{-qt}\rho,$$

where

$$\rho(w, \bar{w}; q) = \sum_{i,j=1}^{\infty} \rho_{i,j} w^{i-1} \bar{w}^{j-1}, \quad \rho_{1,1} = 1 \tag{3.4}$$

is time-independent.

Similarly to the exterior problem, $\rho(w, \bar{w}; q)$ satisfies an integral equation which now reads as follows:

$$L\rho = -q\rho, \tag{3.5}$$

where L is given by (3.2b). A solution of (3.5) which is analytic and non-vanishing at $w = 0$ now corresponds to the moments of derivatives of the interior problem.

The expansion coefficients of the interior problem $\rho_{i,j}$ (see (3.4)) satisfy a recurrence relation of type (3.3) with the following boundary conditions:

$$\rho_{1,1} = 1, \quad \rho_{i<0,j} = \rho_{i,j<0} = 0,$$

and recurrence coefficients ($C_{i,j}^{l,m} = C_{j,i}^{m,l}$)

$$\begin{aligned} C_{i,j}^{0,0} &= -\eta_{i-j} - i - j + 2, & C_{i,j}^{1,1} &= -4(\eta_{i-j} - 2q), \\ C_{i,j}^{2,2} &= -\eta_{i-j} + i + j - 6 + 2q, & C_{i,j}^{0,1} &= 2\left(\eta_{i-j+1} + i - 1 - q\right), \\ C_{i,j}^{0,2} &= -\eta_{i-j+2} + j - i - 2 + q, & C_{i,j}^{1,2} &= 2\left(\eta_{i-j+1} + 3 - j - 2q\right). \end{aligned}$$

4. Universal points, $q = 2$

It has been observed by B. Duplantier et al. in [**5**] that in the interior problem $\rho_{i,i} = i^2$, $\beta = 3$, when $q = 2$ and the evolution is driven by any symmetric Lévy process with $\eta_1 = 3$. The second observation consists of the fact that for $q = 2$ and an evolution driven by any symmetric process with $\eta_1 = 1$ one has $\rho_{i,i} = i^3$, $\beta = 4$.

The first observation has been proved in [**12**] using the coincidence of SLE and LLE solutions of (3.5) when $q = 2$ and $\eta_1 = 3$. The SLE sub-case of the second observation has also been proved in [**12**]. Below we present a complete proof of conjectures by B. Duplantier et al. containing both cases:[3]

THEOREM. *Let us take the interior whole-plane Loewner evolution and fix $q = 2$. Then:*

1) *For an evolution driven by a symmetric Lévy process with $\eta_1 = 3$, we have*

$$\rho_{i,i} = i^2, \quad \beta(2) = 3.$$

2) *For processes with $\eta_1 = 1$ we have*

$$\rho_{i,i} = i^3, \quad \beta(2) = 4.$$

[3]The proof of the sub-case of the second observation with $\eta_1 = 1$, $\eta_2 = 4$ has also been presented in [**6**]. Unfortunately, it has been unnoticed by the authors of [**6**] that such symmetric Lévy processes on the circle different from the Brownian motion do not exist.

PROOF. Representing $\rho(w,\bar{w})$ in the form

$$\rho(w,\bar{w}) = (1-w)(1-\bar{w})\Theta(w,\bar{w}), \tag{4.1}$$

from (3.5) with $q=2$ we get

$$\begin{aligned} &-\hat{\eta}[(1-w)(1-\bar{w})\Theta] + (w+1)(\bar{w}-1)w\,\partial_w\Theta \\ &\quad + (\bar{w}+1)(w-1)\bar{w}\,\partial_{\bar{w}}\Theta + 3(2w\bar{w}-w-\bar{w})\Theta = 0, \end{aligned} \tag{4.2}$$

where $\Theta = \Theta(w,\bar{w})$ is the series

$$\Theta = \theta_0(\xi) + \sum_{i=1}^{\infty}\left(w^i + \xi^i w^{-i}\right)\theta_i(\xi), \quad \xi = w\bar{w}. \tag{4.3}$$

Substituting it into (4.2) we will get a three-term differential recurrence relation for $\theta_i(\xi)$.

Indeed, making the change of variable $\bar{w} = \xi/w$, then taking into account the fact that, according to (2.3), the operator $\hat{\eta}$ commutes with $\xi = w\bar{w}$ and that $\hat{\eta}[w^i] = \eta_i w^i$, after substitution of the series (4.3) into (4.2) we will get a series in w with coefficients depending on ξ. The coefficients of w^i define the following three-term differential recurrence relations:

$$\begin{aligned} &2\xi(\xi-1)\theta_i'(\xi) - (\eta_i + i + (\eta_i - i - 6)\xi)\,\theta_i(\xi) \\ &\quad + \xi(\eta_i + i - 2)\theta_{i+1}(\xi) + (\eta_i - i - 2)\theta_{i-1}(\xi) = 0, \\ &\theta_{-i}(\xi) = \xi^i\theta_i(\xi). \end{aligned} \tag{4.4}$$

It is easy to see that this recurrence relation truncates at $i=N$, i.e. $\theta_i = 0$ for $i \geq N$, when $\eta_N = N+2$ for some N.

Case 1). The simplest truncation happens when $N=1$, i.e. when $\eta_1 = 3$, which corresponds to the case 1) of the Theorem. In this case only θ_0 does not vanish identically and (4.4) reads as follows:

$$(\xi-1)\theta_0'(\xi) + 3\theta_0(\xi) = 0.$$

Therefore $\theta_0(\xi) = 1/(1-\xi^3)$ and $\rho(w,\bar{w}) = (1-w)(1-\bar{w})/(1-w\bar{w})^3$. Then using definitions of $\rho_{i,j}$ and $\beta(q)$ (see (3.4) and (1.9), respectively) we prove the case 1) of the Theorem.

Case 2). Let us now consider case 2) of the Theorem, i.e. $q=2$, $\eta_1 = 1$. Now the truncation of the recurrence relation does not take place. However, it turns out that in this case, two equations defining functions $\theta_0(\xi)$ and $\theta_1(\xi)$ do not involve other $\theta_i(x)$, $i \neq 0, 1$.

In that case of the Theorem the equations for $i=0$ and $i=1$ read as follows:

$$\begin{aligned} &(\xi-1)\theta_0'(\xi) + 3\theta_0(\xi) - 2\theta_1(\xi) = 0, \\ &\xi(\xi-1)\theta_1'(\xi) + (3\xi-1)\theta_1(\xi) - \theta_0(\xi) = 0. \end{aligned}$$

Since the function ρ is analytic at $w=0$, as well as $\rho(w=0)=1$, we have to look for a solution of the above system of two ODEs such that $\theta_0(0)=1$ and $\theta_1(\xi)$ is finite at $\xi=0$. With these conditions we get

$$\theta_0 = \frac{1+\xi}{(1-\xi)^4} \quad \text{and} \quad \theta_1 = -\frac{1}{(1-\xi)^4}.$$

Again taking the definitions of $\rho_{i,j}$ and $\beta(q)$ into account, we prove case 2) of the Theorem. □

5. $\beta(q)$-spectrum of the interior whole-plane SLE

We recall that in the case (1.5)–(1.6), when the driving Lévy process is the Brownian motion, the Loewner evolution results in a continuous curve (SLE-curve). In the interior whole-plane SLE the curve has two exceptional singular points: One of them is the moving tip of the curve, and the second is an immovable "tip" at infinity. The equations for the moments of derivatives (3.5) or (3.2a) become parabolic PDEs of the second order with

$$L = -\frac{\kappa}{2}\left(w\partial_w - \bar{w}\partial_{\bar{w}}\right)^2 + w\frac{w+1}{w-1}\partial_w + \bar{w}\frac{\bar{w}+1}{\bar{w}-1}\partial_{\bar{w}} - \frac{q}{(w-1)^2} - \frac{q}{(\bar{w}-1)^2} + q. \tag{5.1}$$

The β-spectrum of the curve is so far determined by a combination of two methods: For q smaller than the critical value $Q(\kappa)$, where

$$Q(\kappa) = \frac{\kappa^2 + 8\kappa + 12 - 2\sqrt{2\kappa^2 + 16\kappa + 36}}{16\kappa}, \tag{5.2}$$

the blow-up rates of any positive solution of equation (3.5) with L as in (5.1) turn out to depend only on the singularity of the solution in the vicinity of the moving tip of the curve, i.e. at $w = 1$: Near the moving tip, $\epsilon \to 0$, $w = 1 + i\epsilon\zeta$, $\bar{w} = 1 - i\epsilon\bar{\zeta}$, equation (3.5) with (5.1) reduces to

$$\left[-\frac{\kappa}{2}\left(\partial_\zeta + \partial_{\bar{\zeta}}\right)^2 + \frac{2}{\zeta}\partial_\zeta + \frac{2}{\bar{\zeta}}\partial_{\bar{\zeta}} - q\left(\frac{1}{\zeta^2} + \frac{1}{\bar{\zeta}^2}\right)\right]\rho(\zeta, \bar{\zeta}) = 0.$$

The latter has a solution of the form

$$\rho = (\zeta - \bar{\zeta})^{-\kappa\gamma^2/2}(\zeta\bar{\zeta})^{\gamma},$$

where

$$\gamma(q, \kappa) = \frac{\kappa + 4 - \sqrt{(\kappa+4)^2 - 8q\kappa}}{2\kappa}. \tag{5.3}$$

A part of the β-spectrum is then determined by blow-up rates of such a solution, i.e. by the exponents γ and $-\kappa\gamma^2/2$.

This property of positive solutions has been used by D. Belyaev and S. Smirnov in [**2**] for the rigorous calculation of the part of the β-spectrum of the exterior whole-plane SLE.[4] Their approach relies on the maximum principle for parabolic PDEs. In the case of the interior problem it works only for $q < Q(\kappa)$. Therefore, for bigger values of q one has to use different methods. A method based on exact solutions of (3.5)–(5.1) has been proposed in [**13, 14**].

In more detail: Representing a solution of (3.5)–(5.1) which is analytic at $w = 0$ in the form

$$\begin{aligned} \rho(w, \bar{w}) &= \left((1-w)(1-\bar{w})\right)^{\gamma}\Theta(w, \bar{w}), \\ q &= 2\gamma + \frac{1}{2}\kappa\gamma - \frac{1}{2}\kappa\gamma^2, \end{aligned} \tag{5.4}$$

where γ is given by (5.3) and $\Theta = \Theta(w, \bar{w})$, has the following expansion at $w = 0$:

$$\begin{aligned} \Theta &= \sum_{n\in\mathbb{Z}} w^n f_n(\xi), \\ \xi &:= w\bar{w}, \quad f_{-n}(\xi) = \xi^n f_n(\xi), \quad \bar{f}_n = f_n, \end{aligned} \tag{5.5}$$

[4] The idea of using a PDE for the moments of derivatives and singularities of its solution to determine the multi-fractal spectrum of SLE first appeared in the work of M. Hastings [**9**].

we come to the following three-term differential recurrence relation for $f_n(\xi)$:

$$\xi A_{n+1} f_{n+1} + A_{-n+1} f_{n-1} + (B_n + (1-\xi)C_n)\, f_n + 2\xi(\xi-1)\frac{\mathrm{d}f_n}{\mathrm{d}\xi} = 0, \tag{5.6}$$

with

$$\begin{aligned} A_n &= \frac{\kappa}{2}(n-\gamma)^2 + n - 3\gamma - \frac{\kappa}{2}\gamma(1-\gamma), \\ B_n &= -\kappa(n^2+\gamma^2-\gamma) + 6\gamma, \\ C_n &= \kappa\frac{n^2 - 2\gamma + 2\gamma^2}{2} - n - 6\gamma. \end{aligned}$$

It follows from the recurrence relation (5.6) that the expansion (5.5) truncates at $|n| = M$ (i.e. $f_n = 0$ for $|n| > M$) if $A_{-M} = 0$. This truncation condition determines a countable number of curves in the parametric (q,κ)-plane:

$$\begin{aligned} \kappa &= 2\frac{M+3\gamma}{M^2+2M\gamma+2\gamma^2-\gamma}, \\ q &= \frac{\gamma(M+\gamma)(2M+1+\gamma)}{M^2+2M\gamma+2\gamma^2-\gamma}. \end{aligned} \tag{5.7}$$

Along these curves, the relation (5.6) becomes a system of $M+1$ first order linear ODEs for $f_0, \dots, f_M$ (remember that $f_{-n}(\xi) = \xi^n f_n(\xi)$). The blow-up rates of the solution of the PDE (3.5)–(5.1) at the unit circle $\xi = 1$ are determined by the asymptotics of $f_n(\xi)$ as $\xi \to 1$. In these asymptotics the analysis of the system of ODEs is effectively reduced to the analysis of eigenvalues of a $(2M+1)\times(2M+1)$ three-diagonal matrix. The blow-up rates then give the following $\beta(q)$-spectrum along the set of curves (5.7) (see [**13**, **14**]):

$$\beta = \begin{cases} \kappa\frac{\gamma^2}{2} - 2\gamma - 1, & q \le -1 - \frac{3\kappa}{8}, \\ \kappa\frac{\gamma^2}{2}, & -1 - \frac{3\kappa}{8} \le q \le Q(\kappa), \\ 3q - \frac{1}{2} - \frac{1}{2}\sqrt{1+2q\kappa}, & q \ge Q(\kappa), \end{cases} \tag{5.8}$$

where $\gamma = \gamma(q,\kappa)$ and $Q(\kappa)$ are given by (5.3) and (5.2), respectively.

6. Exterior problem and spectrum of the bulk of SLE

A similar picture takes place in the case of the exterior problem. The approach based on the analysis of singularities at $w = 1$ and the maximum principle for the parabolic PDE (3.2a)–(5.1) works for $q > \mathcal{Q}^+(\kappa)$, where

$$\mathcal{Q}^+(\kappa) = -\frac{(\kappa+4)^2(\kappa+8)}{128}.$$

It is worth noting that the rigorous derivation of the β-spectrum of the exterior problem and the bulk spectrum of SLE attempted by D. Belyaev and S. Smirnov in [**2**] based on this approach turns out to be incomplete and valid only for $q > \mathcal{Q}^+(\kappa)$ (see [**14**, Appendix]).

So far the presented method of derivation of the β-spectrum of the exterior problem for $q \le \mathcal{Q}^+(\kappa)$ relies on exact solutions of the PDE for the moments of derivatives along a countable number of curves in the parametric (q,κ)-plane [**14**]: By analogy with the interior problem one seeks a solution of the PDE (3.2a)–(5.1) that is now analytic at $w = \infty$:

$$\rho = \left((1-w^{-1})(1-\bar{w}^{-1})\right)^{\gamma}\Theta,$$

where Θ is a Fourier polynomial of degree M:

$$\Theta(w,\bar{w}) = \sum_{n=-M}^{M} w^{-n} f_n(\xi),$$
$$\xi = 1/(w\bar{w}), \quad f_{-n}(\xi) = \xi^n f_n(\xi), \quad \bar{f}_n = f_n.$$

The condition of polynomiality (i.e. the truncation condition) determines a countable number of curves in the parametric (q,κ)-plane:

$$\kappa = 2\frac{M-\gamma}{M^2+2M\gamma+\gamma},$$
$$q = \frac{\gamma(M+\gamma)(2M+1+\gamma)}{M^2+2M\gamma+\gamma}.$$

Again, along these curves, the derivation of the β-spectrum reduces to the analysis of the eigenvalues of a three-diagonal $(2M+1)\times(2M+1)$ matrix [**14**]. In this version of SLE the β-spectrum has the following form:

$$\beta = \begin{cases} \kappa\frac{\gamma(q,\kappa)^2}{2} - 2\gamma(q,\kappa) - 1, & q \le -1 - \frac{3\kappa}{8}, \\ \kappa\frac{\gamma(q,\kappa)^2}{2}, & -1-\frac{3\kappa}{8} \le q \le \frac{3(\kappa+4)^2}{32\kappa}, \\ q - \frac{(\kappa+4)^2}{16\kappa}, & q \ge \frac{3(\kappa+4)^2}{32\kappa}. \end{cases}$$

Note that, unlike the interior problem, in this version of SLE the transition in the spectrum does not take place at $q = \mathcal{Q}^+(\kappa)$. Although there is a transition in blow-up rates of solutions at the unit circle $|w| = 1$ far from the moving tip $w = 1$, the average β-spectrum is still determined by the behavior of the solution as $w \to 1$.

However, this transition in blow-up rates leads to complications in the derivation of the bulk spectrum of SLE. In more detail: The bulk spectrum is independent of the version of SLE and defined for generic points of the curve away from the tip. For finite-time radial evolution it is determined through integration away from the tip $w = 1$:

$$\beta_{\text{bulk}}(q) = \limsup_{\epsilon\downarrow 0} \frac{\log\int_\delta^{2\pi-\delta}\langle|\tilde{F}'(w=e^{\epsilon+i\varphi})|^q\rangle d\varphi}{-\log\epsilon}. \tag{6.1}$$

On the other hand, in contrast to the finite-time evolution the exterior whole-plane SLE has two tips: one moving tip as well as a tip at the origin of the curve.[5] The conformal mapping has an additional singularity at the origin of the curve that gives its own contribution to the integral. An effect of this contribution results in the above-mentioned transition in the blow-up rates away from $\varphi = 0$ at $q = \mathcal{Q}^+(\kappa)$ (for details see [**14**, Appendix]). Therefore, the definition of (6.1) is not suitable for all values of q in the case of the whole-plane SLE.[6]

Thus, to complete the derivation of the bulk spectrum one should consider blow-up rates away from $\varphi = 0$ for both versions of the whole-plane SLE: the exterior version for $q \ge 0$ and the interior version for $q \le 0$. The bulk spectrum so

[5] An "immovable tip" at the origin $z = 0$ in the exterior problem corresponds to a "tip" at $z = \infty$ of the interior problem.

[6] The simplest example illustrating the above statement is the whole-plane SLE at κ=0: The mapping in this case is $\mathcal{F}(w,t) = e^t(w+1)^2/w$. Integration out of the moving tip $w = 1$ does not remove the contribution at the origin $w = -1$, and as a consequence the "bulk spectrum" so defined does not differ from the tip spectrum for all values of q.

derived has a Legendre transform which corresponds to the multi-fractal harmonic spectrum predicted from quantum gravity by B. Duplantier [**4**].

7. Conclusion

Summary: From elementary considerations it follows that any solution of the equation for moments (3.2a)/(3.5) which is analytic and non-vanishing at $w = \infty/w = 0$ is unique (up to a constant factor). Therefore, it provides a point in the β-spectrum of the exterior/interior whole-plane LLE, respectively.

The analysis of exact solutions with such properties allows one to compute exactly the sets of points of the average β-spectrum for a Loewner evolution driven by certain classes of Lévy processes, as well as find full tip and bulk β-spectra of the whole-plane SLE.

Acknowledgement

The authors would like to acknowledge help received from J.-J. Sansuc. This work was supported by the European Commission 7th framework IEF grants.

References

[1] David Applebaum, *Lévy processes—from probability to finance and quantum groups*, Notices Amer. Math. Soc. **51** (2004), no. 11, 1336–1347. MR2105239 (2005j:60096)

[2] D. Beliaev and S. Smirnov, *Harmonic measure and SLE*, Comm. Math. Phys. **290** (2009), no. 2, 577–595, DOI 10.1007/s00220-009-0864-7. MR2525631 (2011c:60265)

[3] John Cardy, *SLE for theoretical physicists*, Ann. Physics **318** (2005), no. 1, 81–118, DOI 10.1016/j.aop.2005.04.001. MR2148644

[4] Bertrand Duplantier, *Conformally invariant fractals and potential theory*, Phys. Rev. Lett. **84** (2000), no. 7, 1363–1367, DOI 10.1103/PhysRevLett.84.1363. MR1740371 (2001c:82040)

[5] Bertrand Duplantier, Thi Phuong Chi Nguyen, Thi Thuy Nga Nguyen, and Michel Zinsmeister, *Coefficient estimates for whole-plane SLE processes* (2011), available at `http://hal.inria.fr/hal-00609774`.

[6] ———, *The coefficient problem and multifractality of whole-plane SLE and LLE* (2012), available at `http://arxiv.org/pdf/1211.2451v1.pdf`.

[7] Ilya A. Gruzberg, *Stochastic geometry of critical curves, Schramm-Loewner evolutions and conformal field theory*, J. Phys. A **39** (2006), no. 41, 12601–12655, DOI 10.1088/0305-4470/39/41/S01. MR2277452 (2007i:82022)

[8] Thomas C. Halsey, Mogens H. Jensen, Leo P. Kadanoff, Itamar Procaccia, and Boris I. Shraiman, *Fractal measures and their singularities: the characterization of strange sets*, Phys. Rev. A (3) **33** (1986), no. 2, 1141–1151, DOI 10.1103/PhysRevA.33.1141. MR823474 (87h:58125a)

[9] Matthew B. Hastings, *Exact multifractal spectra for arbitrary laplacian random walks*, Phys. Rev. Lett. **88** (2002), no. 5, 055506, 4 pp.

[10] Gregory F. Lawler, *Conformally invariant processes in the plane*, Mathematical Surveys and Monographs, vol. 114, American Mathematical Society, Providence, RI, 2005. MR2129588 (2006i:60003)

[11] ———, *Conformal invariance and 2D statistical physics*, Bull. Amer. Math. Soc. (N.S.) **46** (2009), no. 1, 35–54, DOI 10.1090/S0273-0979-08-01229-9. MR2457071 (2010a:60283)

[12] Igor Loutsenko, *SLE_κ: correlation functions in the coefficient problem*, J. Phys. A **45** (2012), no. 27, 275001, 10 pp., DOI 10.1088/1751-8113/45/27/275001. MR2947222

[13] Igor Loutsenko and Oksana Yermolayeva, *On exact multi-fractal spectrum of the whole-plane SLE* (2012), available at `http://arxiv.org/abs/1203.2756`.

[14] ———, *Average harmonic spectrum of the whole-plane SLE*, J. Stat. Mech. Theory Exp. **2013** (2013), no. 4, P04007, 17 pp. MR3077814

[15] P. Oikonomou, I. Rushkin, I. A. Gruzberg, and L. P. Kadanoff, *Global properties of stochastic Loewner evolution driven by Lévy processes*, J. Stat. Mech. Theory Exp. **2008** (2008), no. 1, P01019, 27 pp. MR2373527

[16] I. Rushkin, P. Oikonomou, L. P. Kadanoff, and I. A. Gruzberg, *Stochastic Loewner evolution driven by Lévy processes*, J. Stat. Mech. Theory Exp. **2006** (2006), no. 1, P01001, 21 pp. (electronic). MR2205677

IMJ-PRG, Université Paris Diderot, bât. Sophie Germain, case 7012, 75205 Paris Cedex 13, France
E-mail address: anne.boutet-de-monvel@imj-prg.fr

IMJ-PRG, Université Paris Diderot, bât. Sophie Germain, case 7012, 75205 Paris Cedex 13, France
E-mail address: loutsenko@math.jussieu.fr
Current address: Centre de recherches mathématiques, Université de Montréal, P.O. Box 6128, Centre-ville Station, H3C 3J7 Montréal (Québec), Canada
E-mail address: loutseni@crm.umontreal.ca

Institut Henri Poincaré, Université Pierre et Marie Curie, 11 rue Pierre et Marie Curie, 75231 Paris Cedex 05, France
E-mail address: yermolay@ihp.jussieu.fr

Amer. Math. Soc. Transl.
Volume **233**, 2014

Asymptotic Behavior of Large Eigenvalues of a Modified Jaynes–Cummings Model

Anne Boutet de Monvel and Lech Zielinski

To Vladimir A. Marchenko, with deep affection, admiration, and gratitude

ABSTRACT. We prove a large n asymptotic formula for the nth eigenvalue of infinite Jacobi matrices whose structure is motivated by the Jaynes–Cummings model with rotating-wave approximation (RWA).

1. Introduction

1.1. Preliminaries. Following A. Boutet de Monvel, S. Naboko, L. O. Silva [**1**] we consider a self-adjoint operator in $l^2 := l^2(\mathbb{N}^*)$, $\mathbb{N}^* = \{1, 2, \dots\}$ defined by the infinite real Jacobi matrix

$$\begin{pmatrix} d(1) & a(1) & 0 & 0 & \dots \\ a(1) & d(2) & a(2) & 0 & \dots \\ 0 & a(2) & d(3) & a(3) & \dots \\ 0 & 0 & a(3) & d(4) & \dots \\ \vdots & \vdots & \vdots & \vdots & \ddots \end{pmatrix}. \tag{1.1}$$

We call it the "modified Jaynes–Cummings model" if

$$\begin{cases} d(k) = k^\alpha + v(k) \\ a(k) = c\, k^\gamma, \end{cases} \tag{1.2}$$

with some constants $\alpha > \gamma > 0$, $c > 0$ and $v\colon \mathbb{N}^* \to \mathbb{R}$ periodic of period $N \geq 1$; i.e.

$$v(k+N) = v(k) \text{ for any } k \in \mathbb{N}^*. \tag{1.3}$$

Our motivation comes from the fact that the interactions of two-level atoms and one-mode field in quantum optics with the rotating-wave approximation can be described by the "Jaynes–Cummings Hamiltonian" defined by (1.1)–(1.2) with $\alpha = 1$, $\gamma = \frac{1}{2}$ and v periodic of period 2 (see, e.g., È. A. Tur [**9**]).

Since (1.2) with $\alpha > \gamma$ ensures the domination of the diagonal and $d(k) \to \infty$ as $k \to \infty$, the corresponding self-adjoint operator J is bounded below and has

2010 *Mathematics Subject Classification.* Primary 47B36; Secondary 47A10, 47A75, 15A42, 47A55.

Key words and phrases. Jacobi matrices, Jaynes–Cummings model, eigenvalue estimates.

compact resolvent (see **[6]** or **[5**, Proposition 2.1]). We denote by

$$\lambda_1(J) \leq \cdots \leq \lambda_n(J) \leq \dots$$

its eigenvalues, enumerated in non-decreasing order, counting multiplicities. The paper **[1]** gives the asymptotic behavior of $\lambda_n(J)$ as $n \to \infty$ for the model (1.1)–(1.2) with $\alpha = 2$ and $\gamma = \frac{1}{2}$ using an approximation based on a method of successive diagonalisations (see also **[7]**). Similar ideas were used in **[3, 8]** to describe the asymptotic behavior of $\lambda_n(J)$ if $1 + \gamma < \alpha$. If $1 + \gamma < \alpha \leq 2$, then one obtains the large n asymptotic formula

$$\lambda_n(J) = n^\alpha + v(n) + \mathrm{o}(1), \tag{1.4}$$

reflecting the oscillations determined by the periodic nature of v.

The main purpose of this paper is to exhibit a radical change of the asymptotic behavior of $\lambda_n(J)$ in the case when $\alpha = 1$ and $0 < \gamma < \frac{1}{2}$. The new phenomenon consists of the absence of a periodic modulation of large eigenvalues. The case of the Jaynes–Cummings model, i.e., $\alpha = 1$ and $\gamma = \frac{1}{2}$ is more complicated to analyze, but a similar phenomenon holds.

1.2. Main result. Let $l^2 = l^2(\mathbb{N}^*)$ denote the Hilbert space of square-summable complex sequences $x\colon \mathbb{N}^* \to \mathbb{C}$ equipped with the scalar product $\langle x, y\rangle := \sum_{k=1}^\infty \overline{x(k)} y(k)$ and the norm $\|x\|_{l^2} := \sqrt{\langle x, x\rangle}$. Let $\{d(k)\}_{k=1}^\infty$, $\{a(k)\}_{k=1}^\infty$ be as in (1.2), with $\alpha > \gamma > 0$ and v as in (1.3). We denote

$$\mathcal{D} := \left\{x \in l^2 : \sum_{k=1}^{\infty} d(k)^2 |x(k)|^2 < \infty\right\} \tag{1.5}$$

and define $J\colon \mathcal{D} \to l^2$ by the formula

$$(Jx)(k) = d(k)x(k) + a(k)x(k+1) + a(k-1)x(k-1), \tag{1.6}$$

where, by convention, $x(0) = 0$ and $a(0) = 0$. Our main result deals with the case $\alpha = 1$ and $0 < \gamma < \frac{1}{2}$.

THEOREM 1.1. *Let J be the self-adjoint operator defined in $l^2(\mathbb{N}^*)$ by (1.6), where*

$$\begin{cases} d(k) = k + v(k), \\ a(k) = c\, k^\gamma. \end{cases} \tag{1.7}$$

We assume

(i) *v is real-valued, periodic of period $N \geq 1$,*
(ii) *$0 < \gamma < \frac{1}{2}$ and $c > 0$,*

and then we denote

$$\rho_N := \max_{1 \leq k \leq N} |v(k) - \langle v\rangle| \quad \textit{where} \quad \langle v\rangle := \frac{1}{N} \sum_{1 \leq k \leq N} v(k), \tag{1.8}$$

and assume that

(iii) *$\rho_N < 1/(\pi\sqrt{N})$ when $N \geq 3$ and $\rho_2 < 1/2$ for $N = 2$.*

If $\lambda_n(J)$ denotes the nth eigenvalue of J, we then have the large n asymptotic formula

$$\lambda_n(J) = n + \langle v\rangle + \mathrm{O}\left(n^{-\gamma/2} \ln n + n^{2\gamma - 1}\right). \tag{1.9}$$

1.3. Plan of the paper. Our proof uses special properties of auxiliary operators J_n and J_n' acting in $\mathcal{H} := l^2(\mathbb{Z})$. Our choice of working in $\mathcal{H}$ instead of l^2 is motivated by the fundamental identity (2.5) and the use of Fourier series (see Section 4).

The operators J_n have frozen off-diagonal entries and the basic ingredient of our approach is a trace estimate for J_n formulated in Proposition 2.1. We begin the proof of Proposition 2.1 in Section 3 by introducing an auxiliary spatial cut-off and reformulating the trace estimate in Proposition 3.1. Next we reduce Proposition 3.1 to Proposition 3.2 by Fourier transform and finally prove Proposition 3.2 in Section 4.

In Section 5 we investigate operators J_n' which differ from J_n by an additional spatial cut-off, allowing us to replace easily $\mathcal{H}$ by l^2. Moreover, it is easy to replace J_n by J_n' in the trace estimate and in Section 6 we use the trace estimate for J_n' to obtain the asymptotic formula for eigenvalues of J_n' formulated in Proposition 2.3.

We complete the proof of Theorem 1.1 in Section 7, where we prove Proposition 2.4, which ensures that the nth eigenvalue of J is close to a suitable eigenvalue of J_n'.

The paper is organized according to the following scheme

$$\begin{array}{l} \text{Lem. } 4.1 \Longrightarrow \text{Prop. } 3.2 \Longrightarrow \text{Prop. } 3.1 \Longrightarrow \text{Prop. } 2.1 \\ \qquad\qquad\qquad\qquad\qquad\quad \text{Lem. } 5.1 \Longrightarrow \text{Prop. } 2.2 \end{array}\Bigg\} \Longrightarrow \begin{array}{l}\text{Prop. } 2.3 \\ \text{Prop. } 2.4\end{array}\Bigg\} \Longrightarrow \text{Th. } 1.1.$$

2. Outline

2.1. Notations. Let $\mathcal{H} := l^2(\mathbb{Z})$ denote the Hilbert space of square-summable sequences $x\colon \mathbb{Z} \to \mathbb{C}$ with the scalar product $\langle x, y\rangle := \sum_{k\in\mathbb{Z}} \overline{x(k)}y(k)$ and the norm

$$\|x\|_{\mathcal{H}} := \Big(\sum_{k\in\mathbb{Z}} |x(k)|^2\Big)^{1/2}. \tag{2.1}$$

We denote by $\mathcal{B}(\mathcal{H})$ the algebra of bounded operators on $\mathcal{H}$ equipped with the operator norm $\|\cdot\|_{\mathcal{B}(\mathcal{H})}$ and by $\mathcal{B}_1(\mathcal{H}) \subset \mathcal{B}(\mathcal{H})$ the ideal of trace class operators with the norm $\|Q\|_{\mathcal{B}_1(\mathcal{H})} := \operatorname{tr}\sqrt{Q^*Q}$, where Q^* is the adjoint of Q. We also use the notations $\operatorname{Re} Q := \frac{1}{2}(Q + Q^*)$ and $\operatorname{Im} Q := \frac{1}{2\mathrm{i}}(Q - Q^*)$.

Let $\{\mathrm{e}_k\}_{k\in\mathbb{Z}}$ denote the canonical basis of $\mathcal{H}$, i.e., $\mathrm{e}_k(k) = 1$, $\mathrm{e}_k(j) = 0$ if $j \neq k$, and let $\Lambda\colon \mathcal{H}_1 \to \mathcal{H}$ be the closed linear operator satisfying $\Lambda \mathrm{e}_k = k\mathrm{e}_k$ $(k \in \mathbb{Z})$ with domain

$$\mathcal{H}_1 := \Big\{x \in \mathcal{H} : \sum_{k\in\mathbb{Z}} k^2|x(k)|^2 < \infty\Big\}. \tag{2.2}$$

If $b\colon \mathbb{Z} \to \mathbb{C}$, then $b(\Lambda)$ is the closed linear operator satisfying $b(\Lambda)\mathrm{e}_k = b(k)\mathrm{e}_k$ $(k \in \mathbb{Z})$. Let $S \in \mathcal{B}(\mathcal{H})$ denote the shift $S\mathrm{e}_k = \mathrm{e}_{k+1}$ $(k \in \mathbb{Z})$.

Assumption on v. Further on $v\colon \mathbb{Z} \to \mathbb{R}$ is periodic of period $N \geq 1$; hence

$$\|v(\Lambda)\|_{\mathcal{B}(\mathcal{H})} = \sup_{j\in\mathbb{Z}}|v(j)| = \max_{1\leq j\leq N}|v(j)|.$$

Since $\lambda_n(J - \langle v\rangle) = \lambda_n(J) - \langle v\rangle$, without loss of generality we can assume that

$$\langle v\rangle = 0. \tag{2.3}$$

Let ρ_N be as in (1.8). Under assumption (2.3) we find

$$\rho_N = \|v(\Lambda)\|_{\mathcal{B}(\mathcal{H})}. \tag{2.4}$$

2.2. The starting point of our analysis is the simple result

LEMMA 2.1. *For every $t \in \mathbb{R}$ one has*

$$\mathrm{e}^{2it\,\mathrm{Im}\,S}(\Lambda + 2t\,\mathrm{Re}\,S)\mathrm{e}^{-2it\,\mathrm{Im}\,S} = \Lambda. \tag{2.5}$$

PROOF. Since (2.5) holds when $t = 0$, it suffices to check that

$$\frac{\mathrm{d}}{\mathrm{d}t}\Big(\mathrm{e}^{2it\,\mathrm{Im}\,S}(\Lambda + 2t\,\mathrm{Re}\,S)\mathrm{e}^{-2it\,\mathrm{Im}\,S}\Big) = 0 \tag{2.6}$$

holds for all $t \in \mathbb{R}$. The left-hand side of (2.6) indeed has the form

$$\mathrm{e}^{t(S-S^{-1})}([S - S^{-1}, \Lambda] + S + S^{-1})\mathrm{e}^{-t(S-S^{-1})}$$

and the direct computation of the commutator

$$[S^k, \Lambda]\mathrm{e}_n = S^k\Lambda\mathrm{e}_n - \Lambda S^k\mathrm{e}_n = S^k n\mathrm{e}_n - (n+k)\mathrm{e}_{n+k} = -kS^k\mathrm{e}_n$$

for $k \in \mathbb{Z}$ gives $[S - S^{-1}, \Lambda] = -S - S^{-1}$; i.e., (2.6) holds. □

2.3. Operators J_n. For $n \geq 1$ we define $J_n : \mathcal{H}_1 \to \mathcal{H}$ by

$$J_n := \Lambda + v(\Lambda) + 2a(n)\,\mathrm{Re}\,S = J_n^0 + v(\Lambda) \tag{2.7}$$

with

$$J_n^0 := \Lambda + 2a(n)\,\mathrm{Re}\,S. \tag{2.8}$$

By Lemma 2.1 we have $J_n^0 = \mathrm{e}^{-2ia(n)\,\mathrm{Im}\,S}\Lambda\mathrm{e}^{2ia(n)\,\mathrm{Im}\,S}$; hence $\sigma(J_n^0) = \sigma(\Lambda) = \mathbb{Z}$. Then, by an elementary perturbation argument, and assuming $\langle v\rangle = 0$ we have

$$\sigma(J_n) \subset \bigcup_{k\in\mathbb{Z}} [k - \rho_N,\, k + \rho_N] \tag{2.9}$$

with $\rho_N = \|v(\Lambda)\|_{\mathcal{B}(\mathcal{H})}$. All eigenvalues of Λ and J_n^0 are simple, i.e., of multiplicity 1, then by assumption $\rho_N < \frac{1}{2}$ all eigenvalues of J_n are simple as well.

PROPOSITION 2.1 (trace estimate for J_n). *If $\langle v\rangle = 0$ we have the large n estimate*

$$\mathcal{G}_n^0 := \mathrm{tr}\,\chi(J_n - n) - \mathrm{tr}\,\chi(\Lambda - n) = \mathrm{O}\big(n^{-\gamma/2}\ln n\big) \tag{2.10}$$

for any function $\chi\colon \mathbb{R} \to \mathbb{C}$ such that

$$\chi(\lambda) = \int_{\mathbb{R}} \hat{\chi}(t)\mathrm{e}^{it\lambda}\mathrm{d}t \tag{2.11}$$

with $\hat{\chi} \in C_0^\infty(\mathbb{R})$.

PROOF. Successive reformulations of the trace estimate (2.10) are given in Propositions 3.1 and 3.2 and a proof of Proposition 3.2 is given in Section 4.

REMARK 2.1. If $\hat{\chi} \in C_0^\infty(\mathbb{R})$, then $\chi \in \mathcal{S}(\mathbb{R})$—the Schwartz class of rapidly decreasing functions on $\mathbb{R}$. Hence, $\hat{\chi}$ is the Fourier transform of χ; i.e.

$$\hat{\chi}(t) = \frac{1}{2\pi}\int_{\mathbb{R}} \mathrm{e}^{-it\lambda}\chi(\lambda)\,\mathrm{d}\lambda. \tag{2.12}$$

Moreover, $\chi(J_n - n)$ and $\chi(\Lambda - n)$ are trace class operators for $\chi \in \mathcal{S}(\mathbb{R})$.

2.4. Operators J'_n. We fix a real-valued cut-off function $\theta^+ \in C^\infty(\mathbb{R})$ satisfying

$$\theta^+(t) = \begin{cases} 0 & \text{when } t \le \frac{1}{3}, \\ 1 & \text{when } t \ge \frac{1}{2}. \end{cases}$$

Then, for $n \ge 1$ and $k \in \mathbb{Z}$ we define

$$a_n(k) := a(n)\, \theta^+\Big(\frac{k}{n}\Big) \tag{2.13}$$

and introduce an operator $J'_n \colon \mathcal{H}_1 \to \mathcal{H}$ which will serve as a link between J_n and J:

$$J'_n := \Lambda + v(\Lambda) + 2\,\mathrm{Re}\,(S\, a_n(\Lambda)) = J'_{n,0} + v(\Lambda) \tag{2.14}$$

with

$$J'_{n,0} := \Lambda + 2\,\mathrm{Re}\,(S\, a_n(\Lambda))\,. \tag{2.15}$$

Since $a_n(k) = 0$ for $k \le 0$, the closed subspace of $\mathcal{H}$ generated by $\{\mathrm{e}_k\}_{k=1}^\infty$ is invariant under J'_n, and J'_n acts on its orthogonal complement as $\Lambda + v(\Lambda)$. Moreover, $a_n(k) = a(n)$ for $k \ge n/2$. Thus J'_n has discrete spectrum and we can write $\sigma(J'_n) = \{\lambda_k(J'_n)\}_{k\in\mathbb{Z}}$ with $\lambda_k(J'_n) = k + v(k)$ for $k \le 0$ and $\lambda_1(J'_n) \le \lambda_2(J'_n) \le \dots \le \lambda_k(J'_n) \le \dots$ for $k \ge 1$.

PROPOSITION 2.2 (rough estimates of the spectrum $\sigma(J'_n)$). *Assume $\langle v\rangle = 0$ and $\rho_N < \frac{1}{2}$ with $\langle v\rangle$ and ρ_N as in (1.8). Let $\varepsilon > 0$ be such that $\varepsilon + \rho_N < \frac{1}{2}$. Then there is $n_\varepsilon \ge 1$ such that for any $n \ge n_\varepsilon$ and all $k \in \mathbb{Z}$ we have*

$$\sigma(J'_n) \cap \big(k - \tfrac{1}{2},\, k + \tfrac{1}{2}\big] = \{\lambda_k(J'_n)\}. \tag{2.16}$$

Hence, for such n's all eigenvalues of J'_n are of multiplicity 1. Moreover, also for $n \ge n_\varepsilon$

$$\sup_{k\in\mathbb{Z}} |\lambda_k(J'_n) - k| \le \rho_N + \varepsilon, \tag{2.17}$$

$$\sup_{k\in\mathbb{Z}} |\lambda_{k+N}(J'_n) - \lambda_k(J'_n) - N| = \mathrm{O}(n^{\gamma-1}). \tag{2.18}$$

PROOF. See Section 5.

PROPOSITION 2.3. *If $\langle v\rangle = 0$, then we have the large n estimate*

$$\lambda_n(J'_n) = n + \mathrm{O}\big(n^{-\gamma/2}\ln n\big). \tag{2.19}$$

PROOF. It's given in Section 6, using Propositions 2.1 and 2.2.

2.5. End of proof of Theorem 1.1. Due to Proposition 2.3 it suffices to prove

PROPOSITION 2.4. *We have the large n estimate*

$$\lambda_n(J) = \lambda_n(J'_n) + \mathrm{O}(n^{2\gamma-1}). \tag{2.20}$$

PROOF. It's given in Section 7.

3. Reformulations of the trace estimate (2.10)

3.1. First reformulation. Inspired by (2.5) we replace J_n by the operators

$$L_n := \mathrm{e}^{2\mathrm{i}a(n)\,\mathrm{Im}\,S}\, J_n\, \mathrm{e}^{-2\mathrm{i}a(n)\,\mathrm{Im}\,S}. \tag{3.1}$$

Further on $\mathcal{G}_n^0$ and χ are as in Proposition 2.1. Then it is clear that

$$\mathcal{G}_n^0 = \mathrm{tr}\,\chi(L_n - n) - \mathrm{tr}\,\chi(\Lambda - n). \tag{3.2}$$

Next we fix a real-valued cut-off function $\theta \in C^\infty(\mathbb{R})$ such that

$$\theta(t) = \begin{cases} 0 & \text{if } t \le \frac{1}{2}\,, \\ 1 & \text{if } \frac{3}{4} \le t \le \frac{3}{2}\,, \\ 0 & \text{if } t \ge 2. \end{cases}$$

It is clear that

$$||(I - \theta(\Lambda/n))\chi(\Lambda - n)||_{\mathcal{B}_1(\mathcal{H})} = \mathrm{O}(n^{-\infty}) \tag{3.3}$$

and in Section 8.3 we prove the following estimate

$$||(I - \theta(\Lambda/n))\chi(L_n - n)||_{\mathcal{B}_1(\mathcal{H})} = \mathrm{O}(n^{\gamma-1}). \tag{3.4}$$

By (3.2), (3.3), and (3.4), Proposition 2.1 is equivalent to

PROPOSITION 3.1. *If χ is as in Proposition 2.1 and θ as above, we have the large n estimate*

$$\mathcal{G}_n := \mathrm{tr}\,(\theta(\Lambda/n)\,[\chi(L_n - n) - \chi(\Lambda - n)]) = \mathrm{O}(n^{-\gamma/2}\ln n). \tag{3.5}$$

3.2. Fourier transform. Taking $L_n - n$ and $\Lambda - n$ instead of λ in (2.11) we find

$$\chi(L_n - n) - \chi(\Lambda - n) = \int_{\mathbb{R}} \hat{\chi}(t)\, \mathrm{e}^{-\mathrm{i}tn} \left(\mathrm{e}^{\mathrm{i}tL_n} - \mathrm{e}^{\mathrm{i}t\Lambda}\right) \mathrm{d}t.$$

We denote

$$U_n(t) := \mathrm{e}^{-\mathrm{i}t\Lambda}\mathrm{e}^{\mathrm{i}tL_n},$$
$$r_n(\Lambda) := -\theta(\Lambda/n)\big(\Lambda - n - \tfrac{1}{2}\big)^{-1}.$$

Then we can write

$$\begin{aligned}\mathcal{G}_n &= \int_{\mathbb{R}} \hat{\chi}(t)\ \mathrm{tr}\Big(\theta(\Lambda/n)\mathrm{e}^{\mathrm{i}t(\Lambda-n)}(U_n(t) - I)\Big)\,\mathrm{d}t \\ &= \int_{\mathbb{R}} \hat{\chi}(t)\mathrm{e}^{\mathrm{i}t/2}\ \mathrm{tr}\left(r_n(\Lambda)\Big(\mathrm{i}\frac{\mathrm{d}}{\mathrm{d}t}\,\mathrm{e}^{\mathrm{i}t(\Lambda-n-1/2)}\Big)(U_n(t) - I)\right)\mathrm{d}t,\end{aligned}$$

and integration by parts gives

$$\begin{aligned}&\mathcal{G}_n = \mathcal{G}_n' - \mathrm{i}\mathcal{G}_n'' \quad \text{with} \\ &\mathcal{G}_n' := \int_{\mathbb{R}} \Big(\frac{1}{2}\hat{\chi}(t) - \mathrm{i}\frac{\mathrm{d}\hat{\chi}}{\mathrm{d}t}(t)\Big)\ \mathrm{tr}\Big(\mathrm{e}^{\mathrm{i}t(\Lambda-n)}r_n(\Lambda)(U_n(t) - I)\Big)\mathrm{d}t, \\ &\mathcal{G}_n'' := \int_{\mathbb{R}} \hat{\chi}(t)\ \mathrm{tr}\Big(\mathrm{e}^{\mathrm{i}t(\Lambda-n)}r_n(\Lambda)\frac{\mathrm{d}}{\mathrm{d}t}U_n(t)\Big)\mathrm{d}t.\end{aligned}$$

3.3. Expansion of $U_n(t)$. We observe that

$$-\mathrm{i}\frac{\mathrm{d}}{\mathrm{d}t}U_n(t) = H_n(t)U_n(t), \qquad U_n(0) = I \tag{3.6}$$

holds with $H_n(t) := \mathrm{e}^{-\mathrm{i}t\Lambda}(L_n - \Lambda)\mathrm{e}^{\mathrm{i}t\Lambda}$ and we have the expansion formula

$$U_n(t) = I + \mathrm{i}\int_0^t \mathrm{d}t_1\, H_n(t_1) + \sum_{\nu=2}^{\infty} \mathrm{i}^\nu \int_0^t \mathrm{d}t_1 \dots \int_0^{t_{\nu-1}} \mathrm{d}t_\nu\, H_n(t_1)\dots H_n(t_\nu). \tag{3.7}$$

For $\nu \in \mathbb{N}^*$, $(t_1, \dots, t_\nu) \in \mathbb{R}^\nu$ we denote

$$g_{n,\nu}(t; t_1, \dots, t_\nu) := \mathrm{i}^\nu \operatorname{tr}\Big(\mathrm{e}^{\mathrm{i}t(\Lambda - n)} r_n(\Lambda) H_n(t_1)\dots H_n(t_\nu)\Big). \tag{3.8}$$

Thus we can write

$$\mathcal{G}_n = \mathcal{G}_n' - \mathrm{i}\mathcal{G}_n'' = \sum_{\nu=1}^{\infty}(\mathcal{G}_{n,\nu}' - \mathrm{i}\mathcal{G}_{n,\nu}'') \quad \text{with}$$

$$\mathcal{G}_{n,1}' := \int_{\mathbb{R}} \mathrm{d}t\left(\frac{1}{2}\hat{\chi}(t) - \mathrm{i}\frac{\mathrm{d}\hat{\chi}}{\mathrm{d}t}(t)\right)\int_0^t \mathrm{d}t_1\, g_{n,1}(t; t_1),$$

$$\mathcal{G}_{n,\nu}' := \int_{\mathbb{R}} \mathrm{d}t\left(\frac{1}{2}\hat{\chi}(t) - \mathrm{i}\frac{\mathrm{d}\hat{\chi}}{\mathrm{d}t}(t)\right)\int_0^t \mathrm{d}t_1 \dots \int_0^{t_{\nu-1}} \mathrm{d}t_\nu\, g_{n,\nu}(t; t_1, \dots, t_\nu) \quad (\nu \geq 2),$$

$$\mathcal{G}_{n,1}'' := \int_{\mathbb{R}} \mathrm{d}t\, \hat{\chi}(t)\, g_{n,1}(t; t),$$

$$\mathcal{G}_{n,2}'' := \int_{\mathbb{R}} \mathrm{d}t\, \hat{\chi}(t) \int_0^t \mathrm{d}t_2\, g_{n,2}(t; t, t_2),$$

$$\mathcal{G}_{n,\nu}'' := \int_{\mathbb{R}} \mathrm{d}t\, \hat{\chi}(t) \int_0^t \mathrm{d}t_2 \dots \int_0^{t_{\nu-1}} \mathrm{d}t_\nu\, g_{n,\nu}(t; t, t_2, \dots, t_\nu) \quad (\nu \geq 3).$$

3.4. Last reformulation. It is now clear that Proposition 3.1 follows from

PROPOSITION 3.2. *Assume* $\operatorname{supp}\hat{\chi} \subset [-\tau_0, \tau_0]$. *Then we can find* $C > 0$ *such that*

$$|g_{n,1}(t; t_1)| \leq C n^{-\gamma/2} \ln n, \tag{3.9}$$

$$\int_{-\tau_0}^{\tau_0} |g_{n,\nu}(t; t_1, \dots, t_\nu)|\, \mathrm{d}t_\nu \leq C^\nu n^{-\gamma/2} \ln n \tag{3.10}$$

hold for all $t, t_1, \dots, t_{\nu-1} \in [-\tau_0, \tau_0]$ *and* $\nu \geq 2$.

4. Proof of Proposition 3.2

4.1. Proof of estimate (3.9). Since v is N-periodic with $\langle v \rangle = 0$ we can write

$$v(k) = \sum_{\omega \in \Omega^*} c_\omega \mathrm{e}^{\mathrm{i}\omega k}, \tag{4.1}$$

where $c_\omega \in \mathbb{C}$ and $\Omega^* = \{2\pi k/N\}_{k=1,\dots,N-1}$. Thus $L_n = \Lambda + \tilde{V}_n$ holds with

$$\tilde{V}_n := \sum_{\omega \in \Omega^*} c_\omega \mathrm{e}^{2\mathrm{i}a(n)\operatorname{Im} S}\mathrm{e}^{\mathrm{i}\omega\Lambda}\mathrm{e}^{-2\mathrm{i}a(n)\operatorname{Im} S} \tag{4.2}$$

and using $\mathrm{e}^{-\mathrm{i}\omega\Lambda}\mathrm{e}^{2\mathrm{i}a(n)\operatorname{Im} S}\mathrm{e}^{\mathrm{i}\omega\Lambda} = \mathrm{e}^{2\mathrm{i}a(n)\operatorname{Im}(\mathrm{e}^{-\mathrm{i}\omega}S)}$ we find the expression

$$\mathrm{e}^{2\mathrm{i}a(n)\operatorname{Im} S}\mathrm{e}^{\mathrm{i}\omega\Lambda}\mathrm{e}^{-2\mathrm{i}a(n)\operatorname{Im} S} = \mathrm{e}^{\mathrm{i}\omega\Lambda}\mathrm{e}^{\mathrm{i}\psi_n^\omega(S)} \tag{4.3}$$

with $\psi_n^\omega(S) := 2a(n)\,\mathrm{Im}\big((\mathrm{e}^{-\mathrm{i}\omega} - 1)S\big)$. Hence

$$H_n(t) = \mathrm{e}^{-\mathrm{i}t\Lambda}\tilde{V}_n\mathrm{e}^{\mathrm{i}t\Lambda} = \sum_{\omega\in\Omega^*} c_\omega \mathrm{e}^{\mathrm{i}\omega\Lambda}\mathrm{e}^{\mathrm{i}\psi_n^\omega(\mathrm{e}^{-\mathrm{i}t}S)}. \tag{4.4}$$

Since $\xi = t + \frac{\omega}{2}$ and $\xi = t + \frac{\omega}{2} + \pi$ are non-degenerate critical points of

$$\psi_n^\omega(\mathrm{e}^{\mathrm{i}(\xi-t)}) = 2a(n)\,\mathrm{Im}\Big((\mathrm{e}^{-\mathrm{i}\omega} - 1)\mathrm{e}^{\mathrm{i}(\xi-t)}\Big) = -4cn^\gamma \sin\frac{\omega}{2}\cos\Big(\xi - t - \frac{\omega}{2}\Big),$$

the stationary phase method ensures

$$h_n^\omega(t) := \int_0^{2\pi} \mathrm{e}^{\mathrm{i}\psi_n^\omega(\mathrm{e}^{\mathrm{i}(\xi-t)})}\,\frac{\mathrm{d}\xi}{2\pi} = \mathrm{O}(n^{-\gamma/2}). \tag{4.5}$$

However, for $b\colon \mathbb{Z} \to \mathbb{C}$ bounded and $q\colon \mathbb{C} \to \mathbb{C}$ continuous we have

$$\langle \mathrm{e}_j, b(\Lambda)q(S)\mathrm{e}_k\rangle = b(j)\int_0^{2\pi} \mathrm{e}^{\mathrm{i}(j-k)\xi}q(\mathrm{e}^{\mathrm{i}\xi})\,\frac{\mathrm{d}\xi}{2\pi}$$

and assuming b has bounded support we find $b(\Lambda)q(S) \in \mathcal{B}_1(\mathcal{H})$ with

$$\mathrm{tr}\,(b(\Lambda)q(S)) = \sum_{k\in\mathbb{Z}}\langle \mathrm{e}_k, b(\Lambda)q(S)\mathrm{e}_k\rangle = \sum_{k\in\mathbb{Z}} b(k)\int_0^{2\pi} q(\mathrm{e}^{\mathrm{i}\xi})\frac{\mathrm{d}\xi}{2\pi}. \tag{4.6}$$

To complete the proof of (3.9) it suffices to write

$$g_{n,1}(t;t_1) = \sum_{\omega\in\Omega^*} \mathrm{i}\,c_\omega \sum_{k\in\mathbb{Z}} r_n(k)\,\mathrm{e}^{\mathrm{i}t(k-n)+\mathrm{i}\omega k}\,h_n^\omega(t_1)$$

and observe that

$$\sum_{k\in\mathbb{Z}}|r_n(k)| \le \sum_{n/2\le k\le 2n} C\Big|k - n - \frac{1}{2}\Big|^{-1} = \mathrm{O}(\ln n). \tag{4.7}$$

4.2. Auxiliary results. We denote $\Psi_n^\omega(t, \mathrm{e}^{\mathrm{i}\xi}) := 2a(n)\big(\mathrm{e}^{-\mathrm{i}\omega} - 1\big)\mathrm{e}^{\mathrm{i}(\xi-t)}$ and

$$H_n^\omega(t) := \mathrm{e}^{-\mathrm{i}t\Lambda}\mathrm{e}^{2\mathrm{i}a(n)\,\mathrm{Im}\,S}\mathrm{e}^{\mathrm{i}\omega\Lambda}\mathrm{e}^{-2\mathrm{i}a(n)\,\mathrm{Im}\,S}\mathrm{e}^{\mathrm{i}t\Lambda}. \tag{4.8}$$

Then $H_n^\omega(t) = \mathrm{e}^{\mathrm{i}\omega\Lambda}\mathrm{e}^{\mathrm{i}\,\mathrm{Im}\,\Psi_n^\omega(t,S)}$ holds due to (4.3) and we claim that for $\nu \in \mathbb{N}^*$, $\underline{t} = (t_1, \dots, t_\nu) \in \mathbb{R}^\nu$, $\underline{\omega} = (\omega_1, \dots, \omega_\nu) \in (\Omega^*)^\nu$,

$$H_n^{\underline{\omega}}(\underline{t}) := H_n^{\omega_1}(t_1)\dots H_n^{\omega_\nu}(t_\nu) = \mathrm{e}^{\mathrm{i}|\underline{\omega}|_1\Lambda}\mathrm{e}^{\mathrm{i}\,\mathrm{Im}\,\Psi_n^{\underline{\omega}}(\underline{t},S)} \tag{4.9}$$

holds with $|\underline{\omega}|_1 = \omega_1 + \dots + \omega_\nu$ and $\Psi_n^{\underline{\omega}}$ satisfying $\Psi_n^{\underline{\omega}}(\underline{t}, \mathrm{e}^{\mathrm{i}\xi}) = \Psi_n^{\underline{\omega}}(\underline{t}, 1)\mathrm{e}^{\mathrm{i}\xi}$. Indeed, reasoning by induction we take $\nu \ge 2$, $\underline{\omega} = (\underline{\omega}', \omega) \in (\Omega^*)^{\nu-1} \times \Omega^*$, $\underline{t} = (\underline{t}', t) \in \mathbb{R}^{\nu-1} \times \mathbb{R}$, assume that $\Psi_n^{\underline{\omega}'}(\underline{t}', \mathrm{e}^{\mathrm{i}\xi}) = \Psi_n^{\underline{\omega}'}(\underline{t}', 1)\mathrm{e}^{\mathrm{i}\xi}$ is such that

$$H_n^{\underline{\omega}'}(\underline{t}') = \mathrm{e}^{\mathrm{i}|\underline{\omega}'|_1\Lambda}\mathrm{e}^{\mathrm{i}\,\mathrm{Im}\,\Psi_n^{\underline{\omega}'}(\underline{t}',S)}, \tag{4.10}$$

and observe that $\mathrm{e}^{-\mathrm{i}\omega\Lambda}\mathrm{e}^{\mathrm{i}\,\mathrm{Im}\,\Psi_n^{\underline{\omega}'}(\underline{t}',S)}\mathrm{e}^{\mathrm{i}\omega\Lambda} = \mathrm{e}^{\mathrm{i}\,\mathrm{Im}\,\Psi_n^{\underline{\omega}'}(\underline{t}',\,\mathrm{e}^{-\mathrm{i}\omega}S)}$ allows us to express

$$H_n^{\underline{\omega}'}(\underline{t}')H_n^\omega(t) = \mathrm{e}^{\mathrm{i}|\underline{\omega}'|_1\Lambda}\mathrm{e}^{\mathrm{i}\,\mathrm{Im}\,\Psi_n^{\underline{\omega}'}(\underline{t}',S)}\mathrm{e}^{\mathrm{i}\omega\Lambda}\mathrm{e}^{\mathrm{i}\,\mathrm{Im}\,\Psi_n^\omega(t,S)} \tag{4.11}$$

in the form (4.9) with $\Psi_n^{\underline{\omega}}(\underline{t}, \mathrm{e}^{\mathrm{i}\xi}) = \Psi_n^{\underline{\omega}}((\underline{t}', t), \mathrm{e}^{\mathrm{i}\xi}) = \Psi_n^{\underline{\omega}'}(\underline{t}', \mathrm{e}^{\mathrm{i}(\xi-\omega)}) + \Psi_n^\omega(t, \mathrm{e}^{\mathrm{i}\xi})$.

LEMMA 4.1. *Let $\Psi_n^{\underline{\omega}}$ be defined as above. Then there exists $c_0 > 0$ and a measurable function $\eta_n^{\underline{\omega}}\colon \mathbb{R}^{\nu-1} \to [0, 2\pi)$ such that one has*

$$|\Psi_n^{\underline{\omega}}(\underline{t}, \mathrm{e}^{\mathrm{i}\xi})| = |\Psi_n^{\underline{\omega}}(\underline{t}, 1)| \ge c_0 n^\gamma |t + \eta_n^{\underline{\omega}}(\underline{t}')|_{\mathrm{mod}\,\pi}, \tag{4.12}$$

where $\underline{t} = (\underline{t}', t) \in \mathbb{R}^{\nu-1} \times \mathbb{R}$ and $|s|_{\mathrm{mod}\,\pi} := \mathrm{dist}(s + \pi\mathbb{Z}, \pi\mathbb{Z})$.

PROOF. The induction formula for $\Psi_n^{\underline{\omega}}$ allows us to express

$$\frac{\Psi_n^{\underline{\omega}}(\underline{t},1)}{\Psi_n^{\omega}(t,1)} = 1 + \frac{\Psi_n^{\underline{\omega}'}(\underline{t}',\mathrm{e}^{-\mathrm{i}\omega})}{\Psi_n^{\omega}(t,1)} = 1 + \Phi_n^{\underline{\omega}}(\underline{t}')\mathrm{e}^{\mathrm{i}t} \tag{4.13}$$

with a certain function $\Phi_n^{\underline{\omega}}\colon \mathbb{R}^{\nu-1} \to \mathbb{C}$. If $\eta_n^{\underline{\omega}}(\underline{t}')$ is the argument of $\Phi_n^{\underline{\omega}}(\underline{t}')$, then

$$\operatorname{Im}\left(\frac{\Psi_n^{\underline{\omega}}(\underline{t},1)}{\Psi_n^{\omega}(t,1)}\right) = |\Phi_n^{\underline{\omega}}(\underline{t}')|\sin(t+\eta_n^{\underline{\omega}}(\underline{t}')) \tag{4.14}$$

ensures

$$|\Phi_n^{\underline{\omega}}(\underline{t}')| > \frac{1}{2} \implies \left|\frac{\Psi_n^{\underline{\omega}}(\underline{t},1)}{\Psi_n^{\omega}(t,1)}\right| > \frac{1}{2}|\sin(t+\eta_n^{\underline{\omega}}(\underline{t}'))|. \tag{4.15}$$

However, $|\Psi_n^{\omega}(t,1)| = 4a(n)\sin(\omega/2)$ and combining (4.15) with

$$|\Phi_n^{\underline{\omega}}(\underline{t}')| \le \frac{1}{2} \implies \left|\frac{\Psi_n^{\underline{\omega}}(\underline{t},1)}{\Psi_n^{\omega}(t,1)}\right| \ge \frac{1}{2} \tag{4.16}$$

we obtain $|\Psi_n^{\underline{\omega}}(\underline{t},1)| \ge \frac{2}{\pi}a(n)\sin(\omega/2)\,|t+\eta_n^{\underline{\omega}}(\underline{t}')|_{\mathrm{mod}\,\pi}$. □

4.3. Proof of estimate (3.10). Let

$$g_n^{\underline{\omega}}(t;\underline{t}) := \operatorname{tr}\Big(\mathrm{e}^{\mathrm{i}t(\Lambda-n)}r_n(\Lambda)H_n^{\underline{\omega}}(\underline{t})\Big). \tag{4.17}$$

Let $\tilde{\eta}_n^{\underline{\omega}}(\underline{t}) \in [0,2\pi)$ denote the argument of $\Psi_n^{\underline{\omega}}(\underline{t},1)$. Then combining

$$\operatorname{Im}\big(\Psi_n^{\underline{\omega}}(\underline{t},\mathrm{e}^{\mathrm{i}\xi})\big) = \operatorname{Im}\big(\Psi_n^{\underline{\omega}}(\underline{t},1)\mathrm{e}^{\mathrm{i}\xi}\big) = |\Psi_n^{\underline{\omega}}(\underline{t},1)|\,\sin(\tilde{\eta}_n^{\underline{\omega}}(\underline{t})+\xi) \tag{4.18}$$

with (4.9) and (4.6) we find the expression

$$g_n^{\underline{\omega}}(t;\underline{t}) = \sum_{k\in\mathbb{Z}} r_n(k)\,\mathrm{e}^{\mathrm{i}t(k-n)+\mathrm{i}k|\underline{\omega}|_1}h_n^{\underline{\omega}}(\underline{t}) \tag{4.19}$$

with

$$h_n^{\underline{\omega}}(\underline{t}) := \int_0^{2\pi} \mathrm{e}^{\mathrm{i}|\Psi_n^{\underline{\omega}}(\underline{t},1)|\,\sin(\xi+\tilde{\eta}_n^{\underline{\omega}}(\underline{t}))}\frac{\mathrm{d}\xi}{2\pi}. \tag{4.20}$$

Then the stationary phase formula allows us to estimate

$$|h_n^{\underline{\omega}}(\underline{t})| \le C_0|\Psi_n^{\underline{\omega}}(\underline{t},1)|^{-1/2} \tag{4.21}$$

and due to (4.7) we can estimate

$$|g_n^{\underline{\omega}}(t;\underline{t})| \le C_0\ln n\,|\Psi_n^{\underline{\omega}}(\underline{t},1)|^{-1/2}. \tag{4.22}$$

However, Lemma 4.1 allows us to estimate the left-hand side of (3.10) by

$$\sum_{\underline{\omega}\in(\Omega^*)^{\nu}} c_{\omega_1}\dots c_{\omega_\nu}\ln n\int_{-\tau_0}^{\tau_0} C_1 n^{-\gamma/2}|t_\nu+\eta_n^{\underline{\omega}}(\underline{t}')|_{\mathrm{mod}\,\pi}^{-1/2}\mathrm{d}t_\nu$$

and we get (3.10) from the fact that $t \to |t|^{-1/2}$ is locally integrable on $\mathbb{R}$.

5. Proof of Proposition 2.2

5.1. Restriction of $J'_{n,0}$ to l^2. We identify $l^2 = l^2(\mathbb{N}^*)$ with the closed subspace of $\mathcal{H} = l^2(\mathbb{Z})$ generated by $\{\mathrm{e}_k\}_{k=1}^{\infty}$ and denote $J^+_{n,0}$ the restriction to l^2 of $J'_{n,0}\colon \mathcal{H}_1 \to \mathcal{H}$ defined by (2.15); i.e.

$$J'_{n,0} := \Lambda + 2\,\mathrm{Re}\,(S\,a_n(\Lambda)),$$

where $a_n(k)$ is given by (2.13). It's easy to check that $J^+_{n,0}$ is bounded below with compact resolvent. Thus it has discrete spectrum and we can write $\sigma(J^+_{n,0}) = \{\lambda_k(J^+_{n,0})\}_{k=1}^{\infty}$ with $\lambda_1(J^+_{n,0}) \le \lambda_2(J^+_{n,0}) \le \dots \le \lambda_k(J^+_{n,0}) \le \dots$.

LEMMA 5.1. *If $J^+_{n,0}$ denotes the restriction to l^2 of $J'_{n,0}$, we have the large n estimate*

$$\sup_{k\in\mathbb{N}^*} |\lambda_k(J^+_{n,0}) - k| = \mathrm{O}(n^{2\gamma-1}). \tag{5.1}$$

PROOF. Let $S^+ \in \mathcal{B}(l^2)$ and $\Lambda^+\colon \mathcal{D} \to l^2$ denote the restriction of Λ and S to l^2 (i.e. $S^+\mathrm{e}_k = \mathrm{e}_{k+1}$ and $\Lambda^+\mathrm{e}_k = k\mathrm{e}_k$ for $k \in \mathbb{N}^*$). Moreover, we denote

$$A^+_n := 2\,\mathrm{Re}\left(S^+ a_n(\Lambda^+)\right), \tag{5.2}$$

$$B^+_n := 2\,\mathrm{Im}(S^+ a_n(\Lambda^+)). \tag{5.3}$$

Then $[S^+, \Lambda^+] = -S^+$ follows similarly as in the proof of (2.6) and we find

$$[\mathrm{i}B^+_n, \Lambda^+] = 2\,\mathrm{Re}[S^+ a_n(\Lambda^+), \Lambda^+] = 2\,\mathrm{Re}[S^+, \Lambda^+]a_n(\Lambda^+) = -A^+_n. \tag{5.4}$$

For $t \in \mathbb{R}$ we introduce

$$J_{0,n}(t) := \Lambda^+ + tA^+_n, \tag{5.5}$$

$$G_n(t) := \mathrm{e}^{\mathrm{i}tB^+_n} J_{0,n}(t)\mathrm{e}^{-\mathrm{i}tB^+_n}. \tag{5.6}$$

Since $G_n(0) = \Lambda^+$, $\lambda_k(G_n(1)) = \lambda_k(J_{0,n}(1))$ and $J_{0,n}(1) = J^+_{n,0}$ we obtain

$$|\lambda_k(J^+_{n,0}) - k| = |\lambda_k(G_n(1)) - \lambda_k(G_n(0))| \le ||G_n(1) - G_n(0)||_{\mathcal{B}(l^2)} \tag{5.7}$$

using the min-max principle to get the inequality. To estimate $G_n(1) - G_n(0)$ we write

$$\frac{\mathrm{d}}{\mathrm{d}t} G_n(t) = \mathrm{e}^{\mathrm{i}tB^+_n}\left(\mathbb{D}_{B^+_n} J_{0,n}(t)\right)\mathrm{e}^{-\mathrm{i}tB^+_n} \tag{5.8}$$

with

$$\mathbb{D}_{B^+_n} J_{0,n}(t) := \frac{\mathrm{d}}{\mathrm{d}t} J_{0,n}(t) + [\mathrm{i}B^+_n, J_{0,n}(t)] = t\,[\mathrm{i}B^+_n, A^+_n], \tag{5.9}$$

where the last equality is due to (5.4). Due to (5.8)–(5.9) the right-hand side of (5.7) can be estimated by the norm of $[B^+_n, A^+_n]$ and to complete the proof we show

$$||[B^+_n, A^+_n]||_{\mathcal{B}(l^2)} = \mathrm{O}(n^{2\gamma-1}). \tag{5.10}$$

In order to estimate the norm of $[\mathrm{i}B^+_n, A^+_n] = 2\,\mathrm{Re}[S^+ a_n(\Lambda^+), A^+_n]$ we observe that

$$\begin{aligned}
[S^+ a_n(\Lambda^+), A^+_n] &= [S^+ a_n(\Lambda^+),\ S^+ a_n(\Lambda^+) + a_n(\Lambda^+)(S^+)^*] \\
&= S^+ a_n(\Lambda^+)^2 (S^+)^* - a_n(\Lambda^+)(S^+)^* S^+ a_n(\Lambda^+) \\
&= a_n(\Lambda^+ - I)^2 - a_n(\Lambda^+)^2.
\end{aligned}$$

Thus we complete the proof of (5.10) using the estimate

$$|a_n(k-1)^2 - a_n(k)^2| = a(n)^2|\theta^+((k-1)/n)^2 - \theta^+(k/n)^2| \leq Cn^{2\gamma-1}.$$

5.2. Proof of estimate (2.17). Let $J^+_{n,0}$, $J'_{n,0}$ be as in Section 5.1. Since $J'_{n,0}\mathrm{e}_k = k\mathrm{e}_k$ when $k \leq 0$ and $J'_{n,0}\mathrm{e}_k = J^+_{n,0}\mathrm{e}_k$ when $k \geq 1$, $J'_{n,0}$ has discrete spectrum and we can write $\sigma(J'_{n,0}) = \{\lambda_k(J'_{n,0})\}_{k\in\mathbb{Z}}$ with $\lambda_k(J'_{n,0}) = k$ when $k \leq 0$ and $\lambda_k(J'_{n,0}) = \lambda_k(J^+_{n,0})$ when $k \geq 1$.

Let $0 < \varepsilon < \frac{1}{2} - \rho_N$ be fixed. Since $\gamma < \frac{1}{2}$ we have $n^{2\gamma-1} \downarrow 0$ as $n \uparrow \infty$, then choosing n_ε large enough we have $Cn^{2\gamma-1} \leq \varepsilon$ for any $n \geq n_\varepsilon$. Hence by Lemma 5.1

$$\sup_{k\in\mathbb{Z}}|\lambda_k(J'_{n,0}) - k| = \sup_{k\in\mathbb{N}}|\lambda_k(J^+_{n,0}) - k| \leq \varepsilon \text{ for } n \geq n_\varepsilon. \tag{5.11}$$

Thus for $n \geq n_\varepsilon$ all eigenvalues of $J'_{n,0}$ are simple and we get (2.17) combining (5.11) with

$$\sup_{k\in\mathbb{N}}|\lambda_k(J'_n) - \lambda_k(J'_{n,0})| \leq \|J'_n - J'_{n,0}\|_{\mathcal{B}(\mathcal{H})} = \rho_N. \tag{5.12}$$

5.3. Proof of estimate (2.18). Since (2.17) holds with $\varepsilon + \rho_N < \frac{1}{2}$, it is clear that all eigenvalues $\lambda_k(J'_n)$ are simple for $n \geq n_\varepsilon$. Next we observe that

$$\begin{aligned} S^{-N}v(\Lambda)S^N &= v(\Lambda + N) = v(\Lambda), \\ S^{-N}\theta^+(\Lambda/n)S^N &= \theta^+((\Lambda+N)/n), \end{aligned}$$

and due to $|\theta^+((\lambda+N)/n) - \theta^+(\lambda/n)| \leq C/n$ we have

$$S^{-N}J'_nS^N = J'_n + N + R_n \tag{5.13}$$

with $\|R_n\|_{\mathcal{B}(\mathcal{H})} = \mathrm{O}(n^{\gamma-1})$. Moreover,

$$\sigma(J'_n) = \sigma(S^{-N}J'_nS^N) = \sigma(J'_n + N + R_n) \subset \bigcup_{j\in\mathbb{Z}} \Delta_{j,n} \tag{5.14}$$

holds with $\Delta_{j,n} := \big[\lambda_j(J'_n) + N - \|R_n\|_{\mathcal{B}(\mathcal{H})}, \lambda_j(J'_n) + N + \|R_n\|_{\mathcal{B}(\mathcal{H})}\big]$. However,

$$\lambda \in \Delta_{j,n} \implies |\lambda - \lambda_j(J'_n) - N| \leq \|R_n\|_{\mathcal{B}(\mathcal{H})}$$

holds by definition of $\Delta_{j,n}$ and (2.17) ensures

$$\lambda \in \Delta_{j,n} \implies |\lambda - j - N| \leq \rho_N + \varepsilon + \|R_n\|_{\mathcal{B}(\mathcal{H})}.$$

Since $\rho_N + \varepsilon < \frac{1}{2}$ and $\|R_n\|_{\mathcal{B}(\mathcal{H})} \to 0$ as $n \to \infty$, we can find $n_0 \in \mathbb{N}^*$ such that

$$n \geq n_0 \implies \rho_N + \varepsilon + \|R_n\|_{\mathcal{B}(\mathcal{H})} < \frac{1}{2}.$$

Therefore, denoting $\Delta_r := (r - \frac{1}{2}, r + \frac{1}{2})$ we find $n \geq n_0 \implies \Delta_{j,n} \subset \Delta_{j+N}$ and we conclude $n \geq n_0 \implies \lambda_{k+N}(J'_n) \in \Delta_{k+N} \cap \bigcup_{j\in\mathbb{Z}} \Delta_{j,n} = \Delta_{k,n}$, which implies

$$|\lambda_{k+N}(J'_n) - \lambda_k(J'_n) - N| \leq \|R_n\|_{\mathcal{B}(\mathcal{H})} = \mathrm{O}(n^{\gamma-1}).$$

□

6. Proof of Proposition 2.3

Step 1. We first claim that the large n estimate

$$\mathcal{G}_n^+ := \operatorname{tr}\chi(J_n' - n) - \operatorname{tr}\chi(\Lambda - n) = \mathrm{O}\big(n^{-\gamma/2}\ln n\big) \tag{6.1}$$

follows from Proposition 2.1. Indeed, in Section 8.2 below we prove the large n estimate

$$\mathcal{G}_n^+ - \mathcal{G}_n^0 = \operatorname{tr}\chi(J_n' - n) - \operatorname{tr}\chi(J_n - n) = \mathrm{O}(n^{\gamma-1}). \tag{6.2}$$

Step 2. We deduce (2.19) from (6.1), (2.17), and (2.18). To begin we write

$$\operatorname{tr}\chi(\Lambda - n) = \sum_{l\in\mathbb{Z}}\chi(l) = \sum_{0\le k\le N-1}\sum_{m\in\mathbb{Z}}\chi(Nm+k)$$

and denote $\chi_k(\lambda) := \chi(N\lambda+k)$ $(0\le k\le N-1)$. Then using (2.12) and the Poisson summation formula we can write

$$\sum_{0\le k\le N-1}\sum_{m\in\mathbb{Z}}\chi_k(m) = \sum_{0\le k\le N-1}\sum_{m\in\mathbb{Z}}\hat{\chi}_k(2m\pi) \tag{6.3}$$

with

$$\hat{\chi}_k(t) = (2\pi)^{-1}\int_{\mathbb{R}}\mathrm{e}^{-\mathrm{i}t\lambda}\,\chi(N\lambda+k)\,\mathrm{d}\lambda = \mathrm{e}^{\mathrm{i}kt/N}\hat{\chi}(t/N)/N.$$

We denote $\rho_N' := \rho_N + \varepsilon_0$ with $\varepsilon_0 > 0$ fixed small enough to ensure $\rho_2' < \frac{1}{2}$ and $\rho_N' < 1/(\pi\sqrt{N})$ if $N\ge 3$. Then (2.17) ensures $|\lambda_k(J_n') - k| \le \rho_N'$ for $n\ge n_0$. Further on we always assume $n\ge n_0$ and consider

$$r_n(k) := \lambda_{n+k}(J_n') - n - k \in [-\rho_N', \rho_N'].$$

However, $r_n(N+k) - r_n(k) = \lambda_{n+k+N}(J_n') - N - \lambda_{n+k}(J_n')$ and we obtain

$$|r_n(mN+k) - r_n(k)| \le Cmn^{\gamma-1} \tag{6.4}$$

due to (2.18). Next we fix $0<\varepsilon\le\frac{1}{8}$ and we observe that

$$\operatorname{tr}\chi(J_n'-n) - \sum_{0\le k\le N-1}\sum_{|m|\le n^\varepsilon}\chi(\lambda_{n+mN+k}(J_n') - n) = \mathrm{O}\left(n^{-\infty}\right). \tag{6.5}$$

Moreover, $|m|\le n^\varepsilon$ ensures

(6.6)

$$\lambda_{n+mN+k}(J_n') - n = mN + k + r_n(mN+k) = mN+k+r_n(k) + \mathrm{O}(n^{\varepsilon+\gamma-1})$$

and

$$\operatorname{tr}\chi(J_n'-n) - \sum_{0\le k\le N-1}\sum_{|m|\le n^\varepsilon}\chi(mN+k+r_n(k)) = \mathrm{O}\left(n^{2\varepsilon+\gamma-1}\right). \tag{6.7}$$

Then we denote $\chi_{n,k}(\lambda) := \chi(\lambda N + k + r_n(k))$ and write

$$\operatorname{tr}\chi(J_n'-n) - \sum_{k=0}^{N-1}\sum_{m\in\mathbb{Z}}\chi_{n,k}(m) = \mathrm{O}\left(n^{2\varepsilon+\gamma-1}\right). \tag{6.8}$$

Using (2.12) and the Poisson summation formula we find

$$\operatorname{tr}\chi(J_n'-n) - \sum_{0\le k\le N-1}\sum_{m\in\mathbb{Z}}\hat{\chi}_{n,k}(2\pi m) = \mathrm{O}(n^{2\varepsilon+\gamma-1}) \tag{6.9}$$

with

$$\hat{\chi}_{n,k}(t) = (2\pi)^{-1}\int_{\mathbb{R}} \mathrm{e}^{-\mathrm{i}t\lambda}\,\chi_{n,k}(\lambda)\,\mathrm{d}\lambda = \mathrm{e}^{\mathrm{i}(k+r_n(k))t/N}\,\hat{\chi}(t/N)/N.$$

We fix $1 \le j \le N$ and $\hat{\chi} \in C_0^\infty(\mathbb{R})$ such that $\hat{\chi}(2\pi m/N) = N\delta_{m,j}$ for $m \in \mathbb{Z}$. Then

$$\sum_{0\le k\le N-1}\sum_{m\in\mathbb{Z}} (\hat{\chi}_{n,k}(2\pi m) - \hat{\chi}_k(2\pi m)) = \sum_{0\le k\le N-1} (z_{k+1}(n)^j - w_{k+1}^j) \tag{6.10}$$

holds with $z_{k+1}(n) := \mathrm{e}^{2\pi\mathrm{i}(k+r_n(k))/N}$ and $w_{k+1} := \mathrm{e}^{2\pi\mathrm{i}k/N}$ for $k = 0, \ldots, N-1$.

Next we introduce $F_j\colon \mathbb{C}^N \to \mathbb{C}$ by the formula

$$F_j(z_1, \ldots, z_N) := (z_1^j + \cdots + z_N^j)/j$$

and denote $z(n) := (z_1(n), \ldots, z_N(n))$, $w := (w_1, \ldots, w_N)$. Then (6.10) and (6.1) give

$$j\big(F_j(z(n)) - F_j(w)\big) = \mathcal{G}_n^+ + \mathrm{O}(n^{2\varepsilon+\gamma-1}) = \mathrm{O}\big(n^{-\gamma/2}\ln n\big) \tag{6.11}$$

for $j = 1, \ldots, N$. Further on we denote $F(z) := (F_1(z), \ldots, F_N(z)) \in \mathbb{C}^N$. Then $F'(z) = (z_l^{j-1})_{j,l=1}^N$ and introducing

$$G(z) := \int_0^1 \big(F'(w + t(z-w)) - M\big)\,\mathrm{d}t$$

with $M := F'(w)$ we find $F(z) - F(w) - M(z-w) = G(z)(z-w)$ and

$$z(n) - w = M^{-1}(F(z(n)) - F(w)) - M^{-1}G(z(n))(z(n)-w).$$

We denote $z(n,t) := w + t(z(n) - w)$ and we want to estimate

$$F'(z(n,t)) - M = \big(z_l(n,t)^{j-1} - w_l^{j-1}\big)_{j,l=1}^N \quad (0 \le t \le 1).$$

However, $|z_l(n,t)^{j-1} - w_l^{j-1}| \le N|z_l(n,t) - w_l| \le Nt|z_l(n) - w_l| \le 2\pi\rho_N' t$ implies $\|F'(z(n,t)) - M\| \le 2\pi N\rho_N' t$ and $\|G(z(n))\| \le \pi N\rho_N'$. Hence

$$|z(n) - w| \le |M^{-1}(F(z(n)) - F(w))| + \mu_N|z(n) - w| \tag{6.12}$$

holds with $\mu_N := \pi N\rho_N'\|M^{-1}\|$. Since $M^*M = N\,I$ we find $\|M\| = \sqrt{N}$ and $M^{-1} = M^*/N$; i.e. $\|M^{-1}\| = 1/\sqrt{N}$ and $\mu_N = \pi\sqrt{N}\rho_N'$. Therefore, we can estimate

$$(1-\mu_N)|z(n) - w| \le |M^{-1}(F(z(n)) - F(w))| \le |F(z(n)) - F(w)| \tag{6.13}$$

and our choice of ρ_N' ensures $\mu_N < 1$ if $N \ge 3$. Hence

$$r_n(k) = \mathrm{O}(|F(z(n)) - F(w)|). \tag{6.14}$$

Thus (6.14) and (6.11) complete the proof when $N \ge 3$.

If $N = 2$ then $(w_1, w_2) = (1, -1)$,

$$M = \begin{pmatrix} 1 & 1 \\ 1 & -1 \end{pmatrix}, \qquad G(z) = \begin{pmatrix} 0 & 0 \\ (z_1 - 1)/2 & (z_2+1)/2 \end{pmatrix}$$

and $\max\{|r_n(0)|, |r_n(1)|\} < \frac{1}{2} \implies \|G(z(n))\| = \frac{1}{2}\left(|z_1(n) - 1|^2 + |z_2(n) + 1|^2\right)^{1/2} < 1$ ensures $\|M^{-1}G(z(n))\| < \|M^{-1}\| = 1/\sqrt{2}$; i.e. (6.13) still holds with $\mu_2 < 1$.

7. Proof of Proposition 2.4

Let $S^+ \in \mathcal{B}(l^2)$ and $\Lambda^+ \colon \mathcal{D} \to l^2$ denote the restriction of Λ and S to l^2 (i.e. $S^+ \mathrm{e}_k = \mathrm{e}_{k+1}$ and $\Lambda^+ \mathrm{e}_k = k\mathrm{e}_k$ for $k \in \mathbb{N}^*$). We want to compare the nth eigenvalues of

$$J_n^+ := \Lambda^+ + 2\,\mathrm{Re}\,\big(S^+ a_n(\Lambda^+)\big) + v(\Lambda^+) \tag{7.1}$$

and

$$\tilde{J}_n := \Lambda^+ + 2\,\mathrm{Re}\,\big(S^+ \tilde{a}_n(\Lambda^+)\big) + v(\Lambda^+) \tag{7.2}$$

where

$$\tilde{a}_n(k) := \begin{cases} a(k) & \text{if } n - C_1 n^\gamma \le k \le n + C_1 n^\gamma, \\ a_n(k) & \text{otherwise,} \end{cases} \tag{7.3}$$

and C_1 is fixed large enough. We choose $n_0 \ge (2C_1)^{1/(1-\gamma)}$ and assume $n \ge n_0$. Then

$$k \ge n - C_1 n^\gamma \ge n/2 \implies a_n(k) = a(n)$$

and due to $a(k+1) - a(k) = \mathrm{O}(k^{\gamma-1})$ we have the estimate

$$\sup_{k\in\mathbb{Z}} |\tilde{a}_n(k) - a_n(k)| \le \sup_{|i|\le C_1 n^\gamma} |a(n+i) - a(n)| \le C_2 n^{2\gamma-1}. \tag{7.4}$$

Hence $|\lambda_n(\tilde{J}_n) - \lambda_n(J_n^+)| \le ||\tilde{J}_n - J_n^+||_{\mathcal{B}(l^2)} = \mathrm{O}(n^{2\gamma-1})$. It remains to show that

$$\lambda_n(J) = \lambda_n(\tilde{J}_n) + \mathrm{O}(n^{-\nu}) \tag{7.5}$$

holds for any $\nu > 0$ under the assumption that C_1 is chosen large enough in (7.3).

For this purpose we will apply [**4**, Theorem 2.3] to the Jacobi matrix J with diagonal entries $k + v(k)$ and off-diagonal entries ck^γ. This form of J corresponds to the choice of parameters $\alpha = 1$, $\beta = \gamma$, $c = 1$ and the use of the auxiliary functions

$$\kappa(\lambda) := \lambda + C_0\lambda^\gamma, \qquad \kappa(\lambda, \lambda') := \lambda' - C_0\lambda^\gamma \tag{7.6}$$

in the statement of [**4**, Theorem 2.3]. Let $\{\lambda_n\}_{n=1}^\infty$ be a fixed real sequence satisfying $\lambda_n = n + \mathrm{O}(n^\gamma)$. If C_1 is large enough in (7.3), then

$$\kappa(\lambda_n, \lambda_n - \lambda_n^{-\nu}) \le k \le \kappa(\lambda_n) \implies J\mathrm{e}_k = \tilde{J}_n \mathrm{e}_k \tag{7.7}$$

for $n \ge n_0$ and [**4**, Theorem 2.3] used with $\lambda = \lambda_n$, $\lambda' = \lambda_n - \lambda_n^{-\nu}$, $J_{\lambda,\lambda'} = \tilde{J}_n$ gives

$$\mathrm{card}\,\big(\sigma(J) \cap (\lambda_n - \lambda_n^{-\nu}, \lambda_n]\big) \le \mathrm{card}\,\Big(\sigma(\tilde{J}_n) \cap \big(\lambda_n - 2\lambda_n^{-\nu}, \lambda_n + \lambda_n^{-\nu}\big]\Big) \tag{7.8}$$

for $n \ge n_0$. However, Lemma 5.1 and $\lambda_n(\tilde{J}_n) = \lambda_n(J_n^+) + \mathrm{O}(n^{2\gamma-1})$ imply

$$n \ge n_0 \implies |\lambda_n(\tilde{J}_n) - n| \le \rho_N + \tilde{\varepsilon} < \frac{1}{2}. \tag{7.9}$$

Hence the cardinal on the right-hand side of (7.8) is at most 1. Let $\lambda_n = \lambda_n(J)$. Then both cardinals are equal to 1 and there is an eigenvalue $\lambda_{k(n)}(\tilde{J}_n) \in \sigma(\tilde{J}_n)$ such that

$$\lambda_{k(n)}(\tilde{J}_n) \in \big(\lambda_n(J) - 2\lambda_n(J)^{-\nu}, \lambda_n(J) + \lambda_n(J)^{-\nu}\big]. \tag{7.10}$$

It remains to check that $k(n) = n$ and due to (7.9)–(7.10) it suffices to know that

$$n \ge n_0 \implies \lambda_n(J) \in \Big(n - \frac{1}{2}, n + \frac{1}{2}\Big). \tag{7.11}$$

However, the operator $J^0 := J - v(\Lambda^+) = \Lambda^+ + 2\,\mathrm{Re}\,(S^+ a(\Lambda^+))$ was investigated in [**2**] (see also [**5**]), where we proved the large n asymptotic formula

$$\lambda_n(J^0) = n + \mathrm{O}(n^{2\gamma-1}). \tag{7.12}$$

Since (7.11) follows from (7.12) and $|\lambda_n(J) - \lambda_n(J^0)| \le \rho_N < \frac{1}{2}$, the proof of Proposition 2.4 is complete.

8. Auxiliary estimates of trace norms

8.1. Auxiliary lemmas.

LEMMA 8.1. *Let $\chi \in C^2(\mathbb{R})$ be such that*

$$|\chi(\lambda)| + |\chi'(\lambda)| + |\chi''(\lambda)| = \mathrm{O}(\lambda^{-2}) \quad (\lambda \in \mathbb{R}). \tag{8.1}$$

Assume that L, $L'\colon \mathcal{D} \to \mathcal{H}$ are self-adjoint operators with the same domain $\mathcal{D}$. Assume also that $\mathcal{D}$ is invariant with respect to $h \in \mathcal{B}(\mathcal{H})$. If $Lh - hL' \in \mathcal{B}(\mathcal{H})$, then

$$\|\chi(L)h - h\chi(L')\|_{\mathcal{B}(\mathcal{H})} \le C_\chi \|Lh - hL'\|_{\mathcal{B}(\mathcal{H})}. \tag{8.2}$$

PROOF. Since (8.1) ensures that $t \to \hat\chi(t)$ and $t \to t\hat\chi(t)$ are integrable on $\mathbb{R}$, we have

$$\big(\chi(L)h - h\chi(L')\big)x = \int_{\mathbb{R}} \hat\chi(t)\,\big(\mathrm{e}^{\mathrm{i}tL}h - h\mathrm{e}^{\mathrm{i}tL'}\big)x\,\mathrm{d}t. \tag{8.3}$$

Let $C_\chi := \int_{\mathbb{R}} |t\hat\chi(t)|\,\mathrm{d}t$. To deduce (8.2) from (8.3) we observe that

$$\big(\mathrm{e}^{\mathrm{i}tL}h - h\mathrm{e}^{\mathrm{i}tL'}\big)x = \int_0^1 \mathrm{e}^{\mathrm{i}stL}\mathrm{i}t(Lh - hL')\mathrm{e}^{\mathrm{i}(1-s)tL'}x\,\mathrm{d}s$$

holds for $x \in \mathcal{D}$ and implies $\|\mathrm{e}^{\mathrm{i}tL}h - h\mathrm{e}^{\mathrm{i}tL'}\|_{\mathcal{B}(\mathcal{H})} \le \|t(Lh - hL')\|_{\mathcal{B}(\mathcal{H})}$. □

LEMMA 8.2. *Let θ be as in Section* 3.1 *and χ as in Lemma* 8.1. *Then*

$$\|(I - \theta(\Lambda/n))\chi(J_n - n)\|_{\mathcal{B}(\mathcal{H})} = \mathrm{O}(n^{\gamma-1}). \tag{8.4}$$

PROOF. Lemma 8.1 with $L = J_n/n$, $L' = \Lambda/n$ and $h = I$ ensures

$$\|\theta(J_n/n) - \theta(\Lambda/n)\|_{\mathcal{B}(\mathcal{H})} \le C\|(J_n - \Lambda)/n\|_{\mathcal{B}(\mathcal{H})} = \mathrm{O}(n^{\gamma-1}). \tag{8.5}$$

Hence, to complete the proof it suffices to observe that

$$\|(1 - \theta(J_n/n))\chi(J_n - n)\|_{\mathcal{B}(\mathcal{H})} = \sup_{\lambda\in\mathbb{R}}|(1 - \theta(\lambda/n))\chi(\lambda - n)| = \mathrm{O}(n^{-2}).$$ □

LEMMA 8.3. *Let χ be as in Lemma* 8.1. *Then*

$$||\chi(J_n' - n) - \chi(J_n - n)||_{\mathcal{B}(\mathcal{H})} = \mathrm{O}(n^{\gamma-1}). \tag{8.6}$$

PROOF. Let θ be as in Lemma 8.2. Then using (8.4) and a similar estimate

$$||\chi(J_n' - n)(I - \theta(\Lambda/n))||_{\mathcal{B}(\mathcal{H})} = \mathrm{O}(n^{\gamma-1}) \tag{8.7}$$

we find that in order to prove (8.6) it suffices to check

$$||\chi(J_n' - n)\theta(\Lambda/n) - \theta(\Lambda/n)\chi(J_n - n)||_{\mathcal{B}(\mathcal{H})} = \mathrm{O}(n^{\gamma-1}). \tag{8.8}$$

Since $(J_n - J_n')\theta(\Lambda/n) = 0$ by definition of J_n, J_n' and θ, using Lemma 8.1 with $L = J_n$, $L' = J_n'$, $h = \theta(\Lambda/n)$ we can estimate the left-hand side of (8.8) by

$$C_\chi||J_n'\theta(\Lambda/n) - \theta(\Lambda/n)J_n||_{\mathcal{B}(\mathcal{H})} = C_\chi||\,[J_n, \theta(\Lambda/n)]\,||_{\mathcal{B}(\mathcal{H})} \tag{8.9}$$

and to complete the proof we observe that the right-hand side of (8.9) is $\mathrm{O}(n^{\gamma-1})$. □

LEMMA 8.4. *Let χ be as in Lemma 8.1. Then $\sup_n \|\chi(J_n - n)\|_{\mathcal{B}_1(\mathcal{H})} < \infty$ and $\sup_{n\geq n_0} \|\chi(J'_n - n)\|_{\mathcal{B}_1(\mathcal{H})} < \infty$, where n_0 is chosen large enough.*

PROOF. Using (2.17) with $\rho_N + \varepsilon < \frac{1}{2}$ we can estimate

$$\begin{aligned} \|\chi(J'_n - n)\|_{\mathcal{B}_1(\mathcal{H})} &= \sum_{k\in\mathbb{Z}} |\chi(\lambda_k(J'_n) - n)| \\ &\leq \sum_{k\in\mathbb{Z}} \frac{C}{1 + |\lambda_k(J'_n) - n|^2} \\ &\leq \sum_{k\in\mathbb{Z}} \frac{C'}{1 + |k - n|^2} \\ &= \sum_{m\in\mathbb{Z}} \frac{C'}{1 + |m|^2} \end{aligned}$$

and a similar estimate holds for J_n instead of J'_n. □

8.2. Proof of estimate (6.2). Let $\chi \in \mathcal{S}(\mathbb{R})$ and $\chi_2(\lambda) = (1 + \lambda^2)^{-1}$. Then $\chi = \chi_1\chi_2$ with $\chi_1 \in \mathcal{S}(\mathbb{R})$ and we can express $\chi(J'_n - n) - \chi(J_n - n)$ in the form

$$\big(\chi_1(J'_n - n) - \chi_1(J_n - n)\big)\chi_2(J'_n - n) + \chi_1(J_n - n)\big(\chi_2(J'_n - n) - \chi_2(J_n - n)\big).$$

Thus the left-hand side of (6.2) can be estimated by the sum of

$$||\chi_1(J'_n - n) - \chi_1(J_n - n)||_{\mathcal{B}(\mathcal{H})}||\chi_2(J'_n - n)||_{\mathcal{B}_1(\mathcal{H})}, \tag{8.10}$$

$$||\chi_1(J_n - n)||_{\mathcal{B}_1(\mathcal{H})}||\chi_2(J'_n - n) - \chi_2(J_n - n)||_{\mathcal{B}(\mathcal{H})}. \tag{8.11}$$

Lemmas 8.3 and 8.4 applied to $\chi = \chi_1$ and $\chi = \chi_2$ allow us to estimate (8.10) and (8.11) by $\mathrm{O}(n^{\gamma-1})$.

8.3. Proof of estimate (3.4). If θ is as in Section 3.1 and $\chi \in \mathcal{S}(\mathbb{R})$ we have

$$||(I - \theta(\Lambda/n))\chi(J_n - n)||_{\mathcal{B}_1(\mathcal{H})} = \mathrm{O}(n^{\gamma-1}). \tag{8.12}$$

Indeed, writing $\chi = \chi_1\chi_2$ with $\chi_2(\lambda) := (1 + \lambda^2)^{-1}$ and applying Lemmas 8.2 and 8.4 for $\chi = \chi_1$ and $\chi = \chi_2$ we can estimate the left-hand side of (8.12) by

$$||(I - \theta(\Lambda/n))\chi_1(J_n - n)||_{\mathcal{B}(\mathcal{H})}||\chi_2(J_n - n)||_{\mathcal{B}_1(\mathcal{H})} = \mathrm{O}(n^{\gamma-1}). \tag{8.13}$$

Next we observe that due to $J_n = \mathrm{e}^{-2ia(n)\operatorname{Im} S}\, L_n\, \mathrm{e}^{2ia(n)\operatorname{Im} S}$ we have

$$||(I - \theta(\Lambda/n))\chi(L_n - n)||_{\mathcal{B}_1(\mathcal{H})} = ||(I - \theta(\tilde{J}^0_n/n))\chi(J_n - n)||_{\mathcal{B}_1(\mathcal{H})}$$

with $\tilde{J}^0_n := \mathrm{e}^{-2ia(n)\operatorname{Im} S}\, \Lambda\, \mathrm{e}^{2ia(n)\operatorname{Im} S}$ and due to (8.12) it remains to show

$$||(\theta(\Lambda/n) - \theta(\tilde{J}^0_n/n))\chi(J_n - n)||_{\mathcal{B}_1(\mathcal{H})} = \mathrm{O}(n^{\gamma-1}). \tag{8.14}$$

Using Lemmas 8.4 and 8.1 we can estimate the left-hand side of (8.14) by

$$C_0||\theta(\tilde{J}^0_n/n)) - \theta(\Lambda/n)||_{\mathcal{B}(\mathcal{H})} \leq C_0C_\theta||(\Lambda - \tilde{J}^0_n)/n||_{\mathcal{B}(\mathcal{H})}. \tag{8.15}$$

Finally, to complete the proof we observe that (2.5) ensures $\Lambda - \tilde{J}^0_n = 2a(n)\operatorname{Re} S$; hence the right-hand side of (8.15) is $\mathrm{O}(n^{\gamma-1})$.

References

[1] A. Boutet de Monvel, S. Naboko, and L. O. Silva, *The asymptotic behavior of eigenvalues of a modified Jaynes-Cummings model*, Asymptot. Anal. **47** (2006), no. 3-4, 291–315. MR2233923 (2007e:81148)

[2] A. Boutet de Monvel and L. Zielinski, *Eigenvalue asymptotics for Jaynes–Cummings type models without modulations*, BiBoS preprint **08-03-278** (2008), available at `http://www.physik.uni-bielefeld.de/bibos/`.

[3] A. Boutet de Monvel and L. Zielinski, *Explicit error estimates for eigenvalues of some unbounded Jacobi matrices*, Spectral theory, mathematical system theory, evolution equations, differential and difference equations, Oper. Theory Adv. Appl., vol. 221, Birkhäuser/Springer Basel AG, Basel, 2012, pp. 189–217, DOI 10.1007/978-3-0348-0297-0_11. MR2954067

[4] A. Boutet de Monvel and L. Zielinski, *Approximation of eigenvalues for unbounded Jacobi matrices using finite submatrices*, Cent. Eur. J. Math. **12** (2014), no. 3, 445–463, DOI 10.2478/s11533-013-0348-z. MR3145902

[5] A. Boutet de Monvel, J. Janas, and L. Zielinski, *Asymptotics of large eigenvalues for a class of band matrices*, Rev. Math. Phys. **25** (2013), no. 8, 1350013, 23 pp., DOI 10.1142/S0129055X1350013X. MR3116930

[6] P. A. Cojuhari and J. Janas, *Discreteness of the spectrum for some unbounded Jacobi matrices*, Acta Sci. Math. (Szeged) **73** (2007), no. 3-4, 649–667. MR2380070 (2009g:47082)

[7] J. Janas and S. Naboko, *Infinite Jacobi matrices with unbounded entries: asymptotics of eigenvalues and the transformation operator approach*, SIAM J. Math. Anal. **36** (2004), no. 2, 643–658 (electronic), DOI 10.1137/S0036141002406072. MR2111793 (2006b:47042)

[8] M. Malejki, *Asymptotics of large eigenvalues for some discrete unbounded Jacobi matrices*, Linear Algebra Appl. **431** (2009), no. 10, 1952–1970, DOI 10.1016/j.laa.2009.06.035. MR2567805 (2010i:15026)

[9] È. A. Tur, *Jaynes–Cummings model: solution without rotating wave approximation*, Optics and Spectroscopy **89** (2000), no. 4, 574–588.

IMJ-PRG, Université Paris Diderot, bâtiment Sophie Germain, case 7012, 75205 Paris Cedex 13, France

E-mail address: `anne.boutet-de-monvel@imj-prg.fr`

LMPA, Université du Littoral, 50 rue Ferdinand Buisson, CS 80699, 62228 Calais Cedex, France

E-mail address: `Lech.Zielinski@lmpa.univ-littoral.fr`

Amer. Math. Soc. Transl.
Volume **233**, 2014

Fourier Integrals and a New Representation of Maslov's Canonical Operator Near Caustics

S. Yu. Dobrokhotov, G. Makrakis, and V. E. Nazaikinskii

Dedicated to Vladimir Aleksandrovich Marchenko

Abstract. We propose a new representation of Maslov's canonical operator in a neighborhood of the caustics using a special class of coordinate systems ("eikonal coordinates") on Lagrangian manifolds.

Introduction

Rapidly oscillating Fourier type integrals are well known in the mathematical literature. One of the main constructions in this area is given by Maslov's canonical operator [**15**] (see also [**18, 12**] and the bibliography therein), which is used to construct short-wave (high-frequency, or rapidly oscillating) asymptotic solutions of a broad class of problems for differential equations with real characteristics.[1] The asymptotics provided by the canonical operator are a far-reaching generalization of ray expansions in problems of optics and electrodynamics as well as of the WKB asymptotics for equations of quantum mechanics. The construction of the canonical operator is based on the fundamental geometric notion of a *Lagrangian manifold.* Assume that the original partial differential equation is defined on the n-dimensional configuration space $\mathbb{R}^n_x$ with coordinates $x = (x_1, \dots, x_n)$. One of the main ideas in the canonical operator theory is to proceed from this equation to a simpler (in fact, ordinary) differential equation naturally induced on an n-dimensional Lagrangian manifold Λ in the phase space $\mathbb{R}^{2n}_{(x,p)}$ with coordinates (x,p), $x = (x_1, \dots, x_n)$. The manifold Λ depends on the problem considered and is usually constructed by solving the canonical equations of classical (Hamiltonian) mechanics. Once we have found an appropriate manifold Λ and a solution ϕ (which is called an *amplitude*) of the induced differential equation on Λ (the choice of a specific solution depends on the original problem as well), we can write out the (asymptotic) solution of the

2010 *Mathematics Subject Classification.* Primary 81Q20; Secondary 35S30.

Supported by RFBR (Grant 11-01-00973-a) and by the Archimedes Center for Modeling, Analysis & Computation (ACMAC), Crete, Greece (grant FP7-REGPDT-2009-1). S. Yu. D. and V. E. N. are grateful to the staff of ACMAC and the Department of Applied Mathematics, University of Crete for support and kind hospitality.

[1]There is also a version of the canonical operator for equations with complex characteristics (see [**16, 17**] and also [**5, 14, 13**]), which we do not discuss here.

original problem in the form

$$u(x) = [K_\Lambda a](x), \tag{0.1}$$

where K_Λ is the canonical operator. Note that

- There exist known recipes or algorithms for constructing Lagrangian manifolds and amplitudes for many types of problems (and for various original differential equations).
- Equation (0.1) is the *answer to the original problem*, automatically including the behavior in caustic regions, passage through the caustics, matching of the asymptotic expansions in various regions, etc.

The standard construction [**15**] of the canonical operator is universal, but it is not the only possible one. For a broad class of interesting problems, one can more conveniently use different representations, and it is these new representations of oscillating solutions that are considered in our paper. We point out that our analysis does not alter the general *concept* of the canonical operator or the *fundamental geometric objects* underlying its construction. We just suggest a new implementation of the canonical operator in a neighborhood of the caustics, which can be more convenient when solving specific physical problems with the use of software such as *Wolfram Mathematica*® [**19**] or *MatLab*® [**20**]. Note also that our formulas are a special case of the general formulas of the theory of Fourier integral operators [**10**], and our main result is the specific form of these formulas and an algorithm for their construction. Finally, note that our formulas may in particular be useful in problems related to the asymptotics of solutions of wave-type equations with localized initial data (e.g., see [**8, 7**]) or in scattering problems and point source problems for equations of Helmholtz type.

The outline of the paper is as follows. The standard construction of the canonical operator is described in Section 1. In Section 2, we describe the main result, our new equations (2.9) and (2.10). Section 3 provides some examples. Section 4 describes the relationship, much discussed at a certain time in the past, between the canonical operator and Fourier integrals. Finally, Section 5 contains auxiliary material: it describes some notation used in the paper and also, for the reader's convenience, reproduces the famous theorem of the stationary phase method. We omit the proofs (which are mostly technical and involve lengthy computations) everywhere except in Section 4, where we think that the short proofs provided might be of interest.

The asymptotics discussed here have the small parameter $h \to 0$, as is customary in the semiclassical approximation in quantum mechanics. In wave problems, one often uses the large parameter $k \to \infty$. To make the formulas fit this case, one should just set $h = 1/k$.

1. Standard construction of the canonical operator

In this section, we recall the construction of Maslov's canonical operator according to [**15, 18, 12**]. Let us start with a very brief overview and then fill in the details.

The input elements of the construction are as follows:

- A *Lagrangian manifold* Λ in the $2n$-dimensional phase space $\mathbf{R}^{2n}_{(x,p)}$.
- A *measure* $d\mu$ on Λ.
- A point $\alpha_0 \in \Lambda$, referred to as the *central point*.

If the *quantization conditions* [1, 15] are satisfied, then these elements uniquely (modulo lower-order terms) determine the canonical operator $K^{1/h}_{(\Lambda,d\mu)}$, which takes each function $a \in C_0^\infty(\Lambda)$ to a rapidly oscillating function $u(x,h) = [K^{1/h}_{(\Lambda,d\mu)}a](x)$ on the configuration space $\mathbf{R}^n_x$. The construction is essentially *local*: first, $[K^{1/h}_{(\Lambda,d\mu)}a](x)$ is defined for functions a supported in certain open sets called the *canonical charts* on Λ; then a partition of unity on Λ is used to paste the local definitions together into the *global canonical operator*. The local expression for $[K^{1/h}_{(\Lambda,d\mu)}a](x)$ in a canonical chart has the form of a rapidly oscillating exponential in the simplest case where the chart is diffeomorphically projected onto a domain in the configuration space $\mathbf{R}^n_x$ (a *nonsingular chart*); in a *singular chart* (a chart containing a *focal point*), the local expression has the form of the Fourier transform of a rapidly oscillating exponential with respect to part of the variables. A change of the central point α_0 results in the multiplication of all local expressions by a unimodular phase factor. In general, the local expressions depend not only on the position of the central point but also on the choice of paths from the central point to the respective canonical charts. The role of the quantization condition is that it guarantees that the local expressions for the canonical operator are independent of the choice of these paths and coincide with each other on functions supported in intersections of canonical charts.

Note that although the canonical operator is an object of function theory significant in its own right, its main applications are related to various problems for partial differential equations, hence the importance of the *commutation formula*, which shows how a differential operator acts on the function $[K^{1/h}_{(\Lambda,d\mu)}a](x)$ and provides conditions (in the form of a geometric condition on Λ and an ordinary differential equation for a) ensuring that this function is a solution of the corresponding differential equation.

Now let us proceed to more detailed explanations.

1.1. Lagrangian manifold, measure, and central point. First, let us discuss the input elements of the construction. A *Lagrangian manifold* in $\mathbf{R}^{2n}_{(x,p)}$ is an n-dimensional submanifold $\Lambda \subset \mathbf{R}^{2n}_{(x,p)}$ such that the symplectic form $\omega^2 = dp \wedge dx \equiv dp_1 \wedge dx_1 + \cdots + dp_n \wedge dx_n$ vanishes on the vectors tangent to Λ. We denote the points of Λ by the letter α and use the notation $\alpha = (\alpha_1, \dots, \alpha_n)$ for various local coordinate systems on Λ. Then the embedding $\Lambda \subset \mathbf{R}^{2n}_{(x,p)}$ is given by equations of the form $x = X(\alpha)$, $p = P(\alpha)$, $\alpha \in \Lambda$.

Next, a *measure* on Λ is understood as a volume form[2] (a nonvanishing differential n-form) $d\mu$. In local coordinates $(\alpha_1, \dots, \alpha_n)$, one has $d\mu = \mu(\alpha)\, d\alpha_1 \wedge \cdots \wedge d\alpha_n$, where the function $\mu(\alpha) \neq 0$ is called the *density* of $d\mu$ in these coordinates.

Finally, a *central point* is an arbitrarily chosen point $\alpha_0 \in \Lambda$. (We assume Λ to be connected; otherwise, we need one central point per connected component.)

1.2. Regular and focal points. Canonical coordinates on Λ. Let $\alpha_* \in \Lambda$. If $\det \frac{\partial X}{\partial \alpha}(\alpha_*) \neq 0$ (this condition is independent of the choice of local coordinates), then the point α_* is said to be *regular*; otherwise, it is said to be *singular*, or *focal*. If α_* is a regular point, then the equation $x = X(\alpha)$ can be solved for

[2]Thus, Λ is orientable; the theory may pretty well be constructed without this assumption, which we make only to simplify the exposition by avoiding the notion of *odd* differential forms.

α in a neighborhood of the point (x_*, α_*), $x_* = X(\alpha_*)$, and hence the variables $x = (x_1, \dots, x_n)$ can be used as local coordinates on Λ in a neighborhood of α_*. For an arbitrary $\alpha_* \in \Lambda$, the *lemma on local coordinates* [**1**] states that there exists a subset $I \subset \{1, \dots, n\}$ such that $\det \frac{\partial(X_I, P_{\overline{I}})}{\partial\alpha}(\alpha_*) \neq 0$ (where $\overline{I} = \{1, \dots, n\} \setminus I$ is the complementary subset); consequently, the equations $x_I = X_I(\alpha)$, $p_{\overline{I}} = P_{\overline{I}}(\alpha)$ can be solved for α in a neighborhood of the point $(x_{I*}, p_{\overline{I}*}, \alpha_*)$, $x_{I*} = X_I(\alpha_*)$, $p_{\overline{I}*} = P_{\overline{I}}(\alpha_*)$, and the variables $(x_I, p_{\overline{I}})$ can be used as local coordinates on Λ in a neighborhood of α_*. These coordinates are called *canonical coordinates.* For a regular point α_*, we can of course take $\overline{I} = \varnothing$, so that $(x_I, p_{\overline{I}}) = x$, but if α_* is a focal point, then $\overline{I}$ is necessarily nonempty.

It follows from the preceding that there exists a *canonical atlas* of Λ in which every chart is given for some $I \subset \{1, \dots, n\}$ by the canonical coordinates $(x_I, p_{\overline{I}})$ defined on an open connected simply connected subset $U \subset \Lambda$; such a chart is denoted by (U, I) and called a *canonical chart* (*nonsingular* if $\overline{I} = \varnothing$ and *singular* otherwise). The equations of Λ in the canonical chart have the form

$$x_{\overline{I}} = X_{\overline{I}}(x_I, p_{\overline{I}}), \quad p_I = P_I(x_I, p_{\overline{I}}) \tag{1.1}$$

(where for brevity we write $X_{\overline{I}}(x_I, p_{\overline{I}})$ instead of $X_{\overline{I}}(\alpha(x_I, p_{\overline{I}}))$ etc.).

1.3. Maslov index of paths and cycles on Λ. Let $\varepsilon \geq 0$ be a given number. The form $d(X_1 - i\varepsilon P_1) \wedge \cdots \wedge d(X_n - i\varepsilon P_n)$ is a differential form of maximal degree n on Λ and hence a multiple of $d\mu$; thus, the function

$$\mathcal{J}^\varepsilon(\alpha) = \frac{d(X_1 - i\varepsilon P_1) \wedge \cdots \wedge d(X_n - i\varepsilon P_n)}{d\mu} \tag{1.2}$$

is well defined; it is called the *Jacobian* of the functions $(X_1 - i\varepsilon P_1, \dots, X_n - i\varepsilon P_n)$ *with respect to the measure* $d\mu$ and can be computed by the formula

$$\mathcal{J}^\varepsilon(\alpha) = \frac{1}{\mu(\alpha)} \det \frac{\partial(X_1 - i\varepsilon P_1, \dots, X_n - i\varepsilon P_n)}{\partial(\alpha_1, \dots, \alpha_n)}, \tag{1.3}$$

where $\mu(\alpha)$ is the density of $d\mu$ in local coordinates $(\alpha_1, \dots, \alpha_n)$. It can be shown that *for* $\varepsilon > 0$ *the Jacobian* (1.2) *vanishes nowhere on* Λ. On the other hand, for $\varepsilon = 0$ the Jacobian (1.2) becomes the Jacobian

$$\mathcal{J}(\alpha) = \frac{dX_1 \wedge \cdots \wedge dX_n}{d\mu} \equiv \frac{1}{\mu(\alpha)} \det \frac{\partial(X_1, \dots, X_n)}{\partial(\alpha_1, \dots, \alpha_n)}, \tag{1.4}$$

which is nonzero at the regular points and vanishes at the focal points.

Let $\gamma \colon [0,1] \longrightarrow \Lambda$ be a path on Λ with regular endpoints $\gamma(0)$ and $\gamma(1)$. The *Maslov index* of γ is defined by[3]

$$\operatorname{ind}\gamma = \frac{1}{\pi} \lim_{\varepsilon \to +0} \operatorname*{var}_{\gamma} \arg \mathcal{J}^\varepsilon(\alpha) = \frac{1}{\pi i} \lim_{\varepsilon \to +0} \int_\gamma \frac{d\mathcal{J}^\varepsilon}{\mathcal{J}^\varepsilon}, \tag{1.5}$$

where $\arg z$ is the argument of a complex number z and $\operatorname{var}_\gamma$ stands for the variation along γ. The index $\operatorname{ind}\gamma$ is an integer depending only on the homotopy class of γ in the set of paths with regular endpoints. Next, if γ is a closed path (*cycle*) on Λ, then the Maslov index of γ is defined by the same equation (1.5). (The limit as $\varepsilon \to +0$ is in fact unnecessary, because the variation of the argument is independent

[3]See [**1**] for the definition of the Maslov index of γ as the intersection number of γ with the cycle of singularities on Λ.

of ε in this case.) The Maslov index of a cycle is an integer homotopy invariant that depends only on the homology class of the cycle in $H_1(\Lambda)$.

1.4. Jacobians and Maslov index of canonical charts. From now on, to avoid unnecessary technical complications, we assume that *the central point* α_0 *is nonsingular and* $\mathcal{J}(\alpha_0) > 0$. Let (U, I) be some canonical chart. Then the Jacobian

$$\mathcal{J}_I(\alpha) = \frac{dX_I \wedge dP_{\overline{I}}}{d\mu} \equiv \frac{1}{\mu(\alpha)} \det \frac{\partial(X_I, P_{\overline{I}})}{\partial(\alpha_1, \ldots, \alpha_n)} \tag{1.6}$$

is nonzero in U. (Note that for $\overline{I} = \varnothing$ this is just the Jacobian (1.4).) The Maslov index $m_{(U,I)}$ of the chart (U, I) is defined as follows. Choose some path $\gamma\colon [0,1] \longrightarrow \Lambda$ with $\gamma(0) = \alpha_0$ and $\gamma(1) \equiv \alpha_* \in U$. Assume momentarily that α_* is a nonsingular point. Then

$$m_{(U,I)} = \operatorname{ind}\gamma + \frac{1}{\pi}\left[\arg \frac{dX_I \wedge d((1-\theta)X_I - i\theta P_{\overline{I}})}{d\mu}(\alpha_*)\right]\Bigg|_{\theta=0}^{\theta=1} + \frac{|\overline{I}|}{2}, \tag{1.7}$$

where $[\arg(\cdot)]_{\theta=0}^{\theta=1}$ is the variation of the argument as θ varies from 0 to 1 on the interval $[0,1]$. If α_* cannot be assumed to be nonsingular, then one can use the slightly more cumbersome equation

$$m_{(U,I)} = \frac{1}{\pi} \operatorname*{var}_{\gamma} \arg \frac{d(X_I - i\varepsilon(t)P_I) \wedge d(\theta(t)X_I - i\varkappa(t)P_{\overline{I}})}{d\mu}(\alpha_*) + \frac{|\overline{I}|}{2}, \tag{1.8}$$

where $e(t)$, $\theta(t)$, and $\varkappa(t)$ are continuous functions on $[0,1]$ positive on $(0,1)$ and satisfy the conditions $\theta(0) = \varkappa(1) = 1$ and $\varkappa(0) = \varepsilon(0) = \varepsilon(1) = \theta(1) = 0$. Note that (i) equations (1.7) and (1.8) agree if α_* is a nonsingular point; (ii) $m_{(U,I)}$ is an integer, and $\pi m_{(U,I)}$ is a branch of $\arg \mathcal{J}_I$ in U; (iii) $m_{(U,I)} = \operatorname{ind}\gamma$ if $\overline{I} = \varnothing$ (i.e., (U, I) is a nonsingular chart); (iv) $m_{(U,I)}$ *depends on the choice of the* (*homotopy class of the*) *path* γ (unless the Maslov index of all cycles on Λ is zero).

1.5. Action (eikonal) in canonical charts. Since Λ is Lagrangian, it follows that the Pfaff equation

$$d\tau(\alpha) = P(\alpha)\, dX(\alpha) \equiv P_1(\alpha)\, dX_1(\alpha) + \cdots + P_n(\alpha)\, dX_n(\alpha) \tag{1.9}$$

is locally solvable on Λ, and the solution is unique up to an additive constant. A solution of Equation (1.9) is called an *eikonal* (or *action*). Let (U, I) be a canonical chart. We define the eikonal in this chart by the formula

$$\tau_{(U,I)}(\alpha) = \int_\gamma P(\alpha)\, dX(\alpha) + \int_{\alpha_*}^{\alpha} P(\alpha)\, dX(\alpha), \tag{1.10}$$

where γ and α_* are the same as in Section 1.4 and the second integral is taken over an arbitrary path entirely lying in U. The eikonal $\tau_{(U,I)}(\alpha)$ depends on the choice of the path γ (unless the cohomology class of the form $P\,dX$ in $H^1(\Lambda)$ is trivial).

1.6. Local canonical operator. Now we are in a position to write out a formula specifying the local canonical operator $K_{(U,I)}$ in the canonical chart (U, I). Let $a \in C_0^\infty(U)$. Then

$$[K_{(U,I)}a](x,h) = \overline{\mathcal{F}}^{1/h}_{p_{\overline{I}} \to x_{\overline{I}}} \left[\frac{e^{\frac{i}{h}[\tau_{(U,I)}(\alpha) - p_{\overline{I}} X_{\overline{I}}(\alpha)] - i\pi m_{(U,I)}/2} a(\alpha)}{\sqrt{|\mathcal{J}_I(\alpha)|}}\right]_{\alpha = \alpha(x_I, p_{\overline{I}})}, \tag{1.11}$$

where $\bar{\mathcal{F}}^{1/h}_{p_{\overline{I}}\to x_{\overline{I}}}$ is the inverse $1/h$-Fourier transform with respect to the variables $p_{\overline{I}}$ (see Section 5). In a nonsingular chart ($|\overline{I}| = \varnothing$) the Fourier transform disappears, and the formula acquires the simpler form

$$[K_{(U,I)}a](x,h) = \frac{e^{\frac{i}{h}\tau_{(U,I)}(\alpha)-i\pi m_{(U,I)}/2}a(\alpha)}{\sqrt{|\mathcal{J}(\alpha)|}}\bigg|_{\alpha=\alpha(x)}. \tag{1.12}$$

1.7. Quantization condition and global canonical operator. Now assume that the *Bohr–Sommerfeld quantization conditions*[4]

$$\frac{2}{\pi h}\oint_\gamma P(\alpha)\,dX(\alpha) \equiv \operatorname{ind}\gamma \pmod 4 \tag{1.13}$$

hold for all cycles γ on Λ. (It suffices to require that (1.13) holds for a basis of independent cycles on Λ.)

THEOREM 1.1. *(cf.* [**15**, p. 253, Theorem 2.3]*.) If the quantization conditions* (1.13) *are satisfied, then the following assertions hold*:

- *The local canonical operators* (1.11) *are independent of the choice of the paths γ in Section* 1.4 *and coincide modulo $O(h)$ on the intersections of the canonical charts.*
- *Let $\{e_{(U,I)}\}$ be a locally finite partition of unity on Λ subordinate to the cover of Λ by the canonical charts (U,I). Define an operator $K^{1/h}_{(\Lambda,d\mu)}$ on $C_0^\infty(\Lambda)$ by the formula*

$$[K^{1/h}_{(\Lambda,d\mu)}a](x,h) = \sum_{(U,I)}[K_{(U,I)}(e_{(U,I)}a)](x,h). \tag{1.14}$$

This operator is modulo $O(h)$ independent of the choice of the canonical atlas and the partition of unity.

The operator $K^{1/h}_{(\Lambda,d\mu)}$ defined in (1.14) is called *Maslov's canonical operator* on the Lagrangian manifold Λ with measure $d\mu$ and central point α_0.

1.8. Commutation formula and asymptotic solutions. Consider a differential or pseudodifferential operator $\widehat{L} = L(\overset{2}{x},\overset{1}{\hat{p}},h) \equiv L\Big(\overset{2}{x}, -ih\overset{1}{\frac{\partial}{\partial x}}, h\Big)$ with smooth symbol $L(x,p,h) = H(x,p) + hL_1(x,p) + \cdots$. The leading term $H(x,p)$ of the expansion of L in powers of h is called the *Hamiltonian.*

THEOREM 1.2. *(cf.* [**15**, p. 397, Lemma 6.13]*.) Let $a \in C_0^\infty(\Lambda)$. Then*

$$\widehat{L}K^{1/h}_{(\Lambda,d\mu)}a = K^{1/h}_{(\Lambda,d\mu)}(H|_\Lambda a) + O(h), \tag{1.15}$$

where $H|_\Lambda = H(X(\alpha),P(\alpha))$ is the restriction of $H(x,p)$ to Λ. Moreover, if $H|_\Lambda \equiv 0$ and the measure $d\mu$ is invariant under shifts along the trajectories of the Hamiltonian vector field $\xi_H = H_p(x,p)\partial_x - H_x(x,p)\partial_p$, then

$$\widehat{L}K^{1/h}_{(\Lambda,d\mu)}a = -ihK^{1/h}_{(\Lambda,d\mu)}\big(\xi_H a - \tfrac{1}{2}(\operatorname{tr} H_{xp})|_\Lambda a + iL_1|_\Lambda a\big) + O(h^2), \tag{1.16}$$

where $\operatorname{tr} H_{xp}$ is the trace of the matrix $H_{xp}(x,p)$.

---

[4]These conditions should be understood as follows: If Λ depends on parameters, then, for each $h > 0$, conditions (1.13) single out the set of admissible values of these parameters.

This theorem suggests a natural way for constructing asymptotic solutions of the equation $\widehat{L}u = 0$: find a Lagrangian manifold Λ with $H|_\Lambda = 0$, equip it with an invariant measure $d\mu$, and solve the *transport equation* $\xi_H a - \frac{1}{2}(\operatorname{tr} H_{xp})|_\Lambda a + iL_1|_\Lambda a = 0$. The desired solutions have the form $u = K^{1/h}_{(\Lambda,d\mu)} a$.

2. New formulas

Now let us present new formulas for Maslov's canonical operator. These formulas differ most dramatically from the standard formulas in the singular charts (although the expression for the nonsingular charts acquires a slightly different form as well), and they can be written out, provided that the Lagrangian manifold (or at least the part of it where we intend to use the new formulas) satisfies Condition 2.1 below. We point out that our formulas give functions with the same asymptotics as the standard canonical operator. (See Theorem 2.5.) Therefore, the counterparts of Theorems 1.1 and 1.2 hold for the new expression of the canonical operator automatically, and that is why we do not bother even to state or mention them in what follows.

Let Λ be a Lagrangian manifold in $\mathbf{R}^{2n}_{(x,p)}$ equipped with a measure $d\mu$ and an initial point α_0. Throughout this section, we assume that the quantization condition (1.13) is satisfied.

2.1. Main condition and eikonal coordinates. From now on, we assume that Λ satisfies the following

CONDITION 2.1. The form $P(\alpha)\,dX(\alpha)$ is nonzero for each $\alpha \in \Lambda$.

Thus, if τ is an eikonal in a neighborhood U of some point of Λ, then $d\tau \neq 0$, and hence (provided that U is sufficiently small) we can supplement τ with some functions $\psi_1, \dots, \psi_{n-1}$ such that $(\tau, \psi) \equiv (\tau, \psi_1, \dots, \psi_{n-1})$ is a coordinate system in U. A coordinate system of this kind will be called an *eikonal coordinate system.* The expressions of the functions $(X(\alpha), P(\alpha))$ via eikonal coordinates will be denoted by[5] $(X(\tau,\psi), P(\tau,\psi))$ or for simplicity by (X, P) with the arguments omitted. The same notation will be used for other functions on Λ. The symbol $\mu = \mu(\tau,\psi)$ will from now on be used to denote the density of the measure $d\mu$ in eikonal coordinates (τ, ψ), so that

$$d\mu = \mu(\tau,\psi)\,d\tau \wedge d\psi_1 \wedge \cdots \wedge d\psi_{n-1}.$$

One can readily prove the following assertion.

PROPOSITION 2.2. *In eikonal coordinates, one has the relations*

$$\langle P, X_\tau \rangle = 1, \qquad \langle P, X_{\psi_j} \rangle = 0, \qquad j = 1, \dots, n-1, \tag{2.1}$$

$$\langle P_{\psi_j}, X_\tau \rangle = \langle P_\tau, X_{\psi_j} \rangle, \quad \langle P_{\psi_j}, X_{\psi_k} \rangle = \langle P_{\psi_k}, X_{\psi_j} \rangle, \quad j,k = 1, \dots, n-1. \tag{2.2}$$

2.2. Canonical operator in a nonsingular chart. In a nonsingular chart (U, I), $\overline{I} = \varnothing$, one still uses equation (1.12). The only refinement is that now we have eikonal coordinates (τ, ψ), where $\tau = \tau_{(U,I)}$, instead of the general coordinates α in U, and so we can partly compute the Jacobian $\mathcal{J}$ in the eikonal coordinates using Proposition 2.2. Namely, the following assertion holds.

[5] Rather than the technically correct $(X(\alpha(\tau,\psi)), P(\alpha(\tau,\psi)))$.

PROPOSITION 2.3. *In the eikonal coordinates, one has*

$$|\mathcal{J}(\tau,\psi)| = \frac{\sqrt{\det(X_\psi^*(\tau,\psi)X_\psi(\tau,\psi))}}{|\mu(\tau,\psi)||P(\tau,\psi)|}, \tag{2.3}$$

where $X_\psi^* X_\psi = \big(\langle X_{\psi_j}, X_{\psi_k}\rangle\big)$ *is the Gram matrix of the vectors* $X_{\psi_1},\dots,X_{\psi_{n-1}}$.

Accordingly, the equaton (1.12) for the local canonical operator becomes

$$[K_{(U,I)}a](x,h) = e^{\frac{i}{h}\tau(x)-i\pi m_{(U,I)}/2} a(\tau,\psi) \frac{\sqrt{|\mu(\tau,\psi)||P(\tau,\psi)|}}{\sqrt[4]{\det(X_\psi^*(\tau,\psi)X_\psi(\tau,\psi))}}\Bigg|_{\substack{\tau=\tau(x)\\ \psi=\psi(x)}}, \tag{2.4}$$

where $\tau=\tau(x)$, $\psi=\psi(x)$ is the expression of the eikonal coordinates (τ,ψ) via the canonical coordinates x in the chart (U,I).

2.3. Canonical operator near focal points. We have defined the action of the canonical operator on functions $a\in C_0^\infty(\Lambda)$ whose support does not meet the set $\Gamma\in\Lambda$ of focal points. Now we should define how the canonical operator acts on functions supported near focal points. This is where our construction differs from the standard one.

Let $\alpha_*\in\Gamma$; i.e., $\mathcal{J}_x(\alpha_*)=0$. We will construct a "new singular chart" in a neighborhood U of α_* and define the "new local canonical operator" on functions $a\in C_0^\infty(U)$. We choose and fix some path γ on Λ with $\gamma(0)=\alpha_0$ and $\gamma(1)=\alpha_*$ and define the eikonal $\tau(\alpha)$ in a sufficiently small neighborhood of α_* by equation (1.10), where the second integral is taken over an arbitrary path lying in that neighborhood. Next, we supplement the eikonal with $n-1$ functions $\psi_1,\dots,\psi_{n-1}$, thus obtaining a system (τ,ψ) of eikonal coordinates on Λ in a neighborhood of α_*. The coordinates of α_* will be denoted by (τ_*,ψ_*). Let $k=\operatorname{rank}X_\psi(\tau_*,\psi_*)$. We have $k<n-1$, because otherwise α_* would not be a focal point. Take k linearly independent columns of the matrix $X_\psi(\tau_*,\psi_*)$ and accordingly divide the variables ψ into two parts ψ' and ψ'', the first part including the variables corresponding to the chosen linearly independent columns and the second part including all the other variables.[6] We assume (renumbering the variables ψ if necessary) that $\psi'=(\psi_1,\dots,\psi_k)$ and $\psi''=(\psi_{k+1},\dots,\psi_{n-1})$. Note that the Gram matrix $X_{\psi'}^*X_{\psi'}$ is invertible in a neighborhood of the point (τ_*,ψ_*).

Consider the system of $k+1$ equations

$$\langle P(\tau,\psi), x-X(\tau,\psi)\rangle = 0, \qquad \langle X_{\psi_j}(\tau,\psi), x-X(\tau,\psi)\rangle = 0, \quad j=1,\dots,k. \tag{2.5}$$

PROPOSITION 2.4. *Equation* (2.5) *defines smooth functions*

$$\tau=\tau(x,\psi''), \qquad \psi'=\psi'(x,\psi'') \tag{2.6}$$

in a neighborhood of the point (x_*,ψ_*''), *where* $x_*=X(\tau_*,\psi_*)$, *such that* $\tau_*=\tau(x_*,\psi_*'')$ *and* $\psi_*'=\psi'(x_*,\psi_*'')$. *Moreover, there exists a neighborhood* W *of the point* $(x_*,\psi_*'')\in\mathbf{R}^{2n-1-k}$ *such that the following conditions hold*:

(i) *The differentials* $d\tau_{\psi_{k+1}},\dots,d\tau_{\psi_{n-1}}$ *are linearly independent at each point of the set*

$$\Pi=\{(x,\psi'')\in W\colon \tau_{\psi''}(x,\psi'')=0\},$$

which is therefore an n*-dimensional submanifold.*

---

[6]If $k=0$, then ψ' is empty and the equations given below will undergo obvious modifications. This is always the case for $n=2$.

(ii) *The image* U *of* Π *under the mapping* $(x,\psi'') \longmapsto (x,\tau_x(x,\psi''))$ *is contained in* Λ *and is a neighborhood of the point* α_* *in* Λ.

(iii) *For* $(x,\psi'') \in W$, *one has* $\det M(\tau(x,\psi''),\psi'(x,\psi''),\psi'') \neq 0$, *where*

$$M = \begin{pmatrix} P & X_{\psi'} & P_{\psi''} - P_{\psi'}(X^*_{\psi'}X_{\psi'})^{-1}X^*_{\psi'}X_{\psi''} \end{pmatrix}. \tag{2.7}$$

The domain $U \subset \Lambda$, together with the eikonal coordinates (τ,ψ) and the functions (2.6), will be called a *new singular chart* on Λ. (Without loss of generality, we can assume that both U and W are connected and simply connected.)

We define the *index* m_U of the new singular chart by setting

$$m_U = \frac{1}{\pi}\left(\arg \mathcal{J}^\varepsilon(\alpha_0)\Big|_{\varepsilon=0}^{\varepsilon=1} + \operatorname*{var}_{\gamma} \arg \mathcal{J}^1(\alpha) - \sum_{s=1}^{2n-k-1} \arg\lambda_s\right), \tag{2.8}$$

where γ is the same path as above, the λ_s are the eigenvalues of the $(2n-k-1)\times(2n-k-1)$ matrix $\begin{pmatrix} E - i\tau_{xx}(x_*,\psi''_*) & -i\tau_{x\psi''}(x_*,\psi''_*) \\ -i\tau_{\psi''x}(x_*,\psi''_*) & -i\tau_{\psi''\psi''}(x_*,\psi''_*)\end{pmatrix}$, and $\arg\lambda_j \in \left[-\frac{\pi}{2},\frac{\pi}{2}\right]$.

For $a \in C_0^\infty(U)$, set

$$[K_U^{1/h}a](x,h) = \frac{e^{-i\pi m_U/2}}{(2\pi h)^{\frac{n-1-k}{2}}} \int \frac{e^{i\frac{\tau}{h}}a\sqrt{|\mu\det M|}}{\sqrt{\det(X^*_{\psi'}X_{\psi'})}}\Bigg|_{\substack{\tau=\tau(x,\psi'') \\ \psi'=\psi'(x,\psi'')}} \chi(x,\psi'')\,d\psi'', \tag{2.9}$$

where M is the matrix (2.7) and $\chi(x,\psi'')$ is a smooth cutoff function on W such that $\chi = 1$ on Π and $\chi(x,\cdot)$ is compactly supported in $W_x = \{\psi'' \colon (x,\psi'') \in W\}$ for each x. For $k=0$ (the coordinates ψ' are absent), equation (2.9) becomes

$$[K_U^{1/h}a](x,h) = \frac{e^{-i\pi m_U/2}}{(2\pi h)^{\frac{n-1-k}{2}}} \int \left[e^{i\frac{\tau}{h}}a\sqrt{|\mu\det(P,P_\psi)|}\right]_{\tau=\tau(x,\psi)} \chi(x,\psi)\,d\psi. \tag{2.10}$$

2.4. Comparison with the standard canonical operator. Let us compare the canonical operator $K_U^{1/h}$ constructed in Section 2.3 with the standard local canonical operator. Without loss of generality, we assume that the domain U is sufficiently small and hence can be covered with a single "old" canonical chart with coordinates $(x_I, p_{\overline{I}})$. Consider the canonical operator $K_{(U,I)}$ defined by equaton (1.11), where the eikonal $\tau_{(U,I)}$ and the index $m_{(U,I)}$ are defined with the use of the same path γ as in the construction of $K_U^{1/h}$. Then the following assertion holds.

THEOREM 2.5. *Under these assumptions, one has*

$$K_U^{1/h}a = K_{(U,I)}a + O(h) \qquad \text{for every } a \in C_0^\infty(U). \tag{2.11}$$

Moreover, for each $a \in C_0^\infty(U)$ *there exist* $a_j \in C_0^\infty(U)$, $j = 1,2,\dots$, *such that*

$$K_U^{1/h}a = K_{(U,I)}\left(a + \sum_{j=1}^{N-1} h^j a_j\right) + O(h^N), N = 1,2,\dots. \tag{2.12}$$

The proof is based on Theorem 4.6 in the Appendix. It is rather technical, and we omit the lengthy computations. □

2.5. Closing remark. Condition 2.1 is actually not restrictive, because if it is violated in a specific problem, then one can introduce an additional variable x_{n+1} (a cyclic variable) that does not occur in the Hamiltonian and consider solutions of the form $v(x, x_{n+1}) = u(x)e^{\frac{i}{h}x_{n+1}}$, where $u(x)$ is the desired solution of the original problem. If Λ_u is the Lagrangian manifold corresponding to u, then the Lagrangian manifold corresponding to v has the form

$$\Lambda_v = \Lambda_u \times \{(x_{n+1}, p_{n+1}) \in \mathbf{R}^2 \colon p_{n+1} = 1,\ x_{n+1} \text{ is arbitrary}\},$$

and one can readily see that the 1-form

$$(p_1\, dx_1 + \cdots + p_{n+1}\, dx_{n+1})|_{\Lambda_v} = dx_{n+1} + (p_1\, dx_1 + \cdots + p_n\, dx_n)|_{\Lambda_u}$$

is nonzero on Λ_v. Thus, this uniformization procedure always permits one to ensure that Condition 2.1 is satisfied. We do not elaborate further on the topic.

3. Examples

3.1. Bessel function of order $1/2$. Consider the Lagrangian manifold

$$\Lambda^3 = \{(x,p) \in \mathbf{R}^6 \colon x = X(\tau,\omega),\ p = P(\tau,\omega),\ \tau \in \mathbf{R},\ \omega \in \mathbf{S}^2\}, \tag{3.1}$$

where $X(\tau,\omega) = \tau\mathbf{n}(\omega)$, $P(\tau,\omega) = \mathbf{n}(\omega)$, and if $\omega \in \mathbf{S}^2$ is represented by the spherical coordinates, $\omega = (\theta,\psi)$, then $\mathbf{n}(\omega) \equiv \mathbf{n}(\theta,\psi) = (\sin\theta\cos\psi, \sin\theta\sin\psi, \cos\theta)$. We equip Λ^3 with the measure $d\mu = d\tau \wedge d\omega$, where $d\omega$ is the surface area element of the unit sphere $\mathbf{S}^2$. In the spherical coordinates,

$$d\mu = \mu(\tau,\theta,\psi)\, d\tau \wedge d\theta \wedge d\psi, \qquad \mu = \sin\theta. \tag{3.2}$$

Obviously,

$$P(\tau,\omega)\, dX(\tau,\omega) = d\tau, \tag{3.3}$$

so that (τ,ω) are eikonal coordinates on Λ. Next, the equation

$$\langle P(\tau,\omega), x - X(\tau,\omega)\rangle = 0 \tag{3.4}$$

is uniquely solvable for τ,

$$\tau(x,\omega) = \langle x, \mathbf{n}(\omega)\rangle, \tag{3.5}$$

and the Jacobian

$$\det(P, P_\omega) = \det(P, P_\theta, P_\psi) = \det\begin{pmatrix} \sin\theta\cos\psi & \cos\theta\cos\psi & -\sin\theta\sin\psi \\ \sin\theta\sin\psi & \cos\theta\sin\psi & \sin\theta\cos\psi \\ \cos\theta & -\sin\theta & 0 \end{pmatrix} = \sin\theta \tag{3.6}$$

is nonzero except for $\theta = 0, \pi$. (These singularities, however, are artificial: they are due to the degeneration of spherical coordinates at these points and would not occur if one used different spherical coordinates to represent ω.) Thus, the entire manifold Λ^3 is covered by one singular chart with coordinates (τ,ω), and the canonical operator on it has the form

$$\begin{aligned}[K_{\Lambda^3}a](x,h) &= -\frac{1}{2\pi h}\int\!\!\int \Big[e^{\frac{i}{h}\tau} a\sqrt{|\mu\det(P,P_\omega)|}\Big]_{\tau=\tau(x,\omega)} \chi(x,\omega)\, d\theta\wedge d\psi \\ &= -\frac{1}{2\pi h}\int\!\!\int e^{\frac{i}{h}\langle x,\mathbf{n}(\omega)\rangle} a(\langle x,\mathbf{n}(\omega)\rangle,\theta,\psi)\sin\theta\, d\theta\wedge d\psi.\end{aligned} \tag{3.7}$$

Take $a \equiv 1$; then the integral depends only on $|x|$, and by taking $x = (0, 0, |x|)$ we obtain

(3.8) $$[K_{\Lambda^3} 1](x, k) = -\frac{1}{h} \int_0^\pi e^{\frac{i}{h}|x|\cos\psi} \sin\psi \, d\psi = \frac{1}{i|x|} e^{\frac{i}{h}|x|\cos\psi}\Big|_{\psi=0}^{\psi=\pi} = -\frac{2\sin(|x|/h)}{|x|} = -\sqrt{\frac{2\pi}{h|x|}} J_{\frac{1}{2}}(|x|/h),$$

where

(3.9) $$J_{\frac{1}{2}}(z) = \sqrt{\frac{2}{\pi z}} \sin z$$

is the Bessel function of order $1/2$.

3.2. Bessel type beams in three-dimensional space. Consider the Lagrangian manifold

(3.10) $$\Lambda_0^3 = \{(x, p) \in \mathbf{R}^6 : x = X_0(\alpha, \psi, \phi),\ p = P_0(\alpha, \psi, \phi),\ \ \alpha, \phi \in \mathbf{R},\ \psi \in \mathbf{S}^1\},$$

where

(3.11) $$X_0(\alpha, \psi, \phi) = \begin{pmatrix} \mathbf{n}(\psi)\alpha \\ \phi \end{pmatrix}, \ P_0(\alpha, \psi, \phi) = \begin{pmatrix} \lambda(\phi)\mathbf{n}(\psi) \\ \alpha\lambda'(\phi) + k \end{pmatrix}, \quad \mathbf{n}(\psi) = \begin{pmatrix} \cos\psi \\ \sin\psi \end{pmatrix},$$

$\lambda(\phi) > 0$ is a smooth function (for example, $\lambda = a(1 + \tanh\phi) + b$, $a, b > 0$), and k is a constant. In what follows, we use the notation $x_\perp = (x_1, x_2)$ and $p_\perp = (p_1, p_2)$.

One can readily see that the cycle $\Gamma_0 \subset \Lambda_0^3$ of singularities (the set where the projection $\Lambda_0^3 \to \mathbf{R}_x^3$ degenerates) is given by the equation $\alpha = 0$; thus, Γ_0 is a two-dimensional cylinder, and the caustic—the projection γ_0 of Γ_0 onto the configuration (physical) space $\mathbb{R}_x^3$—is the x_3-axis. It is *degenerate* in that to each point in γ_0 there corresponds a continuum (namely, a circle) of focal points. The manifold Λ_0^3 is diffeomorphic to the product of the circle $\mathbf{S}^1$ by the plane $\mathbf{R}^2$, and the projection $\Lambda_0^3 \backslash \Gamma_0 \to \mathbb{R}_x^3 \backslash \gamma_0$ is a double covering. We equip Λ_0^3 with the measure

(3.12) $$d\mu_0 = \lambda^{-1}(\phi) \, d\alpha \wedge d\phi \wedge d\psi.$$

Let us construct the canonical operator on the manifold Λ_0^3 with the measure $d\mu_0$. First, let us find eikonal coordinates. One has

$$P_0 \, dX_0 = \lambda(\phi) \, d\alpha + \alpha\lambda'(\phi) \, d\phi + k \, d\phi = d(\lambda(\phi)\alpha + k\phi),$$

and we can define the eikonal to be

(3.13) $$\tau = \lambda(\phi)\alpha + k\phi$$

and the eikonal coordinates to be (τ, ϕ, ψ). These coordinates are in fact global on Λ_0^3, because the passage to the original coordinates is always possible by the formula $\alpha = (\tau - k\phi)/\lambda(\phi)$. We have

(3.14) $$\frac{\partial\alpha}{\partial\tau} = \frac{1}{\lambda(\phi)}, \qquad \frac{\partial\alpha}{\partial\phi} = -\frac{k + \alpha\lambda'(\phi)}{\lambda(\phi)}.$$

The measure in the eikonal coordinates is

(3.15) $$d\mu_0 = \lambda^{-2}(\phi) \, d\tau \wedge d\phi \wedge d\psi.$$

We claim that Λ_0^3 is covered by a single singular chart with the functions $\tau = \tau(x, \psi)$ and $\phi(x, \psi)$ computed as follows. (Thus, the role of ψ' and ψ'' in (2.6)

is played by ϕ and ψ, respectively, in our example.) Indeed, let us write out Equation (2.5) for our case. It reads:

$$\langle \mathbf{n}(\psi), x_\perp \rangle = \frac{\tau - k\phi}{\lambda(\phi)}, \qquad x_3 = \phi, \tag{3.16}$$

and we obtain the global solutions

$$\tau(x,\psi) = \lambda(x_3)\langle \mathbf{n}(\psi), x_\perp \rangle + kx_3, \qquad \phi(x,\psi) = x_3. \tag{3.17}$$

Let us compute the determinant of the matrix (2.7) and other objects occurring in equation (2.9) as applied to our case. In the eikonal coordinates, one can readily prove the orthogonality relations

$$\langle X_{0\phi}, X_{0\psi} \rangle = \langle X_{0\phi}, P_0 \rangle = \langle X_{0\phi}, P_{0\psi} \rangle = \langle P_0, P_{0\psi} \rangle = 0, \tag{3.18}$$

and hence, after some computations, we obtain

$$\det M = \det\bigl(P_0, X_{0\phi}, P_{0\psi}\bigr) = \lambda^2(x_3) + (\lambda'(x_3)\langle \mathbf{n}(\psi), x_\perp \rangle + k)^2,$$
$$\det(X_{0\phi}^* X_{0\phi}) = \langle X_{0\phi}, X_{0\phi} \rangle = \frac{1}{\lambda^2(x_3)}\bigl[\lambda^2(x_3) + (\lambda'(x_3)\langle \mathbf{n}(\psi), x_\perp \rangle + k)^2\bigr].$$

Thus, the expression (2.9) for the canonical operator becomes

$$[K_{\Lambda^3} a](x,h) = \sqrt{\frac{i}{2\pi h}} e^{\frac{i}{h}kx_3} \int_0^{2\pi} e^{\lambda(x_3)\langle x_\perp, \mathbf{n}(\psi)\rangle} a(\lambda(x_3)\langle \mathbf{n}(\psi), x_\perp \rangle + kx_3, x_3, \psi)\, d\psi. \tag{3.19}$$

If $a(\tau, \phi, \psi)$ actually has the form $a = a(\alpha, \phi)$ and is even in α, then we obtain

$$[K_{\Lambda^3} a](x,h) = \sqrt{\frac{2\pi i}{h}} a(|x_\perp|, x_3) e^{\frac{i}{h}kx_3} \mathcal{J}_0\left(\frac{\lambda(x_3)|x_\perp|}{h}\right) + O(h), \tag{3.20}$$

where J_0 is the Bessel function of zero order. If $a = \text{const}$ and $\lambda = \text{const}$, then this function, which acquires the form $\text{const} \cdot e^{\frac{i}{h}kx_3} \mathcal{J}_0\bigl(\frac{\lambda|x_\perp|}{h}\bigr)$, is known as the *Bessel beam* in optics (see [**11**], where further references can be found). If a is a compactly supported function, then the function (3.19) is a pulse oscillating in the x_3-direction and having the shape of a Bessel function in the variables (x_1, x_2) with scale $\lambda(x_3)$ depending on x_3.

Now consider the problem of the beam evolution according to the three-dimensional wave equation

$$\frac{1}{c^2}\frac{\partial^2 u}{\partial t^2} = \triangle u \equiv \left(\frac{\partial^2 u}{\partial x_1^2} + \frac{\partial^2 u}{\partial x_2^2} + \frac{\partial^2 u}{\partial x_3^2}\right). \tag{3.21}$$

More precisely, we wish to study the solution of the corresponding Cauchy problem for this equation. (For $\lambda, \alpha = \text{const}$, the function $e^{\frac{i}{h}(ct\sqrt{\lambda^2+k^2}+kx_3)} \mathcal{J}_0\bigl(\frac{\lambda|x_\perp|}{h}\bigr)$ is the exact solution of this equation.) It is well known that the solution of the wave equation splits into two parts that describe waves running in opposite directions. We restrict ourselves to a beam propagating in one direction. To this end, we factorize the wave operator,

$$-h^2\frac{\partial^2}{\partial t^2} + h^2c^2\triangle = \left(-ih\frac{\partial}{\partial t} - c\sqrt{-h^2\triangle}\right)\left(-ih\frac{\partial}{\partial t} + c\sqrt{-h^2\triangle}\right),$$

and consider the Cauchy problem for an equation of first order in time,

$$-ih\frac{\partial u}{\partial t} + c\sqrt{-h^2\triangle}u = 0, u|_{t=0} = u_0, \tag{3.22}$$

where u_0 has the form (3.19). This problem can be solved by means of the Fourier transform, which gives the answer in a form of rapidly oscillating integral. (One can also pass to the cylindrical coordinates in the equation, but these coordinates produce a fictitious singularity on the x_3-axis, which is inconvenient.) The study of this integral is not trivial, and in our opinion it is much easier to use the Maslov canonical operator to describe this solution. By applying Maslov's general theory, we see that the asymptotic solution,

$$u = K_{\Lambda_t}[a(t)], \tag{3.23}$$

of (3.22) is based on the family of the Lagrangian manifolds $\Lambda_t = g_H^t \Lambda_0^3$ obtained by the shift of the manifold Λ_0^3 with the help of the phase flow g_H^t corresponding to the Hamiltonian $H = c|p| \equiv c\sqrt{p_1^2+p_2^2+p_3^2}$. Since this Hamiltonian does not depend on x, we can readily integrate the corresponding Hamilton system and obtain

$$\begin{gathered}\Lambda_t^3 = \{p_\perp = \mathbf{n}(\psi)\mathcal{P},\, p_3 = P_3,\, x_\perp = \mathbf{n}(\psi)\mathcal{X},\, x_3 = X_3,\},\\ \mathcal{P} = \lambda(\phi), P_3 = \alpha\lambda'(\phi)+k, |P| = \sqrt{\lambda^2+(\alpha\lambda'(\phi)+k)^2},\\ \mathcal{X} = \alpha + t\frac{\mathcal{P}}{|P|} = \alpha + tc\frac{\lambda(\phi)}{\sqrt{\lambda^2+(\alpha\lambda'(\phi)+k)^2}}, X_3 = \phi + tc\frac{P_3}{|P|}.\end{gathered} \tag{3.24}$$

Since the Hamiltonian is of first order in the momenta, it follows that the eikonal is constant along the trajectories and is still given by equation (3.13) if we use the coordinates (α,ϕ,ψ) brought to Λ_t from Λ_0^3 by the Hamiltonian flow. In particular, we still have the global eikonal coordinates (τ,ϕ,ψ). In these coordinates,

$$X_\psi = \begin{pmatrix}\mathbf{n}_\perp\mathcal{X}\\ 0\end{pmatrix}, \qquad X_\phi = \begin{pmatrix}\mathbf{n}(\mathcal{X}_\phi+\mathcal{X}_\alpha\alpha_\phi)\\ X_{3\phi}+X_{3\alpha}\alpha_\phi)\end{pmatrix},$$

where the derivatives on the right-hand side are taken in the coordinate system (α,ϕ,ψ); moreover, the orthogonality relations similar to (3.18) hold, and so we have

$$\begin{pmatrix}X_\psi & X_\phi\end{pmatrix}^* \begin{pmatrix}X_\psi & X_\phi\end{pmatrix} = \begin{pmatrix}\mathcal{X}^2 & 0\\ 0 & X_\phi^2\end{pmatrix}$$

and

$$\det\Big(\begin{pmatrix}X_\psi & X_\phi\end{pmatrix}^* \begin{pmatrix}X_\psi & X_\phi\end{pmatrix}\Big) = \mathcal{X}^2\big((\mathcal{X}_\phi+\mathcal{X}_\alpha\alpha_\phi)^2+(X_{3\phi}+X_{3\alpha}\alpha_\phi)^2\big). \tag{3.25}$$

Hence according to (2.3) the focal points are defined by the equations

$$\mathcal{X} = 0 \tag{3.26}$$

$$(\mathcal{X}_\phi+\mathcal{X}_\alpha\alpha_\phi) = 0, \quad X_{3\phi}+X_{3\alpha}\alpha_\phi = 0. \tag{3.27}$$

After some transformations, we find that the two equations in (3.27) are equivalent to the single equation

$$\big(\lambda^2(\phi)+(\alpha\lambda'(\phi)+k)^2\big)^2 - 2ct\Big(\lambda'(\phi)k+\alpha\Big(\lambda'(\phi)^2-\frac{\lambda(\phi)\lambda''(\phi)}{2}\Big)\Big) = 0. \tag{3.28}$$

The projection of the singularities defined by (3.26) is the x_3-axis $x_1 = x_2 = 0$. We assume for simplicity that Equation (3.28) does not hold on the support of the function $a(\tau,\phi,\psi)$ for $t\in[0,T]$. By virtue of the orthogonality relations (3.18) and the formula $|P_\psi| = \mathcal{P} = \lambda$, we have

$$|\det M| = |P||P_\psi||X_\phi| = \left||P|^2 - \frac{ct\mathcal{P}}{|P|}\Big(2\frac{P_3}{|P|}\lambda' - \alpha\frac{\mathcal{P}}{|P|}\lambda''\Big)\right|. \tag{3.29}$$

Equation (2.5) reads

$$(3.30)\quad \mathcal{P}(\langle \mathbf{n}(\psi), x_\perp\rangle - \mathcal{X}) + P_3(x_3 - X_3) = \mathcal{X}_\phi(\langle \mathbf{n}(\psi), x_\perp\rangle - \mathcal{X}) + X_{3\phi}(x_3 - X_3) = 0.$$

Since the vectors P and X_ϕ are orthogonal, it follows that so are the vectors $\binom{\mathcal{P}}{P_3}$ and $\binom{\mathcal{X}_\phi}{X_{3\phi}}$. Hence the determinant $\mathcal{P}X_{3\phi} - P_3\mathcal{X}_\phi$ of equation (3.30) is up to the sign equal to the product of norms of the latter vectors. This means that the determinant of equation (3.30) is zero if and only if (3.27) is true, which contradicts our assumption. Thus, equation (3.30) is equivalent to the equations

$$\mathcal{X} = q, \quad X_3 = x_3, \qquad \text{where} \quad q = \langle \mathbf{n}(\psi),\, x_\perp\rangle.$$

Let us treat these equations as a system for the unknown variables (α, ϕ),

$$(3.31)\qquad \alpha + ct\frac{\mathcal{P}}{|P|} = q, \quad \phi + ct\frac{P_3}{|P|} = x_3.$$

For small t, this system is obviously solvable by the implicit function theorem. We denote the solution by $\phi = \varphi(q, x_3, t)$, $\alpha = \alpha(q, x_3, t)$, $\tau = \lambda\alpha + k\phi \equiv \tau(q, x_3, t)$. Of course, we cannot write out the solution explicitly, but we still can obtain useful relations for it. For example, by eliminating $1/|P|$, we obtain

$$(q-\alpha)^2 + (x_3 - \phi)^2 = t^2c^2.$$

From this equation and (3.31), we find that

$$(3.32)\qquad \alpha = q - \sqrt{R}, \qquad |P| = \frac{ct\lambda}{\sqrt{R}}, \qquad R(\phi, x_3, t) = t^2c^2 - (x_3 - \phi)^2,$$

$$(3.33)\qquad \tau = \lambda(\phi)(q - \sqrt{R(\phi, x_3, t)}) + k\phi,$$

which can be helpful. Now we can write the solution using the canonical operator:

$$(3.34)\qquad u(x,t,h) = \left(\frac{i}{2\pi h}\right)^{1/2} \int \left[e^{i\tau/h} a(\tau,\phi,\psi)\sqrt{\frac{\lambda|P|}{|X_\phi|}}\,\right]_{\substack{\phi=\varphi(q,x_3,t)\\ \tau=\tau(q,x_3,t)\\ q=\langle \mathbf{n}(\psi),\mathbf{x}_\perp\rangle}} d\psi.$$

Note that it can be proved that for small $x_\perp$ this solution still has the asymptotics equal to the product of an oscillating exponential in the x_3-direction and the Bessel function in the $x_\perp$-direction. The proof of this fact, as well as further simplification of the integral (3.34), requires much space, and we do not present it here. Details can be found in our forthcoming paper [**23**].

4. Appendix. Fourier integrals and the canonical operator

In this short Appendix, we discuss oscillatory integrals of the form (4.6) (see below) with parameter $h \to 0$ and show how such integrals (which are the counterpart of the Fourier integral distributions introduced by Hörmander [**10**]) are related to Maslov's canonical operator (see [**15**] and also [**12**]). The relationship between Fourier integral operators and Maslov's canonical operator was also discussed at length in [**6, 21, 22**].

4.1. Nondegenerate phase functions.

DEFINITION 4.1. A real-valued function $\Phi(x,\theta)$ defined on an open set $V \subset \mathbf{R}^n_x \times \mathbf{R}^m_\theta$ is called a *nondegenerate phase function* if the differentials $d(\Phi_{\theta_1}),\dots,d(\Phi_{\theta_m})$ are linearly independent on the set

$$C_\Phi = \big\{(x,\theta)\in V\colon \Phi_\theta(x,\theta)=0\big\}, \tag{4.1}$$

or, equivalently,

$$\operatorname{rank}\big(\Phi_{\theta x}(x,\theta)\quad \Phi_{\theta\theta}(x,\theta)\big)=m, \qquad (x,\theta)\in C_\Phi. \tag{4.2}$$

By the implicit function theorem, C_Φ is a smooth n-dimensional manifold, and the functions Φ_θ can serve as coordinates in the directions transversal to C_Φ.

PROPOSITION 4.2. *The mapping*

$$j_\Phi\colon C_\Phi \longrightarrow \mathbf{R}^{2n}_{(x,p)}, \quad (x,\theta)\longmapsto \big(x,\Phi_x(x,\theta)\big), \tag{4.3}$$

is a local diffeomorphism of C_Φ onto its image $\Lambda_\Phi = j_\Phi(C_\Phi)\subset \mathbf{R}^{2n}_{(x,p)}$, which is a Lagrangian submanifold (possibly with self-intersections).

PROOF. Let $v={}^t(\eta,\xi)$ be a vector tangent to C_Φ; thus, $\Phi_{\theta x}\eta+\Phi_{\theta\theta}\xi=0$. Let $j_{\Phi*}(v)\equiv(\eta,\Phi_{x\theta}\xi)=0$. Then $\Phi_{\theta\theta}\xi=-\Phi_{\theta x}\eta=0$, and ${}^t\xi\,\big(\Phi_{\theta x}(x,\theta)\quad \Phi_{\theta\theta}(x,\theta)\big)=0$, whence $\xi=0$ by (4.2) and so $v=0$. This proves that j_Φ is a local diffeomorphism onto the image of itself. Next,

$$\begin{aligned}dx\wedge dp = dx\wedge(\Phi_{x\theta}d\theta+\Phi_{xx}dx) = (\Phi_{\theta x}dx)\wedge d\theta = (\Phi_{\theta x}dx+\Phi_{\theta\theta}d\theta)\wedge d\theta\\ = d(\Phi_\theta)\wedge d\theta = 0\end{aligned}$$

on the tangent space to Λ_Φ, because $\Phi_\theta=0$ on C_Φ and the products $dx\wedge(\Phi_{xx}dx)=0$ and $\Phi_{\theta\theta}d\theta\wedge d\theta$ are zero by the antisymmetry of the exterior product. □

DEFINITION 4.3. A nondegenerate phase function $\Phi(x,\theta)$ is called a (*local*) *determining function* of a Lagrangian manifold $\Lambda\subset\mathbf{R}^{2n}$ if $\Lambda_\Phi\subset\Lambda$.

Under the conditions of Definition 4.3, Λ_Φ does not have self-intersections and is (relatively) open in Λ. In what follows, we identify C_Φ with Λ_Φ via the mapping

$$j_\Phi\colon C_\Phi\longrightarrow\Lambda_\Phi, \tag{4.4}$$

so that functions (differential forms) on C_Φ are automatically functions (differential forms) on Λ_Φ and vice versa. The following proposition provides a good example of this.

PROPOSITION 4.4. *The function $\tau=\Phi|_{C_\Phi}$ is an action on Λ_Φ, i.e., satisfies*

$$d\tau = p\,dx\big|_{\Lambda_\Phi}. \tag{4.5}$$

PROOF. One has

$$d\tau = d(\Phi|_{C_\Phi}) = (d\Phi)|_{C_\Phi} = (\Phi_x\,dx+\Phi_\theta\,d\theta)|_{C_\Phi} = p\,dx|_{\Lambda_\Phi},$$

because $\Phi_x=p$ and $\Phi_\theta=0$ on C_Φ. □

4.2. Fourier integrals. Let $\Phi(x,\theta)$, $(x,\theta)\in V\subset \mathbf{R}^n_x\times\mathbf{R}^m_\theta$, be a nondegenerate phase function, and let $a\in C_0^\infty(V)$ be a smooth compactly supported function.

DEFINITION 4.5. The function

$$\mathcal{I}[\Phi,a](x,h)=\frac{e^{i\pi m/4}}{(2\pi h)^{m/2}}\int e^{\frac{i}{h}\Phi(x,\theta)}a(x,\theta)\,d\theta \tag{4.6}$$

is called the *Fourier integral with phase function* Φ *and amplitude* a.

Note that C_Φ is the set of stationary points of the integral (4.6), and hence the asymptotics of $\mathcal{I}[\Phi,a]$ as $h\to 0$ depends only on the behavior of a near C_Φ. In particular, we can assume that a is supported in an arbitrarily small neighborhood of C_Φ. Now if $a|_{C_\Phi}=0$, then a can be represented as a linear combination $a=\sum_{j=1}^m a_j\Phi_{\theta_j}$ of the derivatives Φ_{θ_j} with coefficients $a_j\in C_0^\infty(V)$. Integration by parts gives $\mathcal{I}[\Phi,a]=ih\sum_{j=1}^m\mathcal{I}[\Phi,a_{j\theta_j}]$, and we see that the Fourier integral (4.6) modulo $O(h)$ depends only on the restriction of the amplitude to C_Φ. *In what follows, we are interested only in the leading term of the asymptotics, and accordingly we describe only the restriction of the amplitude to* C_Φ; *its continuation outside* C_Φ *can be chosen arbitrarily.*[7]

4.3. Canonical operator as a Fourier integral. Let Λ be a Lagrangian manifold in $\mathbf{R}^{2n}_{(x,p)}$, and let (U,I), $I\subset\{1,\dots,n\}$, be a canonical chart on Λ. In particular, U is given by (1.1). The local canonical operator in (U,I) is given by (1.11), where $\bar{\mathcal{F}}^{1/h}_{p_{\overline{I}}\to x_{\overline{I}}}$ is the inverse $1/h$-Fourier transform (5.2) with respect to the variables $p_{\overline{I}}$. By substituting (5.2) in (1.11), we see that the local canonical operator has the form (4.6) with

$$\Phi(x,\theta)=\tau_{(U,I)}(\alpha(x_I,\theta))+\theta(x_{\overline{I}}-X_{\overline{I}}(x_I,\theta)), \tag{4.7}$$

where we have denoted $p_{\overline{I}}$ by θ to conform in notation with (4.6). We have

$$d\Phi=P_I\,dx_I+\theta\,dX_{\overline{I}}+\theta(dx_{\overline{I}}-dX_{\overline{I}})+(x_{\overline{I}}-X_{\overline{I}})d\theta=P_I\,dx_I+\theta\,dx_{\overline{I}}+(x_{\overline{I}}-X_{\overline{I}})d\theta$$

(the arguments (x_I,θ) are omitted), $\Phi_x(x,\theta)=(P_I(x,\theta),\theta)$, and $\Phi_\theta(x,\theta)=x_{\overline{I}}-X_{\overline{I}}(x_I,\theta)$. In particular, $\Phi_{\theta x_{\overline{I}}}$ is the $|\overline{I}|\times|\overline{I}|$ identity matrix, and so Φ is a nondegenerate phase function. Next, C_Φ is given by the equations $x_{\overline{I}}=X_{\overline{I}}(x_I,\theta)$, and Λ_Φ is given by the equations $p_I=P_I(x_I,\theta)$, $p_{\overline{I}}=\theta$, $x_{\overline{I}}=X_{\overline{I}}(x_I,\theta)$, or (eliminating the variables θ) by (1.1). We conclude that $\Lambda_\Phi=U\subset\Lambda$, and hence the standard representation (1.11) of the canonical operator in the chart (U,I) is none other than a special case of the general Fourier integrals (4.6).

4.4. Jacobians in Fourier integrals. The amplitude of the Fourier integral (1.11) representing the local canonical operator has the form of the product

$$a(x,p_{\overline{I}})=\frac{\varphi(\alpha(x_I,p_{\overline{I}}))}{\sqrt{\mathcal{J}_I(\alpha(x_I,p_{\overline{I}}))}}.$$

The factor $1/\sqrt{\mathcal{J}_I(\alpha(x_I,p_{\overline{I}}))}$ plays an important role when we compare local canonical operators in different canonical charts on the same Lagrangian manifold. It is convenient to introduce a similar factor in the amplitude of the general Fourier integral (4.6). Thus, let $\Phi(x,\theta)$, $(x,\theta)\in V\subset\mathbf{R}^n_x\times\mathbf{R}^m_\theta$, be a nondegenerate phase

[7]The subsequent terms of the asymptotics can be studied as well, but we will not dwell on the topic.

function, and let $d\mu$ be a measure on Λ_Φ. Using the identification (4.4) of Λ_Φ and C_Φ, we can treat $d\mu$ as a differential n-form on C_Φ. Next, let $\widetilde{d\mu}$ be an arbitrary differential n-form defined in a neighborhood of C_Φ in V such that

$$\imath^*(\widetilde{d\mu}) = d\mu, \qquad \text{where } \imath\colon C_\Phi \longrightarrow V \text{ is the embedding.} \tag{4.8}$$

The product $\widetilde{d\mu} \wedge (-d\Phi_\theta) \overset{\text{def}}{=} \widetilde{d\mu} \wedge d(-\Phi_{\theta_1}) \wedge \dots \wedge d(-\Phi_{\theta_m})$ is a differential form of maximum degree $n+m$, and hence

$$\widetilde{d\mu} \wedge (-d\Phi_\theta) = F(x,\theta)\, dx \wedge d\theta \equiv F(x,\theta)\, dx_1 \wedge \dots \wedge dx_n \wedge d\theta_1 \wedge \dots \wedge d\theta_m$$

for some function $F(x,\theta)$. We write

$$F(x,\theta) \equiv F[\Phi, d\mu](x,\theta) = \frac{\widetilde{d\mu} \wedge (-d\Phi_\theta)}{dx \wedge d\theta}. \tag{4.9}$$

The restriction of $F[\Phi, d\mu](x,\theta)$ to C_Φ is independent of the choice of the form $\widetilde{d\mu}$ satisfying (4.8). Indeed, let $\xi_1, \dots, \xi_{n+m} \in \mathbf{R}^{n+m}$ be linearly independent vectors such that $\xi_1, \dots, \xi_n$ form a basis in the tangent space to C_Φ at some point (x_*, θ_*). Then

$$(\widetilde{d\mu} \wedge (-d\Phi_\theta))(\xi_1, \dots, \xi_{n+m}) = -d\mu(\xi_1, \dots, \xi_n) d\Phi_\theta(\xi_{n+1}, \dots, \xi_{n+m})$$

depends only on $d\mu$. Moreover, it is nonzero, because $d\mu$ is nondegenerate on the tangent space to C_Φ and $d\Phi_\theta$ is nondegenerate in the transversal directions. From now on, we consider Fourier integrals with amplitude

$$a = \varphi\sqrt{F[\Phi, d\mu]}, \tag{4.10}$$

where φ is some function on C_Φ (or, equivalently, on Λ_Φ).

4.5. Fourier integral as the canonical operator. Let $\Phi(x,\theta)$, $(x,\theta) \in V \subset \mathbf{R}^n_x \times \mathbf{R}^m_\theta$, be a nondegenerate phase function, and let $d\mu$ be a measure on Λ_Φ. We claim that the Fourier integral $\mathcal{I}[\Phi, \varphi\sqrt{F[\Phi, d\mu]}]$ is none other than the canonical operator on Λ_Φ applied to the function φ,

$$\mathcal{I}[\Phi, \varphi\sqrt{F[\Phi, d\mu]}] = K^{1/h}_{(\Lambda_\Phi, d\mu)}\varphi + O(h), \tag{4.11}$$

for an appropriate choice of the action and arguments of Jacobians on Λ_Φ. Without loss of generality, we assume that Λ_Φ is covered by a single canonical chart (U, I), $U = \Lambda_\Phi$.

THEOREM 4.6. *Let the canonical operator $K^{1/h}_{(\Lambda_\Phi, d\mu)}$ be defined by* (1.11), *where the eikonal $\tau_{(U,I)}$ coincides with the eikonal τ defined in Proposition* 4.4 *and the index $m_{(U,I)}$ is chosen according to the rule*

$$m_{(U,I)} = -\frac{1}{\pi} \arg F[\Phi, d\mu] - \sigma_-\left[\begin{pmatrix} -\Phi_{\theta\theta} & -\Phi_{\theta x_{\overline{I}}} \\ -\Phi_{x_{\overline{I}}\theta} & -\Phi_{x_{\overline{I}} x_{\overline{I}}} \end{pmatrix}\right] + |\overline{I}|, \tag{4.12}$$

$\sigma_-(A)$ being the number of negative eigenvalues of a symmetric matrix A. Then relation (4.11) *holds.*

COROLLARY 4.7. *Two phase functions are equivalent* (*i.e., define the same space of Fourier integrals*) *if and only if the corresponding Lagrangian manifolds are the same.*

Indeed, Theorem 4.6 reduces an arbitrary Fourier integral to the canonical operator on the corresponding Lagrangian manifold.

PROOF of Theorem 4.6. Let us apply the $1/h$-Fourier transform from the variables $x_{\overline{I}}$ to the variables $p_{\overline{I}}$ (see Section 5) to equation (4.11). We see that it suffices to prove that

$$(4.13)\quad \frac{e^{i\pi(m-|\overline{I}|)/4}}{(2\pi h)^{(m+|\overline{I}|)/2}} \int e^{\frac{i}{h}[\Phi(x,\theta)-p_{\overline{I}}x_{\overline{I}}]}\varphi(x,\theta)\sqrt{F[\Phi,d\mu](x,\theta)}\,d\theta\,dx_{\overline{I}}$$
$$= e^{\frac{i}{h}[\tau(\alpha(x_I,p_{\overline{I}}))-p_{\overline{I}}X_{\overline{I}}(x_I,p_{\overline{I}})]}\frac{\varphi(\alpha(x_I,p_{\overline{I}}))}{\sqrt{\mathcal{J}_I(\alpha(x_I,p_{\overline{I}}))}} + O(h).$$

To this end, we use the stationary phase method (see Theorem 5.1 below). The stationary point equations for the phase function

$$\Psi(x,\theta,p_{\overline{I}}) = \Phi(x,\theta) - p_{\overline{I}}x_{\overline{I}}$$

of the integral on the left-hand side in (4.13) read

$$(4.14)\qquad \Phi_\theta(x,\theta)=0, \qquad \Phi_{x_{\overline{I}}}(x,\theta)-p_{\overline{I}}=0.$$

In particular, if a point $(x_{\overline{I}},\theta)$ is a stationary point of the integral for given $(x_I,p_{\overline{I}})$, then $(x,\theta)\in\Lambda_\Phi$. Let us compute $F[\Phi,d\mu](x,\theta)$ at the stationary points. We can take

$$\widetilde{d\mu} = \mu_I(x_I,\Phi_{x_{\overline{I}}}(x,\theta))\,dx_I\wedge d(\Phi_{x_{\overline{I}}}(x,\theta)), \quad \mu_I(x_I,p_{\overline{I}}) = \frac{1}{\mathcal{J}_I(\alpha(x_I,p_{\overline{I}}))},$$

and so (4.9) gives (all computations are carried out for $(x,\theta)\in C_\Phi$)

(4.15)
$$F[\Phi,d\mu] = \mu_I\frac{dx_I\wedge d\Phi_{x_{\overline{I}}}\wedge(-d\Phi_\theta)}{dx\wedge d\theta} = (-1)^{|\overline{I}|}\mu_I\frac{dx_I\wedge d'(-\Phi_{x_{\overline{I}}})\wedge(-d'\Phi_\theta)}{dx_I\wedge dx_{\overline{I}}\wedge d\theta}$$
$$= (-1)^{|\overline{I}|}\mu_I\frac{d'(-\Phi_{x_{\overline{I}}})\wedge(-d'\Phi_\theta)}{dx_{\overline{I}}\wedge d\theta} = (-1)^{|\overline{I}|}\mu_I\det\begin{pmatrix}-\Phi_{\theta\theta} & -\Phi_{\theta x_{\overline{I}}}\\ -\Phi_{x_{\overline{I}}\theta} & -\Phi_{x_{\overline{I}}x_{\overline{I}}}\end{pmatrix}.$$

Here d' is the differential with respect to the variables $(x_{\overline{I}},\theta)$, the variables x_I being treated as parameters. Since $F[\Phi,d\mu]\neq 0$ on C_Φ, we see that the determinant is nonzero and the stationary point equations (4.14) are nondegenerate, so that their solution is given by smooth functions $x_{\overline{I}} = x_{\overline{I}}(x_I,p_{\overline{I}})$, $\theta=\theta(x_I,p_{\overline{I}})$. Since the point $(x_I,x_{\overline{I}}(x_I,p_{\overline{I}}),\theta(x_I,p_{\overline{I}}))$ lies in C_Φ, it follows that

$$(x_I,x_{\overline{I}}(x_I,p_{\overline{I}}),\Phi_x(x_I,x_{\overline{I}}(x_I,p_{\overline{I}}),\theta(x_I,p_{\overline{I}})))\in\Lambda_\Phi.$$

By the second equation in (4.14), we can replace $\Phi_{x_{\overline{I}}}$ by $p_{\overline{I}}$ here and obtain

$$(x_I,x_{\overline{I}}(x_I,p_{\overline{I}}),\Phi_{x_I}(x_I,x_{\overline{I}}(x_I,p_{\overline{I}}),\theta(x_I,p_{\overline{I}})),p_{\overline{I}})\in\Lambda_\Phi,$$

or, in view of (1.1),

$$x_{\overline{I}}(x_I,p_{\overline{I}}) = X_{\overline{I}}(x_I,p_{\overline{I}}), \qquad \Phi_{x_I}(x_I,x_{\overline{I}}(x_I,p_{\overline{I}}),\theta(x_I,p_{\overline{I}})) = P_I(x_I,p_{\overline{I}}).$$

Thus, the point $(x_I,x_{\overline{I}}(x_I,p_{\overline{I}}),\theta(x_I,p_{\overline{I}}))\in C_\Phi$ corresponds via the mapping j_Φ to the point $(x_I,X_{\overline{I}}(x_I,p_{\overline{I}}),P_I(x_I,p_{\overline{I}}),p_{\overline{I}})\in\Lambda_\Phi$, and accordingly (see Proposition 4.4) we have $\Phi(x_I,x_{\overline{I}}(x_I,p_{\overline{I}}),\theta(x_I,p_{\overline{I}})) = \tau(\alpha(x_I,p_{\overline{I}}))$. The phase function at the stationary point is

$$\Psi(x_I,x_{\overline{I}}(x_I,p_{\overline{I}}),\theta(x_I,p_{\overline{I}}),p_{\overline{I}}) = \Phi(x_I,x_{\overline{I}}(x_I,p_{\overline{I}}),\theta(x_I,p_{\overline{I}})) - p_{\overline{I}}X_{\overline{I}}(x_I,p_{\overline{I}})$$
$$= \tau(\alpha(x_I,p_{\overline{I}})) - p_{\overline{I}}X_{\overline{I}}(x_I,p_{\overline{I}}).$$

Now we apply Theorem 5.1 and obtain

$$(4.16)\quad \frac{e^{i\pi(m-|\overline{I}|)/4}}{(2\pi h)^{(m+|\overline{I}|)/2}}\int e^{\frac{i}{h}(\Phi(x,\theta)-p_{\overline{I}}x_{\overline{I}})}\varphi(x,\theta)\sqrt{F[\Phi,d\mu](x,\theta)}\,d\theta\,dx_{\overline{I}}$$

$$= e^{-i\pi|\overline{I}|/2}e^{\frac{i}{h}[\tau(\alpha(x_I,p_{\overline{I}}))-p_{\overline{I}}X_{\overline{I}}(x_I,p_{\overline{I}})]}\left[\varphi(x,\theta)\sqrt{\frac{F[\Phi,d\mu](x,\theta)}{\det\begin{pmatrix}-\Phi_{\theta\theta} & -\Phi_{\theta x_{\overline{I}}}\\ -\Phi_{x_{\overline{I}}\theta} & -\Phi_{x_{\overline{I}}x_{\overline{I}}}\end{pmatrix}}}\right]+O(h),$$

where the expression in square brackets is taken at the stationary point and the argument of the determinant is chosen as indicated in Theorem 5.1. Now we take into account (4.12), (4.15), and the fact that $\mu_I(x_I,p_{\overline{I}})=\mathcal{J}(\alpha(x_I,p_{\overline{I}}))^{-1}$ and arrive at (4.13). The proof of the theorem is complete. □

5. Auxiliary information

5.1. Notation. All vectors are understood as column vectors. If ξ and η are n-vectors, then we write $\langle\xi,\eta\rangle$ for the bilinear form $\langle\xi,\eta\rangle=\sum_{j=1}^n\xi_j\eta_j$. Sometimes, however, we just write $\xi\eta$ instead. Partial derivatives are denoted by subscripts; for example, $\Phi_x=\partial\Phi/\partial x$.

If I is a subset of $\{1,\dots,n\}$, then by $\overline{I}$ we denote the complementary subset $\overline{I}=\{1,\dots,n\}\setminus I$. By $|\overline{I}|$ we denote the number of elements in $\overline{I}$. The $|I|$-vector with components x_j, $j\in I$, is denoted by x_I. The product $dx_I\wedge dp_{\overline{I}}$ is understood as the exterior product of all differentials dx_j, $j\in I$, and dp_j, $j\in\overline{I}$, all factors being arranged in ascending order of subscripts from 1 to n. Similar intuitively clear notation is used as well. Next, $A_{I\overline{I}}$ is an $|I|\times|\overline{I}|$ matrix with entries A_{jk}, $j\in I$, $k\in\overline{I}$. For example, if $I=\{1,3\}$ and $\overline{I}=\{2,4\}$, then

$$dx_I=dx_1\wedge dx_3,\quad dp_{\overline{I}}=dp_2\wedge dp_4,\quad dx_I\wedge dp_{\overline{I}}=dx_1\wedge dp_2\wedge dx_3\wedge dp_4,$$

$$(x_I,p_{\overline{I}})=(x_1,p_2,x_3,p_4),\qquad \frac{\partial S_I}{\partial x_I\partial p_{\overline{I}}}=\begin{pmatrix}\frac{\partial S_I}{\partial x_1\partial p_2} & \frac{\partial S_I}{\partial x_1\partial p_4}\\ \frac{\partial S_I}{\partial x_3\partial p_2} & \frac{\partial S_I}{\partial x_3\partial p_4}\end{pmatrix}.$$

5.2. Fourier transform. Recall that the $1/h$-Fourier transform from the variables $x_{\overline{I}}$ to the variables $p_{\overline{I}}$ and the inverse transform are defined as

$$(5.1)\qquad [\mathcal{F}^{1/h}_{x_{\overline{I}}\to p_{\overline{I}}}u](x_I,p_{\overline{I}})=\frac{e^{-i\pi|\overline{I}|/4}}{(2\pi h)^{|\overline{I}|/2}}\int e^{-\frac{i}{h}p_{\overline{I}}x_{\overline{I}}}u(x)\,dx_{\overline{I}},$$

$$(5.2)\qquad [\bar{\mathcal{F}}^{1/h}_{p_{\overline{I}}\to x_{\overline{I}}}v](x)=\frac{e^{i\pi|\overline{I}|/4}}{(2\pi h)^{|\overline{I}|/2}}\int e^{\frac{i}{h}p_{\overline{I}}x_{\overline{I}}}v(x_I,p_I)\,dp_{\overline{I}}.$$

5.3. Asymptotics of oscillatory integrals. Here we reproduce the statement of the theorem on the stationary phase method used in the preceding subsection.

THEOREM 5.1 (e.g., see [**9, 12**]). *Let the phase function $\Phi(x,\theta)$ have a unique stationary point $\theta=\theta(x)$ on the support of the amplitude, and assume that this stationary point is nondegenerate,* $\det\Phi_{\theta\theta}(x,\theta(x))\neq 0$. *Then the integral* (4.6) *has the asymptotics*

$$(5.3)\qquad I[\Phi,a](x)=\frac{e^{\frac{i}{h}\Phi(x,\theta(x))}a(x,\theta(x))}{\sqrt{\det(-\Phi_{\theta\theta}(x,\theta(x)))}}+O(h),$$

where the branch of the square root in the denominator is chosen according to the rule

$$\arg\det(-\Phi_{\theta\theta}(x,\theta(x))) = -\pi\sigma_-(-\Phi_{\theta\theta}(x,\theta(x))). \tag{5.4}$$

Here $\sigma_-(A)$ is the negative index of inertia (the number of negative eigenvalues) of a self-adjoint matrix A.

References

[1] V. I. Arnold, *On a characteristic class entering into quantization conditions*, Funktsional. Analiz i Prilozhen. **1**, no. 1 (1967), 1–14. English transl. in Funct. Anal. Appl. **1**, 1967. MR0211415 (35:2296)

[2] V. I. Arnold, *Mathematical Methods of Classical Mechanics*, Graduate Texts in Mathematics, vol. 60, Springer–Verlag, Berlin–Heidelberg–New York, 1989. MR997295 (90c:58046)

[3] V. I. Arnold, *Singularities of caustics and wave fronts*, Mathematics and its Applications (Soviet Series), vol. 62, Kluwer Academic Publishers Group, Dordrecht, 1990. MR1151185 (93b:58019)

[4] V. I. Arnold, S. M. Gussein-Zade and A. N. Varchenko, *Singularities of differentiable maps*, vol. 1, Birkhäuser, 1985.

[5] V. V. Belov and S. Yu. Dobrokhotov, *Semiclassical Maslov asymptotics with complex phases. I. General approach*, Teoret. Mat. Fiz. **92**, no. 2, 1992, 215–254. MR1226013 (94g:58229)

[6] V. G. Danilov and Le Vu Ahn, *On Fourier integral operators*, Matem. Sb. **110 (152)**, no. 3 (11), 1979, 323–368. English transl. in Math. USSR Sb. **38**, no. 3, 1981, 293–334. MR554119 (81f:58034)

[7] S. Dobrokhotov, A. Shafarevich, and B. Tirozzi, *Localized wave and vortical solutions to linear hyperbolic systems and their application to linear shallow water equations*, Russ. J. Math. Phys. **15**, no. 2, 2008, 192–221. MR2410830 (2009c:35271)

[8] S. Yu. Dobrokhotov, B. Tirozzi, and A. I. Shafarevich, *Representations of rapidly decreasing functions by the Maslov canonical operator*, Mat. Zametki **82**, no. 5, 2007, 792–796. MR2399958 (2009d:58051)

[9] M. V. Fedoryuk, *The Saddle-Point Method*, Nauka, Moscow, 1977 [Russian]. MR0507923 (58:22580)

[10] L. Hörmander, *Fourier integral operators* I, Acta Math. **127**, 1971, 79–183. MR0388463 (52:9299)

[11] A. P. Kiselev, *Localized light waves: Paraxial and exact solutions of the wave equation (a review)*, Optics and Spectroscopy **102** (4), 2007, 603–622.

[12] A. S. Mishchenko, B. Yu. Sternin, and V. E. Shatalov, *Lagrangian Manifolds and the Method of Canonical Operator*, Nauka, Moscow, 1978. [Russian]. English transl.: *Lagrangian Manifolds and the Maslov Operator*, Springer–Verlag, Berlin–Heidelberg, 1990. MR511792 (83f:58074)

[13] A. S. Mishchenko, B. Yu. Sternin, and V. E. Shatalov, *Maslov's Canonical Operator Method. Complex Theory*, MIEM, Moscow, 1974. [Russian].

[14] V. Maslov and V. Nazaikinskii, *Algebras with general commutation relations and their applications. I. Pseudodifferential equations with increasing coefficients*, J. Soviet Math. **15**, no. 3, 1981, 176–273.

[15] V. P. Maslov, *Perturbation Theory and Asymptotic Methods*, Moscow State University, Moscow, 1965. (Russian). French transl.: *Théorie des Perturbations et Méthod Asymptotiques*, Dunod, Paris, 1972.

[16] V. P. Maslov, *Operator Methods*, Nauka, Moscow, 1973. English transl.: *Operational Methods*, Mir, Moscow, 1976. MR0445305 (56:3647)

[17] V. P. Maslov, *The Complex WKB Method for Nonlinear Equations*, Nauka, Moscow, 1977. [Russian]. English transl.: *The Complex WKB Method for Nonlinear Equations: I. Linear Theory*, Birkhäuser, Basel–Boston–Berlin, 1994. MR1306505 (95m:58126)

[18] V. P. Maslov and M. V. Fedoryuk, *Semiclassical Approximation for Quantum-Mechanical Equations*, Nauka, Moscow, 1976. [Russian]. MR0461590 (57:1575)

[19] `http://www.wolfram.com/mathematica/`

[20] `http://www.mathworks.com/`

[21] V. Nazaikinskii, V. Oshmyan, B. Sternin, and V. Shatalov, *Fourier integral operators and the canonical operator*, Russ. Math. Surveys **36**:2 (1981), 93–161. MR616798 (82h:58049)

[22] A. Yoshikawa, *On Maslov's canonical operator*, Hokkaido Math. J. **4**:3 (1975), 8–38. MR0385938 (52:6797)

[23] S. Yu Dobrokhotov, G. Makrakis, and V. E. Nazaikinskii, *Maslov's canoncial operator, Hörmander's formula, and localization of the Berry–Balazs solution in wave beam theory*, Teor. Mat. Fiz. **180**:2 (2014), 162–188.

A. Ishlinsky Institute for Problems in Mechanics, Moscow; Moscow Institute of Physics and Technology, Dolgoprudny, Moscow District
E-mail address: dobr@ipmnet.ru

Department of Applied Mathematics, University of Crete; Institute of Applied and Computational Mathematics, Foundation for Research and Technology-Hellas, Heraklion, Crete, Greece
E-mail address: makrakg@iacm.forth.gr

A. Ishlinsky Institute for Problems in Mechanics, Moscow; Moscow Institute of Physics and Technology, Dolgoprudny, Moscow District
E-mail address: nazay@ipmnet.ru

Amer. Math. Soc. Transl.
Volume **233**, 2014

Inverse Uniqueness Results for One-Dimensional Weighted Dirac Operators

Jonathan Eckhardt, Aleksey Kostenko, and Gerald Teschl

To Vladimir Aleksandrovich Marchenko with deep admiration

Abstract. Given a one-dimensional weighted Dirac operator we can define a spectral measure by virtue of singular Weyl–Titchmarsh–Kodaira theory. Using the theory of de Branges spaces we show that the spectral measure uniquely determines the Dirac operator up to a gauge transformation. Our result applies in particular to radial Dirac operators and extends the classical results for Dirac operators with one regular endpoint. Moreover, our result also improves the currently known results for canonical (Hamiltonian) systems. If one endpoint is in the limit circle case, then we also establish corresponding two-spectra results.

1. Introduction

In this paper we are concerned with inverse uniqueness results for one-dimensional weighted Dirac operators. Such operators include the case of one-dimensional Dirac operators which play an important role as a toy model in relativistic quantum mechanics [**4, 46**] as well as canonical (Hamiltonian) systems which are of independent interest. Moreover, such operators also arise from the three-dimensional Dirac equation with an (e.g.) radially symmetric potential. In this latter case both endpoints of the resulting radial operator will be singular and classical Weyl–Titchmarsh–Kodaira theory leads to a definition of two by two Weyl–Titchmarsh matrices. In the case of one-dimensional spherical Schrödinger operators this has led to the development of a singular version of Weyl–Titchmarsh–Kodaira theory which allows to introduce a scalar function, the so-called singular Weyl function; see [**17, 18, 20, 23, 24, 25, 26, 27, 29**] and the references therein. Moreover, in [**10, 11, 13**] this theory has been combined with the theory of de Branges to obtain new powerful inverse uniqueness results for Sturm–Liouville equations which have important applications, for example, to the Camassa–Holm equation [**15**]. The basic theory in the case of one-dimensional Dirac operators has been established in [**6**]

2010 *Mathematics Subject Classification.* Primary 34L40, 34B20; Secondary 46E22, 34A55.
Key words and phrases. Dirac operators, canonical systems, inverse spectral theory, de Branges spaces.
The authors' research was supported by the Austrian Science Fund (FWF) under Grant No. Y330 and M1309 as well as by the AXA Research Fund under the Mittag-Leffler Fellowship Project.

(see also [**14**]) and the aim of the present paper is to combine this basic theory with the theory of de Branges spaces and to extend the aforementioned results [**10, 11**] to the case of Dirac operators. As our main result we will show that the spectral measure uniquely determines the Dirac operator up to a gauge transformation and extends the classical results for the case of a regular endpoint. If one endpoint is in the limit circle case, such that we can vary the boundary condition at this endpoint, then we can also establish corresponding two-spectra results.

We apply our findings to the case of radial Dirac operators which have attracted significant interest recently [**1, 2, 3, 45**]. In particular, we establish a Borg–Marchenko [**5, 37**] result for this case, extending the results from [**9**]. We also extend some of the results for canonical systems from [**48**].

For closely related research we also refer to [**12, 13, 16, 19, 42**] as well as the recent monograph [**44**].

2. Dirac operators with strongly singular coefficients

Let (a, b) be a bounded or unbounded interval and consider the differential expression τ given by

$$\tau f(x) = R(x)^{-1} \left(J f'(x) + Q(x) f(x) \right), \quad x \in (a, b), \tag{2.1}$$

where J is the symplectic matrix

$$J = \begin{pmatrix} 0 & -1 \\ 1 & 0 \end{pmatrix} \tag{2.2}$$

and the coefficients are presumed to satisfy the following set of assumptions.

HYPOTHESIS 2.1. *Suppose that the functions Q, $R : (a, b) \to \mathbb{R}^{2\times 2}$ are measurable and satisfy the following additional conditions:*

(i) The functions $\|Q\|$ and $\|R\|$ are locally integrable on (a, b).
(ii) The matrix $Q(x)$ is self-adjoint for almost all $x \in (a, b)$.
(iii) The matrix $R(x)$ is positive definite for almost all $x \in (a, b)$.

REMARK 2.2. *The assumption (iii) is too restrictive since it excludes a number of interesting cases (for instance, Krein's strings with mass distributions having jumps provide examples of R with* $\det(R) = 0$ *on some intervals* [**21**, Chapter VI.8]*; see also* [**30, 31, 48**]*). However, the results of the paper remain true under much more general assumptions on the coefficients. Namely, it suffices to require that R is non-zero a.e. on (a, b) and that the spectral problem*

$$J f'(x) + Q(x) f(x) = z R(x) f(x), \quad x \in (a, b), \tag{2.3}$$

is definite in the sense of [**32**, §2.5]*, that is, the system*

$$J f'(x) + Q(x) f(x) = 0, \quad R(x) f(x) = 0, \quad x \in (a, b), \tag{2.4}$$

has only a trivial solution. Note that in the case $Q = 0$ on (a, b), the latter is equivalent to the fact that R is of positive type, i.e., the matrix $\int_c^d R(x)dx$ is invertible for some subinterval $(c, d) \subset (a, b)$ (cf. [**21, 32**]*).*

We decided to restrict our considerations to the case of positive definite R since, on one hand, our main motivation is the Dirac equation (and in this case $R = I$ on (a, b)). On the other hand, a rigorous definition of the operator (linear relation) associated with the spectral problem (2.3) *in this case is lengthy (cf., e.g.,* [**22, 32, 48**]*); however, the proofs of our main results remain the same.*

For given functions $f,\ g : (a,b) \to \mathbb{C}^2$, we introduce the Wronskian $W_x(f,g)$ via

$$W_x(f,g) = \langle Jf(x)^*, g(x)\rangle = f_1(x)g_2(x) - f_2(x)g_1(x), \quad x \in (a,b). \tag{2.5}$$

Here and henceforth, subscripts will denote the respective component of a vector-valued function. In particular, the scalar product in (2.5) is the usual one in $\mathbb{C}^2$. As long as the functions f and g are at least locally absolutely continuous, we have the following Lagrange identity

$$W_\beta(f,g) - W_\alpha(f,g) = \int_\alpha^\beta \langle \tau f(x)^*, R(x)g(x)\rangle - \langle f(x)^*, R(x)\tau g(x)\rangle dx \tag{2.6}$$

for all $\alpha,\ \beta \in (a,b)$ with $\alpha < \beta$; see [**47**, Theorem 2.3].

It is well known that the differential expression τ gives rise to self-adjoint operators in the Hilbert space $L^2((a,b);R(x)dx)$ associated with the inner product

$$\langle f,g\rangle = \int_a^b \langle f(x), R(x)g(x)\rangle dx, \quad f,\ g \in L^2((a,b);R(x)dx). \tag{2.7}$$

Although we use the same notation for the scalar products in $L^2((a,b);R(x)dx)$ and in $\mathbb{C}^2$, it will be clear from the context which one is meant. In the following, let S be a self-adjoint realization of the differential expression τ with separated boundary conditions. This means that the operator S is given by

$$\begin{aligned}\operatorname{dom}(S) = \{f \in{} & L^2((a,b);R(x)dx) \,|\, f_1,\ f_2 \in AC_{\mathrm{loc}}(a,b), \\ & \tau f \in L^2((a,b);R(x)dx),\ W_a(f,u_a) = W_b(f,u_b) = 0\},\end{aligned} \tag{2.8}$$

and $Sf = \tau f$ for $f \in \operatorname{dom}(S)$. Hereby, the functions u_a, u_b can, for example, be chosen as real-valued solutions of $\tau u = 0$ which lie in $L^2((a,b);R(x)dx)$ near a, b, respectively. In this case, the limits

$$W_a(f,u_a) = \lim_{x\to a} W_x(f,u_a) \quad \text{and} \quad W_b(f,u_b) = \lim_{x\to b} W_x(f,u_b) \tag{2.9}$$

in (2.8) are guaranteed to exist in view of the Lagrange identity (2.6).

More precisely, the need for boundary conditions in (2.8) actually depends on whether the limit circle case or the limit point case prevails at the respective endpoint. Thereby, an endpoint is said to be in the *limit circle* case if for every $z \in \mathbb{C}$, all solutions of $(\tau - z)u = 0$ lie in $L^2((a,b);R(x)dx)$ near the respective endpoint. Otherwise, for each $z \in \mathbb{C}$ there is a solution of $(\tau - z)u = 0$ which does not lie in $L^2((a,b);R(x)dx)$ near the respective endpoint [**47**, Theorem 5.6] and the endpoint is said to be in the *limit point* case. It is known that if an endpoint is in the limit point case, then the boundary condition there is superfluous, that is, the respective function u_a or u_b can be chosen to be identically zero.

In this article, we will consider quite singular endpoints to the extent that the spectrum of S remains simple. To this end, we say that a function $\Phi : \mathbb{C}\times(a,b) \to \mathbb{C}$ is a *real entire solution* of $(\tau - z)u = 0$ if $\Phi(z,\cdot\,)$ is a solution of $(\tau - z)u = 0$ for every $z \in \mathbb{C}$ and both of the functions $\Phi_1(\,\cdot\,,c)$ and $\Phi_2(\,\cdot\,,c)$ are real entire for one (and hence for all) $c \in (a,b)$. Moreover, in this case, we say that Φ lies in $\operatorname{dom}(S)$ near a if for every $z \in \mathbb{C}$, the function $\Phi(z,\cdot\,)$ lies in $L^2((a,b);R(x)dx)$ near a and satisfies the boundary condition at a if τ is in the limit circle case there.

HYPOTHESIS 2.3. *There is a nontrivial real entire solution Φ of $(\tau - z)u = 0$ which lies in* $\operatorname{dom}(S)$ *near a.*

It is known that in general, the essential spectrum of S is made up of a part arising from the left endpoint and another part from the right endpoint. The assumption in Hypothesis 2.3 is equivalent to presuming that there is no contribution coming from the left endpoint (see [**6**, Section 2], [**20**], [**24**]). In this case, the spectrum of S turns out to be simple, as we will see below.

Under the presumption of Hypothesis 2.3, one may introduce the transformation

$$\hat{f}(z) = \int_a^b \langle f(x)^*, R(x)\Phi(z,x)\rangle dx, \quad z \in \mathbb{C}, \tag{2.10}$$

for all functions $f \in L^2((a,b); R(x)dx)$ which vanish almost everywhere near the right endpoint b. Given this, it is possible to introduce a scalar spectral measure ρ on $\mathbb{R}$ (cf. [**6**, Section 2], [**24**], [**18**]) such that

$$\int_{\mathbb{R}} |\hat{f}(\lambda)|^2 d\rho(\lambda) = \int_a^b \langle f(x), R(x)f(x)\rangle dx \tag{2.11}$$

for all functions f vanishing almost everywhere near b. Moreover, the transformation in (2.10) uniquely extends to a unitary operator $\mathcal{F}$ from $L^2((a,b); R(x)dx)$ onto $L^2(\mathbb{R}; d\rho)$, which maps the operator S onto multiplication with the independent variable in $L^2(\mathbb{R}; d\rho)$. Hereby, the inverse of $\mathcal{F}$ is given by

$$\mathcal{F}^{-1}F(x) = \int_{\mathbb{R}} \Phi(\lambda,x)F(\lambda)d\rho(\lambda), \quad x \in (a,b), \tag{2.12}$$

for functions $F \in L^2(\mathbb{R}; d\rho)$ with compact support. Because of this, the measure ρ is called the spectral measure of S associated with the real entire solution Φ.

One way to obtain the spectral measure ρ is by introducing a singular Weyl–Titchmarsh–Kodaira function [**24**], [**6**]. Therefore, one requires an additional real entire solution Θ of $(\tau - z)u = 0$ such that $W(\Theta, \Phi) = 1$. Given such a fundamental system Θ, Φ of solutions, one may introduce the corresponding singular Weyl–Titchmarsh–Kodaira function M on $\mathbb{C}\backslash\mathbb{R}$ by requiring that the solution

$$\Psi(z,x) = \Theta(z,x) + M(z)\Phi(z,x), \quad x \in (a,b), \tag{2.13}$$

lies in $\operatorname{dom}(S)$ near b for every $z \in \mathbb{C}\backslash\mathbb{R}$. With this definition, the spectral measure ρ of S associated with the real entire solution Φ is given by

$$\rho((\lambda_1,\lambda_2]) = \lim_{\delta\to 0}\lim_{\varepsilon\to 0}\frac{1}{\pi}\int_{\lambda_1+\delta}^{\lambda_2+\delta} \operatorname{Im}(M(\lambda - \mathrm{i}\varepsilon))d\lambda, \tag{2.14}$$

for all λ_1, $\lambda_2 \in \mathbb{R}$ with $\lambda_1 < \lambda_2$.

The purpose of the present article is to investigate to which extent a spectral measure ρ determines the coefficients as well as the boundary conditions of the underlying operator S. This will in general only be possible up to a so-called Liouville transformation. In order to present the concept of a Liouville transformation, let η be a locally absolutely continuous, increasing bijection from (a,b) onto some other interval $(\tilde{a},\tilde{b})$ and $\Gamma : (a,b) \to \mathbb{R}^{2\times 2}$ be locally absolutely continuous such that $\det(\Gamma(x)) = 1$ for all $x \in (a,b)$. Then it is simple to check that the transformation

$$\tilde{f} \mapsto f(x) = \Gamma(x)\tilde{f}(\eta(x)), \quad x \in (a,b) \tag{2.15}$$

maps solutions of the Dirac equation $\tilde{\tau}\tilde{u} = \tilde{g}$ on the interval $(\tilde{a}, \tilde{b})$, with the coefficients $\tilde{Q}$, $\tilde{R}$ of the differential expression $\tilde{\tau}$ given by

$$\eta'\tilde{R} \circ \eta = \Gamma^* R \Gamma, \qquad \eta'\tilde{Q} \circ \eta = \Gamma^* Q \Gamma + \Gamma^* J \Gamma', \tag{2.16}$$

onto solutions of the Dirac equation $\tau u = g$. Moreover, the transformation (2.15) defines a unitary operator from $L^2((\tilde{a}, \tilde{b}); \tilde{R}(x)dx)$ onto $L^2((a, b); R(x)dx)$ which maps the operator S onto a self-adjoint realization $\tilde{S}$ of $\tilde{\tau}$ in $L^2((\tilde{a}, \tilde{b}); \tilde{R}(x)dx)$. In particular, the Liouville transform maps any real entire solution $\tilde{\Phi}$ which lies in $\operatorname{dom}(\tilde{S})$ near the left endpoint to a corresponding real entire solution Φ, as in Hypothesis 2.3. Since the Wronskian is invariant under this transformation, the corresponding spectral measures are identical (that is, $\rho = \tilde{\rho}$).

A particular kind of these Liouville transformations can be performed when the weight matrix R is assumed to be locally absolutely continuous. In this case, we may choose locally absolutely continuous functions η and Γ on (a, b) such that

$$\eta'(x) = \sqrt{\det(R(x))} \quad \text{and} \quad \Gamma(x) = \sqrt{\eta'(x) R(x)^{-1}}, \quad x \in (a, b). \tag{2.17}$$

Now the inverse of the transformation (2.15) maps the differential expression τ to the differential expression $\tilde{\tau}$ which now has the constant weight $\tilde{R} = I$.

Upon performing another Liouville transform, we can even normalize the trace of the potential matrix to zero. Therefore, we choose $\eta(x) = x$ and Γ such that

$$\varphi'(x) = \frac{\operatorname{tr}(Q(x))}{2}, \quad \text{and} \quad \Gamma(x) = \mathrm{e}^{\varphi(x) J}, \quad x \in (a, b). \tag{2.18}$$

Then the inverse of the transformation (2.15) maps the differential expression τ to the differential expression $\tilde{\tau}$ whose potential matrix now has trace zero almost everywhere. Hereby also note that one has $\tilde{R} = I$ as long as initially $R = I$.

Similarly, one can also perform a Liouville transformation which reduces the potential matrix to zero. Therefore, we simply choose $\eta(x) = x$ and let Γ be a solution of $J\Gamma' + Q\Gamma = 0$ with $\det(\Gamma) = 1$. (Hereby note that the determinant of such a solution is constant because $\operatorname{tr}(JQ)$ is zero almost everywhere.) With this choice, the inverse of the transformation (2.15) maps the differential expression τ to the differential expression $\tilde{\tau}$ which now has no potential matrix $\tilde{Q} = 0$.

Finally, we may also perform a Liouville transform which normalizes the determinant of the weight matrix to one. To this end, we choose Γ to be constant and η such that

$$\eta'(x) = \sqrt{\det(R(x))}, \quad x \in (a, b). \tag{2.19}$$

In this case, the inverse of the transformation (2.15) maps the differential expression τ to the differential expression $\tilde{\tau}$ whose weight matrix has determinant one almost everywhere. Also note that under this transformation, $\tilde{\tau}$ has no potential term if and only if τ has none.

3. Dirac operators and de Branges spaces

The proof of our inverse uniqueness result relies on the de Branges subspace ordering theorem. Hence in this section, we will first introduce a chain of de Branges spaces associated with a self-adjoint Dirac operator S with separated boundary conditions, which satisfies Hypothesis 2.1 and Hypothesis 2.3.

To this end, we fix some point $c \in (a,b)$ and consider the entire function

$$E(z,c) = \Phi_1(z,c) - \mathrm{i}\Phi_2(z,c), \quad z \in \mathbb{C}. \tag{3.1}$$

A simple calculation, using the Lagrange identity and the fact that

$$W_a(\Phi(\zeta^*,\cdot\,), \Phi(z,\cdot\,)) = 0, \quad \zeta,\, z \in \mathbb{C}, \tag{3.2}$$

shows that one has for all ζ, $z \in \mathbb{C}$

$$\frac{E(z,c)E(\zeta,c)^* - E(\zeta^*,c)E(z^*,c)^*}{2\mathrm{i}(\zeta^* - z)} = \int_a^c \langle \Phi(\zeta,x), R(x)\Phi(z,x)\rangle dx. \tag{3.3}$$

In particular, choosing $\zeta = z$ shows that $E(\,\cdot\,,c)$ is a de Branges function, that is, $|E(z,c)| > |E(z^*,c)|$ for all z in the open upper complex half-plane $\mathbb{C}^+$. Moreover, one observes that $E(\,\cdot\,,c)$ does not have any real zero λ; otherwise, both components of $\Phi(\lambda,c)$ would vanish.

The de Branges space associated with the de Branges function $E(\,\cdot\,,c)$ will be denoted with $B(c)$ (see [**8**, Section 19] for details). It consists of all entire functions F such that the integral

$$\int_{\mathbb{R}} \frac{|F(\lambda)|^2}{|E(\lambda,c)|^2} d\lambda \tag{3.4}$$

is finite and such that $F/E(\,\cdot\,,c)$ as well as $F^\#/E(\,\cdot\,,c)$ are of bounded type in $\mathbb{C}^+$ (that is, they can be written as the quotient of bounded analytic functions) with non-negative mean type. Hereby, $F^\#$ denotes the entire function

$$F^\#(z) = F(z^*)^*, \quad z \in \mathbb{C}, \tag{3.5}$$

and the mean type of a function N which is of bounded type in $\mathbb{C}^+$ is the number

$$\limsup_{y\to\infty} \frac{\ln|N(\mathrm{i}y)|}{y} \in [-\infty,\infty). \tag{3.6}$$

Equipped with the inner product

$$[F,G]_{B(c)} = \frac{1}{\pi}\int_{\mathbb{R}} \frac{F(\lambda)^* G(\lambda)}{|E(\lambda,c)|^2} d\lambda, \quad F,\, G \in B(c), \tag{3.7}$$

the space $B(c)$ turns into a reproducing kernel Hilbert space. In view of [**8**, Theorem 19]) and Equation (3.3), the reproducing kernel $K(\,\cdot\,,\cdot\,,c)$ such that

$$[K(\zeta,\cdot\,,c), F]_{B(c)} = F(\zeta), \quad F \in B(c), \tag{3.8}$$

for every $\zeta \in \mathbb{C}$ is simply given by

$$K(\zeta,z,c) = \int_a^c \langle \Phi(\zeta,x), R(x)\Phi(z,x)\rangle dx, \quad \zeta,\, z \in \mathbb{C}. \tag{3.9}$$

In what follows, we will always identify $L^2((a,c);R(x)dx)$ with the subspace of functions in $L^2((a,b);R(x)dx)$ which vanish almost everywhere on (c,b).

THEOREM 3.1. *For every $c \in (a,b)$ the transformation $f \mapsto \hat{f}$ defined by* (2.10) *is unitary from $L^2((a,c);R(x)dx)$ onto $B(c)$. In particular,*

$$B(c) = \big\{\hat{f} \,\big|\, f \in L^2((a,c);R(x)dx)\big\}. \tag{3.10}$$

PROOF. The proof is (almost) literally the same as the ones of [**10**, Theorem 3.2], [**13**, Theorem 3.1]. □

It remains to collect several useful properties of these de Branges spaces.

COROLLARY 3.2. *The de Branges spaces $B(c)$ have the following properties:*

(*i*) *The de Branges spaces $B(c)$ are isometrically embedded in $L^2(\mathbb{R}, d\rho)$;*

$$\int_{\mathbb{R}} |F(\lambda)|^2 d\rho(\lambda) = \|F\|^2_{B(c)}, \quad F \in B(c). \tag{3.11}$$

(*ii*) *The union of all de Branges spaces $B(c)$ is dense in $L^2(\mathbb{R}, d\rho)$;*

$$\overline{\bigcup_{c \in (a,b)} B(c)} = L^2(\mathbb{R}, d\rho). \tag{3.12}$$

(*iii*) *The de Branges spaces $B(c)$ are strictly increasing;*

$$B(c_1) \subsetneq B(c_2) \quad \text{for } c_1 < c_2. \tag{3.13}$$

(*iv*) *The de Branges spaces $B(c)$ are continuous;*

$$\overline{\bigcup_{x \in (a,c)} B(x)} = B(c) = \bigcap_{x \in (c,b)} B(x). \tag{3.14}$$

PROOF. Items (i) and (ii) are an immediate consequence of Theorem 3.1 and the fact that the transformation from (2.10) extends to a unitary map from $L^2((a,b); R(x)dx)$ onto $L^2(\mathbb{R}, d\rho)$. The inclusion in (iii) follows from the similar fact that

$$L^2((a,c_1); R(x)dx) \subsetneq L^2((a,c_2); R(x)dx).$$

In much the same manner, the remaining claim is due to

$$\overline{\bigcup_{s \in (a,c)} L^2((a,s); R(x)dx)} = L^2((a,c); R(x)dx) = \bigcap_{s \in (c,b)} L^2((a,s); R(x)dx).$$

□

4. Inverse uniqueness results

In order to state our inverse uniqueness results, let τ and $\tilde{\tau}$ be two differential expressions of the form (2.1), both of them satisfying Hypothesis 2.1. Furthermore, we will denote with S and $\tilde{S}$ two self-adjoint realizations with separated boundary conditions and suppose Hypothesis 2.3 to hold. All of the quantities associated with τ are denoted as in the previous sections and, in obvious notation, the corresponding quantities for $\tilde{\tau}$ are equipped with an additional twiddle.

THEOREM 4.1. *Suppose that the function*

$$E(z,c)\tilde{E}(z,\tilde{c})^{-1}, \quad z \in \mathbb{C}^+ \tag{4.1}$$

is of bounded type for some $c \in (a,b)$ and $\tilde{c} \in (\tilde{a}, \tilde{b})$. If $\rho = \tilde{\rho}$, then there is a locally absolutely continuous bijection η from (a,b) onto $(\tilde{a}, \tilde{b})$ and a locally absolutely continuous function $\Gamma : (a,b) \to \mathbb{R}^{2\times 2}$ with $\det(\Gamma) = 1$ such that

$$\eta' \tilde{R} \circ \eta = \Gamma^* R \Gamma, \qquad \eta' \tilde{Q} \circ \eta = \Gamma^* Q \Gamma + \Gamma^* J \Gamma'. \tag{4.2}$$

Moreover, the operator U from $L^2((\tilde{a}, \tilde{b}); \tilde{R}(x)dx)$ to $L^2((a,b); R(x)dx)$, given by

$$U\tilde{f}(x) = \Gamma(x)\tilde{f}(\eta(x)), \quad x \in (a,b), \tag{4.3}$$

is unitary with $S = U\tilde{S}U^{-1}$.

PROOF. First of all, note that because of the definition of de Branges spaces, the function in (4.1) is of bounded type for all $c \in (a,b)$ and $\tilde{c} \in (\tilde{a},\tilde{b})$. Now fix some arbitrary point $x \in (a,b)$. Since for every $\tilde{x} \in (\tilde{a},\tilde{b})$, both spaces $B(x)$ and $\tilde{B}(\tilde{x})$ are isometrically embedded in $L^2(\mathbb{R},d\rho)$, we infer from the subspace ordering theorem [**8**, Theorem 35] that $B(x)$ is contained in $\tilde{B}(\tilde{x})$ or $\tilde{B}(\tilde{x})$ is contained in $B(x)$. We claim that the infimum $\eta(x)$ of all $\tilde{x} \in (\tilde{a},\tilde{b})$ such that $B(x) \subseteq \tilde{B}(\tilde{x})$ lies in $(\tilde{a},\tilde{b})$. In fact, otherwise we either had $\tilde{B}(\tilde{x}) \subseteq B(x)$ for all $\tilde{x} \in (\tilde{a},\tilde{b})$ or $B(x) \subseteq \tilde{B}(\tilde{x})$ for all $\tilde{x} \in (\tilde{a},\tilde{b})$. The first case would imply that $B(x)$ is dense in $L^2(\mathbb{R},d\rho)$, which is not possible in view of Corollary 3.2 *(*ii*)*. In the second case, this would mean that for every function $F \in B(x)$ and $\zeta \in \mathbb{C}$ we have

$$\begin{aligned}|F(\zeta)|^2 &\le \left|[\tilde{K}(\zeta,\cdot\,,\tilde{x}),F]_{\tilde{B}(\tilde{x})}\right|^2 \le \|F\|^2_{\tilde{B}(\tilde{x})}[\tilde{K}(\zeta,\cdot\,,\tilde{x}),\tilde{K}(\zeta,\cdot\,,\tilde{x})]_{\tilde{B}(\tilde{x})}\\ &= \|F\|^2_{B(x)}\tilde{K}(\zeta,\zeta,\tilde{x})\end{aligned}$$

for every $\tilde{x} \in (\tilde{a},\tilde{b})$. But since $\tilde{K}(\zeta,\zeta,\tilde{x}) \to 0$ as $\tilde{x} \to \tilde{a}$ by (3.9), this would imply $B(x) = \{0\}$ contradicts Theorem 3.1. Now from (3.14) we infer that

$$\tilde{B}(\eta(x)) = \overline{\bigcup_{\tilde{x}<\eta(x)} \tilde{B}(\tilde{x})} \subseteq B(x) \subseteq \bigcap_{\tilde{x}>\eta(x)} \tilde{B}(\tilde{x}) = \tilde{B}(\eta(x))$$

and hence $B(x) = \tilde{B}(\eta(x))$, including norms. In particular, from this we infer that there is a matrix $\Gamma(x) \in \mathbb{R}^{2\times 2}$ with $\det(\Gamma(x)) = 1$ such that

$$\Phi(z,x) = \Gamma(x)\tilde{\Phi}(z,\eta(x)), \quad z \in \mathbb{C}, \tag{4.4}$$

in view of [**7**, Theorem I].

The function $\eta : (a,b) \to (\tilde{a},\tilde{b})$ defined above is strictly increasing and continuous by Corollary 3.2. Moreover, since for every $\zeta \in \mathbb{C}$ we have

$$\tilde{K}(\zeta,\zeta,\eta(x)) = K(\zeta,\zeta,x) \to 0,$$

as $x \to a$, we infer that $\eta(x) \to \tilde{a}$ as $x \to a$. Furthermore, the function η even has to be a bijection because of (3.12). Next, since the reproducing kernels are given by (3.9), we get for each $z \in \mathbb{C}$

$$\int_a^x \langle \Phi(z,s), R(s)\Phi(z,s)\rangle ds = \int_{\tilde{a}}^{\eta(x)} \langle \tilde{\Phi}(z,s), \tilde{R}(s)\tilde{\Phi}(z,s)\rangle ds, \quad x \in (a,b).$$

Thus η turns out to be locally absolutely continuous in view of [**39**, Chapter IX; Exercise 13] and [**39**, Chapter IX; Theorem 3.5] with

$$\langle \Phi(z,x), R(x)\Phi(z,x)\rangle = \eta'(x)\langle \tilde{\Phi}(z,\eta(x)), \tilde{R}(\eta(x))\tilde{\Phi}(z,\eta(x))\rangle \tag{4.5}$$

for almost all $x \in (a,b)$. In order to prove that Γ is locally absolutely continuous as well, fix some $z \in \mathbb{C}\backslash\mathbb{R}$ and note that from the Lagrange identity (2.6) one gets

$$\begin{aligned}&\tilde{\Phi}_1(z,\eta(x))\tilde{\Phi}_2(z^*,\eta(x)) - \tilde{\Phi}_2(z,\eta(x)\tilde{\Phi}_1(z^*,\eta(x))\\ &\qquad = 2\mathrm{i}\,\mathrm{Im}(z)\int_{\tilde{a}}^{\eta(x)} \langle \tilde{\Phi}(z,s), R(s)\tilde{\Phi}(z,s)\rangle ds \neq 0\end{aligned} \tag{4.6}$$

for each $x \in (a,b)$. Then from (4.4) we infer that

$$\Gamma(x) = \begin{pmatrix}\Phi_1(z,x) & \Phi_1(z^*,x)\\ \Phi_2(z,x) & \Phi_2(z^*,x)\end{pmatrix}\begin{pmatrix}\tilde{\Phi}_1(z,\eta(x)) & \tilde{\Phi}_1(z^*,\eta(x))\\ \tilde{\Phi}_2(z,\eta(x)) & \tilde{\Phi}_2(z^*,\eta(x))\end{pmatrix}^{-1}, \quad x \in (a,b).$$

Since the right-hand side is locally absolutely continuous by [**39**, Chapter IX; Section 1], so is Γ.

Next, we will show that the relations in (4.2) hold. To this end, one plugs (4.4) into (4.5) and notes that for $z \in \mathbb{C}\backslash\mathbb{R}$, the vectors $\tilde{\Phi}(z,\eta(x))$ and $\tilde{\Phi}(z^*,\eta(x))$ are linearly independent by (4.6), which then shows that the first relation in (4.2) holds almost everywhere. Subsequently, one differentiates (4.4) to obtain the second relation in (4.2) almost everywhere.

Finally, consider the unitary operator $\mathcal{F}^{-1}\tilde{\mathcal{F}}$ which maps $L^2((\tilde{a},\tilde{b});\tilde{R}(x)dx)$ onto $L^2((a,b);R(x)dx)$ and satisfies $S = \mathcal{F}^{-1}\tilde{\mathcal{F}}\tilde{S}\tilde{\mathcal{F}}^{-1}\mathcal{F}$. In order to identify this operator with U, one simply observes that for every $F \in L^2(\mathbb{R};d\rho)$ with compact support

$$\mathcal{F}^{-1}F(x) = \Gamma(x)\int_{\mathbb{R}} \tilde{\Phi}(\lambda,\eta(x))F(\lambda)d\rho(\lambda) = \Gamma(x)\tilde{\mathcal{F}}^{-1}F(\eta(x)),$$

for almost all $x \in (a,b)$, which finishes the proof. □

In view of applications, the condition that the function (4.1) is of bounded type is somewhat inconvenient. However, let us mention that this condition always holds if the solutions Φ and $\tilde{\Phi}$ satisfy certain growth conditions. Therefore, recall that some entire function F belongs to the Cartwright class if it is of finite exponential type and the logarithmic integral

$$\int_{\mathbb{R}} \frac{\ln^+|F(\lambda)|}{1+\lambda^2}d\lambda \tag{4.7}$$

is finite, where $\ln^+$ is the positive part of the natural logarithm. Now if it is known that the functions $E(\,\cdot\,,c)$ and $\tilde{E}(\,\cdot\,,\tilde{c})$ belong to the Cartwright class for some $c \in (a,b)$ and $\tilde{c} \in (\tilde{a},\tilde{b})$, then a theorem of Krein [**41**, Theorem 6.17], [**33**, Section 16.1] guarantees that they are of bounded type in $\mathbb{C}^+$. Since the quotient of two functions of bounded type is of bounded type itself, this immediately implies that (4.1) is of bounded type as well. In particular, we will see in the next section that it is always possible to choose the real entire solutions Φ and $\tilde{\Phi}$ such that this holds, provided that the limit circle case prevails at the left endpoint.

In general, one can not deduce more information about the Dirac operator from the spectral measure than from what is claimed in Theorem 4.1. However, if one has further a priori information on the differential expression, then it is possible indeed to strengthen the results. For example, in the case of Dirac operators with trivial weight matrices we get the following refinement of Theorem 4.1.

COROLLARY 4.2. *Suppose that* $R = \tilde{R} = I$ *and that the function*

$$E(z,c)\tilde{E}(z,\tilde{c})^{-1}, \quad z \in \mathbb{C}^+ \tag{4.8}$$

is of bounded type for some $c \in (a,b)$ *and* $\tilde{c} \in (\tilde{a},\tilde{b})$. *If* $\rho = \tilde{\rho}$, *then there is an* $\eta_0 \in \mathbb{R}$ *and a locally absolutely continuous real-valued function* φ *on* (a,b) *such that*

$$\tilde{Q}(\eta_0 + x) = \mathrm{e}^{-\varphi(x)J}Q(x)\mathrm{e}^{\varphi(x)J} - \varphi'(x)I \tag{4.9}$$

for almost all $x \in (a,b)$. *Moreover, the operator* U *from* $L^2((\tilde{a},\tilde{b});Idx)$ *to* $L^2((a,b);Idx)$ *given by*

$$U\tilde{f}(x) = \mathrm{e}^{\varphi(x)J}\tilde{f}(\eta_0 + x), \quad x \in (a,b) \tag{4.10}$$

is unitary with $S = U\tilde{S}U^{-1}$.

PROOF. Using Theorem 4.1, we infer from our additional assumptions that

$$\Gamma(x)^*\Gamma(x) = \eta'(x)I, \quad x \in (a,b).$$

Taking $\det(\Gamma) = 1$ into account, we conclude that $\eta(x) = \eta_0 + x$ for some $\eta_0 \in \mathbb{R}$. In particular, this shows that $\Gamma(x)$ is unitary for each $x \in (a,b)$ and we may write

$$\Gamma(x) = \begin{pmatrix} \cos\varphi(x) & -\sin\varphi(x) \\ \sin\varphi(x) & \cos\varphi(x) \end{pmatrix} = \mathrm{e}^{\varphi(x)J}, \quad x \in (a,b),$$

for some real-valued function φ on (a,b). Since the entries of Γ are locally absolutely continuous, we infer that φ may be chosen locally absolutely continuous as well. Now evaluating the second equation in (4.2) establishes the claim. □

If, in addition to the assumptions of Corollary 4.2, we additionally require that the traces of the potentials Q and $\tilde{Q}$ are equal almost everywhere (e.g. if both are normalized to the same constant), then we may conclude from (4.9) that the function φ has to be a real constant. Under various additional a priori assumptions (for example, prescribing the boundary condition at a regular left endpoint), it is furthermore even possible to determine this constant (at least modulo 2π).

As an application of Corollary 4.2, which is of particular physical interest, we would like to single out the case of radial Dirac operators. Namely, we may consider differential expressions of the form

$$\tau f(x) = Jf'(x) + \begin{pmatrix} q_{\mathrm{sc}}(x) & \frac{\kappa}{x} + q_{\mathrm{am}}(x) \\ \frac{\kappa}{x} + q_{\mathrm{am}}(x) & -q_{\mathrm{sc}}(x) \end{pmatrix} f(x), \quad x \in (0,\infty), \tag{4.11}$$

with q_{sc}, $q_{\mathrm{am}} \in L^1_{loc}([0,\infty),\mathbb{R})$ and $\kappa \geq 0$. In the case $\kappa = \frac{1}{2}$ we also assume that

$$\int_0^c \left(|q_{\mathrm{sc}}(x)| + |q_{\mathrm{am}}(x)|\right) |\log(x)| dx < \infty \tag{4.12}$$

for some $c \in (0,\infty)$. Here, q_{sc} and q_{am} are interpreted as a scalar potential and an anomalous magnetic moment, respectively (see [**46**, Chapter 4]). For simplicity we assume that the electrostatic potential is zero which can always be achieved by a suitable gauge transform. Moreover, the case $\kappa < 0$ can be reduced to the case $\kappa > 0$ by performing a Liouville transform with $\Gamma = J$. If $\kappa > 0$, then this differential expression is singular at the left endpoint as $\frac{\kappa}{x}$ is not integrable there and it is also singular at ∞ because the endpoint itself is not finite. Note that τ is always limit point at ∞ (see [**34**, Theorem 8.6.1] and [**47**, Theorem 6.8]). We also refer to the monographs [**34**], [**47**] for background and also to [**46**] for further information about Dirac operators and their applications.

Under the above assumptions, there is a particular real entire solution Φ associated with the differential expression (4.11) which satisfies the spatial asymptotics

$$\Phi_1(z,x) = o(x^\kappa), \qquad \Phi_2(z,x) = x^\kappa(1+o(1)), \tag{4.13}$$

as $x \to 0$ (see [**45**] in the case $\kappa \in \mathbb{N}_0$ and [**6**, Section 8] for the general case). In the case $0 \leq \kappa < \frac{1}{2}$, in which the differential expression is in the limit circle case at the left endpoint, this implies that we impose the boundary condition

$$\lim_{x\to 0} x^\kappa f_1(x) = 0. \tag{4.14}$$

THEOREM 4.3. *Suppose that τ, $\tilde{\tau}$ are differential expressions of the form* (4.11) *and that the boundary conditions of S, $\tilde{S}$ at the left endpoint are given by* (4.14) *if the limit circle case prevails there. Furthermore, let Φ, $\tilde{\Phi}$ be the real entire solutions* *which satisfy* (4.13) *and let* ρ, $\tilde{\rho}$ *be the corresponding spectral measures. If* $\rho = \tilde{\rho}$, *then* $\kappa = \tilde{\kappa}$ *and* $q_{\mathrm{sc}} = \tilde{q}_{\mathrm{sc}}$, $q_{\mathrm{am}} = \tilde{q}_{\mathrm{am}}$ *almost everywhere, as well as* $S = \tilde{S}$.

PROOF. It follows from [**6**, Lemma 8.3] that the solutions $\Phi(\,\cdot\,,x)$ and $\tilde{\Phi}(\,\cdot\,,x)$ belong to the Cartwright class for every fixed $x > 0$. Therefore, so do the functions $E(\,\cdot\,,x)$ and $\tilde{E}(\cdot,x)$ defined by (3.1) and hence the function (4.8) is of bounded type. Thus we may apply Corollary 4.2 and infer that $\eta_0 = 0$ because the left endpoints are the same. Moreover, since the traces of the potential matrices are zero almost everywhere, the function φ is constant. However, the particular form of the potential matrices immediately implies that $\sin\varphi = 0$ (also taking into account that κ, $\tilde{\kappa} \geq 0$). Therefore, $\Gamma = \pm I$, and we infer from (4.9) that

$$\begin{pmatrix} q_{\mathrm{sc}}(x) & \frac{\kappa}{x} + q_{\mathrm{am}}(x) \\ \frac{\kappa}{x} + q_{\mathrm{am}}(x) & -q_{\mathrm{sc}}(x) \end{pmatrix} = \begin{pmatrix} \tilde{q}_{\mathrm{sc}}(x) & \frac{\tilde{\kappa}}{x} + \tilde{q}_{\mathrm{am}}(x) \\ \frac{\tilde{\kappa}}{x} + \tilde{q}_{\mathrm{am}}(x) & -\tilde{q}_{\mathrm{sc}}(x) \end{pmatrix}$$

for almost all $x \in (0,\infty)$, which finishes the proof. □

Clearly, in view of (2.14), we can rephrase this as the following result.

COROLLARY 4.4. *Suppose that* τ, $\tilde{\tau}$ *are differential expressions of the form* (4.11) *and that the boundary conditions of* S, $\tilde{S}$ *at the left endpoint are given by* (4.14) *if the limit circle case prevails there. Furthermore, let* Φ, $\tilde{\Phi}$ *be real entire solutions which satisfy* (4.13) *and let* M, $\tilde{M}$ *be corresponding singular Weyl–Titchmarsh–Kodaira functions. If the function* $M - \tilde{M}$ *is entire, then* $\kappa = \tilde{\kappa}$ *and* $q_{\mathrm{sc}} = \tilde{q}_{\mathrm{sc}}$, $q_{\mathrm{am}} = \tilde{q}_{\mathrm{am}}$ *almost everywhere, as well as* $S = \tilde{S}$.

REMARK 4.5. *Note that various inverse spectral problems for radial Dirac operators in dimension three (i.e., the case* $\kappa \in \mathbb{Z}$*) inside the unit ball have been studied in* [**2, 3, 45**] *(see also the references therein) in the case* q_{am}, $q_{\mathrm{sc}} \in L^p(0,1)$, $p \in (1,\infty)$. *In particular,* [**2, 3, 45**] *provide a complete solution of the inverse spectral problems by one spectrum and norming constants and by two spectra in the special case* q_{am}, $q_{\mathrm{sc}} \in L^2(0,1)$.

The case $\kappa = 0$ *is of course classical and we refer to the monograph* [**34**] *as well as the recent papers* [**1, 35, 36, 38, 40, 43**]. *For a local Borg–Marchenko result in the case* $\kappa = 0$ *and* $q_{\mathrm{am}}, q_{\mathrm{sc}} \in L^\infty(a,b)$, *see* [**9**]. *For related uniqueness results, see* [**19**].

Similarly, one can improve Theorem 4.1 in the case when the weight matrix is not constant but the potential vanishes identically.

COROLLARY 4.6. *Suppose that* $Q = \tilde{Q} = 0$ *and that the function*

$$E(z,c)\tilde{E}(z,\tilde{c})^{-1}, \quad z \in \mathbb{C}^+ \tag{4.15}$$

is of bounded type for some $c \in (a,b)$ *and* $\tilde{c} \in (\tilde{a},\tilde{b})$. *If* $\rho = \tilde{\rho}$, *then there is a locally absolutely continuous bijection* η *from* (a,b) *onto* $(\tilde{a},\tilde{b})$ *and a matrix* $\Gamma_0 \in \mathbb{R}^{2\times 2}$ *with* $\det(\Gamma_0) = 1$ *such that*

$$\eta'(x)\tilde{R}(\eta(x)) = \Gamma_0^* R(x)\,\Gamma_0 \tag{4.16}$$

for almost all $x \in (a,b)$. *Moreover, the operator* U *from* $L^2((\tilde{a},\tilde{b});\tilde{R}(x)dx)$ *to* $L^2((a,b);R(x)dx)$ *given by*

$$U\tilde{f}(x) = \Gamma_0\tilde{f}(\eta(x)), \quad x \in (a,b) \tag{4.17}$$

is unitary with $S = U\tilde{S}U^{-1}$.

PROOF. From Theorem 4.1, we infer from our additional assumptions that

$$\Gamma(x)^* J \Gamma'(x) = 0, \quad x \in (a,b).$$

Since $\det(\Gamma) = 1$ and J is invertible, we conclude that Γ is constant, which establishes the claim. □

If, in addition to the assumptions of Corollary 4.6, we also require that the determinant of the weights R and $\tilde{R}$ are equal to one almost everywhere; then we may conclude from (4.16) that η is linear with gradient one.

REMARK 4.7. *Let us mention that Corollary* 4.6 *was obtained in* [**48**, Proposition 4.9] *under the assumption that* $\operatorname{tr}(R) \in L^1(a,c)$ *for some* $c \in (a,b)$*, that is, the left endpoint* $x = a$ *is in the limit circle case. However, the case of Hamiltonians* R *that may vanish on sets of positive Lebesgue measure has also been studied in* [**48**].

5. Two-spectra uniqueness results

In this section, we will show to which extent the (purely discrete) spectra of two self-adjoint realizations determine the underlying differential expression as well as boundary conditions. Therefore, one of the endpoints has to be in the limit circle case. If this is the case for the left endpoint, then Hypothesis 2.3 holds for any self-adjoint realization with separated boundary conditions, and we may even choose a particular real entire fundamental system of solutions.

PROPOSITION 5.1. *Suppose Hypothesis* 2.1*, that* τ *is in the limit circle case at* a *and that* S *is self-adjoint realization of* τ *with separated boundary conditions holds. Then there is a real entire solution* Φ *of* $(\tau - z)u = 0$ *which lies in* $\operatorname{dom}(S)$ *near* a *as well as a real entire solution* Θ *of* $(\tau - z)u = 0$ *such that*

(5.1)
$$W_a(\Theta(z_1,\cdot),\Theta(z_2,\cdot)) = W_a(\Phi(z_1,\cdot),\Phi(z_2,\cdot)) = 0, \quad W_a(\Theta(z_1,\cdot),\Phi(z_2,\cdot)) = 1,$$

for all z_1*,* $z_2 \in \mathbb{C}$*. In this case, the corresponding Weyl–Titchmarsh–Kodaira function is a Herglotz–Nevanlinna function.*

This result can be proven along the lines of the corresponding result for Sturm–Liouville equations; see, e.g., [**24**, Appendix A], [**6**, Section 5]. Note that hereby, the measure appearing in the Herglotz–Nevanlinna representation

$$M(z) = \operatorname{Re}(M(\mathrm{i})) + m_c z + \int_{\mathbb{R}} \left(\frac{1}{\lambda - z} - \frac{\lambda}{1+\lambda^2} \right) d\rho(\lambda), \quad z \in \mathbb{C}\backslash\mathbb{R} \tag{5.2}$$

of the singular Weyl–Titchmarsh–Kodaira function is precisely the spectral measure of S associated with the real entire solution Φ in view of (2.14). For the purpose of this section, we will now derive a growth restriction for these particular solutions.

COROLLARY 5.2. *Suppose Hypothesis* 2.1*, that* τ *is in the limit circle case at* a *and that* S *is a self-adjoint realization of* τ *with separated boundary conditions holds. If* Φ *is a real entire solution of* $(\tau - z)u = 0$ *as in Proposition* 5.1*, then the entire functions* $\Phi_1(\,\cdot\,,c)$ *and* $\Phi_2(\,\cdot\,,c)$ *belong to the Cartwright class for every* $c \in (a,b)$.

PROOF. This follows essentially from an application of a result by Krein [**28**]. Alternatively, note that according to Proposition 5.1 the fractions on the right-hand side of

$$\frac{1}{\Phi_1(z,c)^2} = \frac{\Theta_1(z,c)}{\Phi_1(z,c)}\left(\frac{\Phi_2(z,c)}{\Phi_1(z,c)} - \frac{\Theta_2(z,c)}{\Theta_1(z,c)}\right), \quad z\in\mathbb{C}\backslash\mathbb{R}$$

are Herglotz–Nevanlinna functions (up to the sign), which are bounded by

$$C\frac{1+|z|^2}{|\mathrm{Im}(z)|} \leq K\exp\left(L\frac{1+\sqrt{|z|}}{\sqrt[4]{|\mathrm{Im}(z)|}}\right), \quad z\in\mathbb{C}\backslash\mathbb{R},$$

for some constants C, K, $L\in\mathbb{R}^+$. Thus also

$$\left|\frac{1}{\Phi_1(z,c)^2}\right| \leq 2K^2\exp\left(2L\frac{1+\sqrt{|z|}}{\sqrt[4]{|\mathrm{Im}(z)|}}\right), \quad z\in\mathbb{C}\backslash\mathbb{R},$$

which guarantees that $\Phi_1(\,\cdot\,,c)$ belongs to the Cartwright class in view of Matsaev's theorem [**33**, Theorem 26.4.4]). In much the same manner one shows that the same holds true for $\Phi_2(\,\cdot\,,c)$. □

In order to state our inverse uniqueness results, let τ and $\tilde\tau$ be two differential expressions of the form (2.1), both of them satisfying Hypothesis 2.1. Again, all of the quantities associated with τ and $\tilde\tau$ will be denoted as in the preceding sections and distinguished with a twiddle.

THEOREM 5.3. *Suppose that τ and $\tilde\tau$ are in the limit circle case at the left endpoint and that S, T as well as $\tilde S$, $\tilde T$ are two distinct self-adjoint realizations with purely discrete spectra and the same boundary condition at the right endpoint (if any). If*

$$\sigma(S)=\sigma(\tilde S) \quad \textit{and} \quad \sigma(T)=\sigma(\tilde T), \tag{5.3}$$

then there is a locally absolutely continuous bijection η from (a,b) onto $(\tilde a,\tilde b)$ and a locally absolutely continuous function $\Gamma:(a,b)\to\mathbb{R}^{2\times 2}$ with $\det(\Gamma)=1$ such that

$$\eta'\tilde R\circ\eta = \Gamma^* R\,\Gamma, \qquad \eta'\tilde Q\circ\eta = \Gamma^* Q\,\Gamma + \Gamma^* J\,\Gamma'. \tag{5.4}$$

Moreover, the operator U from $L^2((\tilde a,\tilde b);\tilde R(x)dx)$ to $L^2((a,b);R(x)dx)$ given by

$$U\tilde f(x) = \Gamma(x)\tilde f(\eta(x)), \quad x\in(a,b) \tag{5.5}$$

is unitary with $S=U\tilde S U^{-1}$ and $T=U\tilde T U^{-1}$.

PROOF. Let Θ, Φ be a real entire fundamental system of $(\tau-z)u=0$, as in Proposition 5.1, associated with the operators S. Due to our assumptions, the corresponding singular Weyl–Titchmarsh–Kodaira function M is meromorphic and has poles precisely at the spectrum of S. Furthermore, note that the value $h=M(\lambda)$ is independent of $\lambda\in\sigma(T)$ and conversely, each $\lambda\in\mathbb{C}$ for which $h=M(\lambda)$ is an eigenvalue of T. Thus we may assume (upon replacing Θ by $\Theta+h\Phi$) that the set of zeros of M is precisely the spectrum of T. If $\tilde\Theta$, $\tilde\Phi$ is a similar real entire fundamental system of $(\tilde\tau-z)u=0$, then we infer from our assumptions (5.3) and a theorem by Krein [**33**, Theorem 27.2.1], that $M=C^2\tilde M$ for some positive constant C. Moreover, upon replacing Φ with $C\Phi$ as well as Θ with $C^{-1}\Theta$, we may even assume that $M=\tilde M$. From this we obtain $\rho=\tilde\rho$ and hence, except for the very last part, the claim follows from Theorem 4.1 (taking Corollary 5.2 into account).

In order to show that U maps $\tilde{T}$ to T, note that U maps $\tilde{T}_{\max}$ given by

$$\operatorname{dom}(\tilde{T}_{\max}) = \{f \in L^2((\tilde{a},\tilde{b}); \tilde{R}(x)dx) \mid f_1,\ f_2 \in AC_{\mathrm{loc}}(\tilde{a},\tilde{b}), \\ \tilde{\tau} f \in L^2((\tilde{a},\tilde{b}); \tilde{R}(x)dx)\},$$

with $\tilde{T}_{\max} f = \tilde{\tau} f$ for $f \in \operatorname{dom}(\tilde{T}_{\max})$, to the corresponding operator $T_{\max}$ given by

$$\operatorname{dom}(T_{\max}) = \{f \in L^2((a,b); R(x)dx) \mid f_1,\ f_2 \in AC_{\mathrm{loc}}(a,b), \\ \tau f \in L^2((a,b); R(x)dx)\},$$

with $T_{\max} f = \tau f$ for $f \in \operatorname{dom}(T_{\max})$. Thus, the operator $U\tilde{T}U^{-1}$ is a self-adjoint realization of τ with separated boundary conditions. Furthermore, from the first part of the proof we know that the boundary condition at the right endpoint is the same as the one for T. In order to show that the boundary condition at a is the same as well, pick some $\lambda \in \sigma(T)$ and note that the function

$$\Gamma(x)\tilde{\Theta}(\lambda, \eta(x)), \quad x \in (a,b),$$

is a solution of $(\tau - \lambda)u = 0$. Moreover, it is an eigenfunction of $U\tilde{T}U^{-1}$ as well as of T, which establishes the remaining claim. □

Of course, as in Section 4, it is possible to improve on Theorem 5.3 if one has further a priori information on the coefficients of the underlying differential expressions. For example, one obtains the following two analogues of Corollary 4.2 and Corollary 4.6 without changing their respective proofs.

COROLLARY 5.4. *Suppose that $R = \tilde{R} = I$, that τ and $\tilde{\tau}$ are in the limit circle case at the left endpoint and that S, T as well as $\tilde{S}$, $\tilde{T}$ are two distinct self-adjoint realizations with discrete spectra and the same boundary condition at the right endpoint (if any). If*

$$\sigma(S) = \sigma(\tilde{S}) \quad \text{and} \quad \sigma(T) = \sigma(\tilde{T}), \tag{5.6}$$

then there is an $\eta_0 \in \mathbb{R}$ and a locally absolutely continuous real-valued function φ on (a,b) such that

$$\tilde{Q}(\eta_0 + x) = \mathrm{e}^{-\varphi(x)J} Q(x) \mathrm{e}^{\varphi(x)J} - \varphi'(x) I \tag{5.7}$$

for almost all $x \in (a,b)$. Moreover, the operator U from $L^2((\tilde{a},\tilde{b}); I dx)$ to $L^2((a,b); I dx)$ given by

$$U\tilde{f}(x) = \mathrm{e}^{\varphi(x)J} \tilde{f}(\eta_0 + x), \quad x \in (a,b), \tag{5.8}$$

is unitary with $S = U\tilde{S}U^{-1}$ and $T = U\tilde{T}U^{-1}$.

Of course, as we remarked after Corollary 4.2, one may infer that φ is a constant if we require that the traces of Q and $\tilde{Q}$ are equal almost everywhere. Also, one can apply this result to radial Dirac operators (4.11) in the case when $|\kappa| < \frac{1}{2}$.

COROLLARY 5.5. *Suppose that $Q = \tilde{Q} = 0$, that τ and $\tilde{\tau}$ are in the limit circle case at the left endpoint and that S, T as well as $\tilde{S}$, $\tilde{T}$ are two distinct self-adjoint realizations with discrete spectra and the same boundary condition at the right endpoint (if any). If*

$$\sigma(S) = \sigma(\tilde{S}) \quad \text{and} \quad \sigma(T) = \sigma(\tilde{T}), \tag{5.9}$$

then there is a locally absolutely continuous bijection η *from* (a,b) *onto* $(\tilde{a},\tilde{b})$ *and a matrix* $\Gamma_0 \in \mathbb{R}^{2\times 2}$ *with* $\det(\Gamma_0) = 1$ *such that*

$$\eta'(x)\tilde{R}(\eta(x)) = \Gamma_0^* R(x)\,\Gamma_0 \tag{5.10}$$

for almost all $x \in (a,b)$. *Moreover, the operator* U *from* $L^2((\tilde{a},\tilde{b});\tilde{R}(x)dx)$ *to* $L^2((a,b);R(x)dx)$ *given by*

$$U\tilde{f}(x) = \Gamma_0\tilde{f}(\eta(x)), \quad x \in (a,b), \tag{5.11}$$

is unitary with $S = U\tilde{S}U^{-1}$ *and* $T = U\tilde{T}U^{-1}$.

Again, we may conclude that η is linear with gradient one if we require the determinant of the weight matrices R and $\tilde{R}$ to be equal to one almost everywhere. Finally, let us also mention that if the boundary conditions of S, T and $\tilde{S}$, $\tilde{T}$ are known a priori, then one may even say more about the matrix Γ_0 from evaluating (4.4) and its analog for $\Theta(z,x)$ at $z=0$. (Note that τ and $\tilde{\tau}$ are necessarily regular at the left endpoint in this case.)

Acknowledgments. We thank Fritz Gesztesy and Alexander Sakhnovich for hints with respect to the literature. Jonathan Eckhardt and Gerald Tesch gratefully acknowledge the stimulating atmosphere during spring 2013 at the *Institut Mittag-Leffler*, where parts of this paper were written during the international research program on *Inverse Problems and Applications*.

References

[1] S. Albeverio, R. Hryniv, and Ya. Mykytyuk, *Inverse spectral problems for Dirac operators with summable potentials*, Russ. J. Math. Phys. **12** (2005), 406–423. MR2201307 (2006i:34021)

[2] S. Albeverio, R. Hryniv, and Ya. Mykytyuk, *Reconstruction of radial Dirac operators*, J. Math. Phys. **48** (2007), 043501, 14 pp. MR2319895 (2008e:81038)

[3] S. Albeverio, R. Hryniv, and Ya. Mykytyuk, *Reconstruction of radial Dirac and Schrödinger operators from two spectra*, J. Math. Anal. Appl. **339** (2008), 45–57. MR2370631 (2008k:81127)

[4] A. A. Balinsky and W. D. Evans, *Spectral Analysis of Relativistic Operators*, Imperial College Press, London 2011. MR2779257 (2012g:81063)

[5] G. Borg, *Uniqueness theorems in the spectral theory of* $y'' + (\lambda - q(x))y = 0$, in "Den 11te Skandinaviske Matematikerkongress, Trondheim, 1949", pp. 276–287, Johan Grundt Tanums Forlag, Oslo, 1952. MR0058063 (15:315a)

[6] R. Brunnhuber, J. Eckhardt, A. Kostenko, and G. Teschl, *Singular Weyl–Titchmarsh–Kodaira theory for one-dimensional Dirac operators*, Monatsh. Math. **174** (2014), 515–547, doi:10.1007/s00605-013-0563-5.

[7] L. de Branges, *Some Hilbert spaces of entire functions*, Trans. Amer. Math. Soc. **96** (1960), 259–295. MR0133455 (24:A3289a)

[8] L. de Branges, *Hilbert Spaces of Entire Functions*, Prentice-Hall, Inc., Englewood Cliffs, N.J. 1968. MR0229011 (37:4590)

[9] S. Clark and F. Gesztesy, *Weyl–Titchmarsh M-function asymptotics, local uniqueness results, trace formulas, and Borg-type theorems for Dirac operators*, Trans. Amer. Math. Soc. **354** (2002), 3475–3534. MR1911509 (2003i:34191)

[10] J. Eckhardt, *Inverse uniqueness results for Schrödinger operators using de Branges theory*, Complex Anal. Oper. Theory **8** (2014), no. 1, 37–50. MR3147711

[11] J. Eckhardt, *Direct and inverse spectral theory of singular left-definite Sturm–Liouville operators*, J. Differential Equations **253** (2012), 604–634. MR2921208

[12] J. Eckhardt, F. Gesztesy, R. Nichols, and G. Teschl, *Supersymmetry and Schrödinger-type operators with distributional matrix-valued potentials*, J. Spectr. Theory (to appear), arXiv1206.4966.

[13] J. Eckhardt, F. Gesztesy, R. Nichols and G. Teschl, *Inverse spectral theory for Sturm–Liouville operators with distributional potentials*, J. Lond. Math. Soc. (2) **88** (2013), no. 3, 801–828. MR3145132

[14] J. Eckhardt, A. Kostenko and G. Teschl, *Spectral asymptotics for one-dimensional weighted Dirac operators*, in preparation.

[15] J. Eckhardt and G. Teschl, *On the isospectral problem of the dispersionless Camassa-Holm equation*, Adv. Math. **235** (2013), 469–495. MR3010066

[16] J. Eckhardt and G. Teschl, *Singular Weyl–Titchmarsh–Kodaira theory for Jacobi operators*, Oper. Matrices **7** (2013), no. 3, 695–712. MR3089355

[17] C. Fulton and H. Langer, *Sturm–Liouville operators with singularities and generalized Nevanlinna functions*, Complex Anal. Oper. Theory **4** (2010), 179–243. MR2652523 (2011i:34053)

[18] C. Fulton, H. Langer, and A. Luger, *Mark Krein's method of directing functionals and singular potentials*, Math. Nachr. **285** (2012), 1791–1798. MR2988006

[19] F. Gesztesy, A. Kiselev, and K. A. Makarov, *Uniqueness results for matrix-valued Schrödinger, Jacobi, and Dirac-type operators*, Math. Nachr. **239-240** (2002), 103–145. MR1905666 (2003i:47047)

[20] F. Gesztesy and M. Zinchenko, *On spectral theory for Schrödinger operators with strongly singular potentials*, Math. Nachr. **279** (2006), 1041–1082. MR2242965 (2007h:47076)

[21] I. Gohberg and M. Krein, *Theory and Applications of Volterra Operators in Hilbert Spaces*, Transl. Math. Monographs **24**, Amer. Math. Soc., Providence, RI, 1970. MR0264447 (41:9041)

[22] I. S. Kac, *Linear relations generated by canonical differential equations*, Funct. Anal. Appl. **17** (1983), 315–317. MR725424 (85k:47064)

[23] A. Kostenko, A. Sakhnovich, and G. Teschl, *Inverse eigenvalue problems for perturbed spherical Schrödinger operators*, Inverse Problems **26** (2010), 105013, 14 pp. MR2719774 (2011g:34026)

[24] A. Kostenko, A. Sakhnovich, and G. Teschl, *Weyl–Titchmarsh theory for Schrödinger operators with strongly singular potentials*, Int. Math. Res. Not. **2012** (2012), 1699–1747. MR2920828

[25] A. Kostenko, A. Sakhnovich, and G. Teschl, *Commutation methods for Schrödinger operators with strongly singular potentials*, Math. Nachr. **285** (2012), 392–410. MR2899633

[26] A. Kostenko and G. Teschl, *On the singular Weyl–Titchmarsh function of perturbed spherical Schrödinger operators*, J. Differential Equations **250** (2011), 3701–3739. MR2773183 (2012b:34066)

[27] A. Kostenko and G. Teschl, *Spectral asymptotics for perturbed spherical Schrödinger operators and applications to quantum scattering*, Comm. Math. Phys. **322** (2013), 255–275. MR3073165

[28] M. G. Krein, *On the indeterminate case of the Sturm–Liouville boundary problem in the interval* $(0,\infty)$, (Russian) Izvestiya Akad. Nauk SSSR Ser. Mat. **16** (1952), 293–324. MR0052004 (14:558g)

[29] P. Kurasov and A. Luger, *An operator theoretic interpretation of the generalized Titchmarsh–Weyl coefficient for a singular Sturm–Liouville problem*, Math. Phys. Anal. Geom. **14** (2011), 115–151. MR2802457 (2012i:34031)

[30] H. Langer and H. Winkler, *Direct and inverse spectral problems for generalized strings*, Integral Equations Operator Theory **30** (1998), 409–431. MR1622358 (99m:34190)

[31] M. Langer and H. Woracek, *A function space model for canonical systems with an inner singularity*, Acta Sci. Math. (Szeged) **77** (2011), 101–165. MR2841147 (2012h:47093)

[32] M. Lesch and M. Malamud, *On the deficiency indices and self-adjointness of symmetric Hamiltonian systems*, J. Differential Equations **189** (2003), 556–615. MR1964480 (2004f:34128)

[33] B. Ya. Levin, *Lectures on Entire Functions*, Transl. Math. Monographs **150**, Amer. Math. Soc., Providence, RI, 1996. MR1400006 (97j:30001)

[34] B. M. Levitan and I. S. Sargsjan, *Sturm–Liouville and Dirac Operators*, Kluwer Academic Publishers, Dordrecht 1991. MR1136037 (92i:34119)

[35] M. M. Malamud, *Uniqueness questions in inverse problems for systems of differential equations on a finite interval*, Trans. Moscow Math. Soc. **60** (1999), 173–224. MR1702668 (2001h:34026)

[36] M. M. Malamud, *On Borg type theorems for first-order systems on a finite interval*, Funct. Anal. Appl. **33** (1999), 64–68. MR1711902 (2000j:34021)
[37] V. A. Marčenko, *Some questions of the theory of one-dimensional linear differential operators of the second order. I*, Trudy Moskov. Mat. Obšč. **1** (1952), 327–420. MR0058064 (15:315b)
[38] Ya. V. Mykytyuk and D. V. Puyda, *Inverse spectral problems for Dirac operators on a finite interval*, J. Math. Anal. Appl. **386** (2012), 177–194. MR2834876 (2012g:34028)
[39] I. P. Natanson, *Theory of Functions of a Real Variable*, Frederick Ungar Publishing Co., New York, 1955. MR0067952 (16:804c)
[40] D. V. Puyda, *Inverse spectral problems for Dirac operators with summable matrix-valued potentials*, Integral Equations Operator Theory **74** (2012), 417–450. MR2992031
[41] M. Rosenblum and J. Rovnyak, *Topics in Hardy Classes and Univalent Functions*, Birkhäuser Verlag, Basel, 1994. MR1307384 (97a:30047)
[42] A. L. Sakhnovich, *Spectral functions of a second-order canonical system*, Math. USSR-Sb. **71** (1992), 355–369. MR1090913 (92c:47056)
[43] A. Sakhnovich, *Dirac type and canonical systems: spectral and Weyl–Titchmarsh matrix functions, direct and inverse problems*, Inverse Problems **18** (2002), 331–348. MR1910241 (2003f:34164)
[44] A. L. Sakhnovich, L. A. Sakhnovich, and I. Ya. Roitberg, *Inverse Problems and Nonlinear Evolution Equations. Solutions, Darboux Matrices and Weyl–Titchmarsh Functions*, Walter de Gruyter, Berlin, 2013. MR3098432
[45] F. Serier, *Inverse spectral problem for singular Ablowitz–Kaup–Newell–Segur operators on* $[0,1]$, Inverse Problems **22** (2006), 1457–1484. MR2249473 (2007e:34016)
[46] B. Thaller, *The Dirac Equation*, Springer, Berlin, 1991. MR1219537 (94k:81056)
[47] J. Weidmann, *Spectral Theory of Ordinary Differential Operators*, Lecture Notes in Mathematics, **1258**, Springer, Berlin, 1987. MR923320 (89b:47070)
[48] H. Winkler and H. Woracek, *Reparametrizations of non trace-normed Hamiltonians*, Oper. Theory Adv. Appl. **221** (2012), 667–690. MR2954096

INSTITUT MITTAG-LEFFLER, AURAVÄGEN 17, SE-182 60 DJURSHOLM, SWEDEN
E-mail address: `jonathaneckhardt@aon.at`

FACULTY OF MATHEMATICS, UNIVERSITY OF VIENNA, OSKAR-MORGENSTERN-PLATZ 1, 1090 WIEN, AUSTRIA
E-mail address: `duzer80@gmail.com; Oleksiy.Kostenko@univie.ac.at`

FACULTY OF MATHEMATICS, UNIVERSITY OF VIENNA, OSKAR-MORGENSTERN-PLATZ 1, 1090 WIEN, AUSTRIA AND INTERNATIONAL ERWIN SCHRÖDINGER INSTITUTE FOR MATHEMATICAL PHYSICS, BOLTZMANNGASSE 9, 1090 WIEN, AUSTRIA
E-mail address: `Gerald.Teschl@univie.ac.at`
URL: `http://www.mat.univie.ac.at/~gerald/`

Amer. Math. Soc. Transl.
Volume **233**, 2014

Spectral Loci of Sturm–Liouville Operators with Polynomial Potentials

Alexandre Eremenko and Andrei Gabrielov

ABSTRACT. We consider the differential equation $-y'' + P(z,a)y = \lambda y$, where P is a polynomial of the independent variable z depending on a parameter a. The spectral locus is the set of (a, λ) such that the equation has a non-trivial solution tending to zero on two fixed rays in the complex plane. We study the topology of the spectral loci for polynomials P of degree 3 or 4 with respect to z.

We consider the differential operators

$$L : y \mapsto -y'' + Py,$$

where the potential P is a polynomial of degree d. When $d \in \{0, 1, 2\}$, the general solution of $Ly = 0$ can be expressed in terms of special functions (elementary, Airy or Weber functions, respectively). The eigenvalue problem for $d = 2$ (harmonic oscillator) plays an important role in quantum mechanics.

We mostly consider the cases $d = 3$ and $d = 4$.

Cubic and quartic oscillators have been studied extensively from the very beginning of quantum mechanics, mostly by perturbative methods. Cubic oscillators arose in quantum field theory [**24**] and in the theory of Painlevé equations [**15, 16**].

We normalize the equation by an affine transformation of the independent variable to obtain: $P(z) = z^d + O(z^{d-2}),\ z \to \infty$.

Every solution y of $Ly = \lambda y$, for $\lambda \in \mathbf{C}$ is an entire function. Consider the Stokes sectors

$$S_j = \left\{ z : \left| \arg z - \frac{2\pi j}{d+2} \right| < \frac{\pi}{d+2} \right\},\ 0 \leq j \leq d+1.$$

For each $j \in \{0, \ldots, d+1\}$ a solution y either grows exponentially along all rays from the origin in S_j or tends to zero exponentially along every such ray in S_j.

We choose two non-adjacent sectors and impose the boundary condition that a solution y of $Ly = \lambda y$ tends to 0 in these sectors. Those λ for which such $y \neq 0$ exists are called eigenvalues, and the corresponding y's are called eigenfunctions.

2010 *Mathematics Subject Classification.* Primary 81Q05, 34M60, 34A05.
Key words and phrases. One-dimensional Schrödinger operators, quasi-exact solvability, PT-symmetry.
Both authors were supported by NSF Grant DMS-1067886.

The interest of physicists in such eigenvalue problems originated with C. Bender and T.-T. Wu's paper [**4**], where a systematic study of the analytic continuation of eigenvalues as parameters vary in the complex plane was made for the first time.

With such boundary conditions, the problem has an infinite discrete spectrum and eigenvalues tend to infinity [**20**]. A one-dimensional eigenspace corresponds to each eigenvalue.

Suppose that the polynomial potential depends analytically on a parameter $a \in \mathbf{C}^n$. Then the spectral locus Z is defined as the set of all pairs $(a, \lambda) \in \mathbf{C}^{n+1}$ such that the differential equation

$$-y'' + P(z, a)y = \lambda y$$

has a solution y satisfying the boundary conditions. The spectral locus is an analytic hypersurface in $\mathbf{C}^{n+1}$; it is the zero-set of an entire function $F(a, \lambda)$, which is called the spectral determinant [**20**].

The multi-valued function $\lambda(a)$ defined by $F(a, \lambda) = 0$ has the following property: its only singularities are algebraic ramification points, and there are finitely many of them over every compact set in the a-space [**8**].

Next we discuss the connectedness of the spectral locus.

THEOREM 1 ([**9**]). *For the cubic oscillator*

$$-y'' + (z^3 - az)y = -\lambda y, \quad y(\pm i\infty) = 0, \tag{1}$$

the spectral locus is a smooth irreducible curve in $\mathbf{C}^2$.

THEOREM 2 ([**8**]). *For the even quartic oscillator*

$$-y'' + (z^4 + az^2)y = \lambda y, \quad y(\pm\infty) = 0,$$

the spectral locus consists of two disjoint smooth irreducible curves in $\mathbf{C}^2$, *one corresponding to even eigenfunctions, another to odd ones.*

These theorems can be generalized to polynomials of arbitrary degree if we use all coefficients as parameters [**14, 1, 2**].

However, if we consider a subfamily of all polynomials of given degree, then the spectral locus can be reducible in an interesting way [**3**].

One example is the family of quasi-exactly solvable quartics L_J

$$-y'' + (z^4 - 2bz^2 + 2Jz)y = \lambda y, \quad y(re^{\pm\pi i/3}) \to 0, \quad r \to +\infty. \tag{2}$$

When J is a positive integer, this problem has J elementary eigenfunctions of the form $p(z)\exp(z^3/3 - bz)$, with a polynomial p. The (b, λ) corresponding to these eigenfunctions form the quasi-exactly solvable part Z_J^{QES} of the spectral locus Z_J, and Z_J^{QES} is an algebraic curve.

THEOREM 3 ([**9**]). *For every positive integer* J, *the QES spectral locus* Z_J^{QES} *is a smooth irreducible curve in* $\mathbf{C}^2$.

Thus the curve Z_J^{QES} is an irreducible component of Z_J. It is not known whether $Z_J \setminus Z_J^{QES}$ is irreducible.

A similar phenomenon occurs in degree 6: there are one-parametric families of even quasi-exactly solvable sextics [**22, 23**], and for each such family the quasi-exactly solvable part of the spectral locus is a smooth irreducible algebraic curve [**9**].

When $J \to \infty$, an appropriate rescaling of Z_J^{QES} tends to the spectral locus of the one-parametric cubic family (1), and a rescaling of the sextic QES spectral locus tends to the spectral locus of the even quartic family [**23**].

Now we consider Hermitian and PT-symmetric operators. An eigenvalue problem may be invariant under a reflection with respect to a line in the complex z-plane. Without loss of generality, we can take this line to be the real line, and the reflection to be the complex conjugation. Then the coefficients of the potential are real. For the boundary conditions, two cases are possible:

a) Each of the two boundary conditions is preserved by the reflection. In this case the problem is Hermitian.

b) The two boundary conditions are interchanged by the reflection. Such problems are called PT-symmetric. (Physicists prefer to choose the reflection with respect to the imaginary line in this case. PT stands for "parity and time".)

For example, there is a one-parametric family of PT-symmetric cubics (1), with real parameter a. There are two different two-parametric families of PT-symmetric quartics which we call type I and II:

$$-y'' + (-z^4 + az^2 + cz)y = -\lambda y, \quad y(\pm i\infty) = 0, \tag{3}$$

and

$$-y'' + (z^4 - 2bz^2 + 2Jz)y = \lambda y, \quad y(re^{\pm\pi i/3}) \to 0, \quad r \to +\infty. \tag{4}$$

The real spectral locus $Z(\mathbf{R})$ is the subset of the spectral locus Z, which consists of points with real coordinates.

We begin with the cubic PT-symmetric spectral locus (1).

THEOREM 4 ([**10**]). *For every integer $n \geq 0$, there exists a simple curve $\Gamma_n \subset \mathbf{R}^2$, that is, the image of a proper analytic embedding of a line, and which has these properties:*
(i) For every $(a, \lambda) \in \Gamma_n$ problem (1) has an eigenfunction with $2n$ non-real zeros.
(ii) The curves Γ_n are disjoint and the real spectral locus of (1) is $\bigcup_{n\geq 0} \Gamma_n$
(iii) The map

$$\Gamma_n \cap \{(a, \lambda) : a \geq 0\} \to \mathbf{R}_{\geq 0}, \quad (a, \lambda) \mapsto a$$

is a 2-to-1 covering.
(iv) For $a \geq 0$, $(a, \lambda) \in \Gamma_n$ and $(a, \mu) \in \Gamma_{n+1}$ imply $\mu > \lambda$.

The following computer-generated plot of the real spectral locus of (1) is taken from Trinh's thesis [**21**]. Theorem 4 rigorously establishes some features of this picture. At the points whose abscissas are marked in Figure 1, pairs of real eigenvalues collide and escape to the complex plane. Theorem 4 proves rigorously that infinitely many such points exist. Consider the PT-symmetric quartic family (3) of type I: It is equivalent to the PT-symmetric family

$$-w'' + (z^4 + az^2 + icz)w = \lambda w, \quad w(\pm\infty) = 0,$$

studied by Bender, *et al* [**3**] and Delabaere and Pham [**7**].

THEOREM 5 ([**10**]). *The real spectral locus of (3) consists of disjoint smooth analytic properly embedded surfaces $S_n \subset \mathbf{R}^3$, $n \geq 0$, homeomorphic to a punctured disk. For $(a, c, \lambda) \in S_n$, the eigenfunction has exactly $2n$ non-real zeros. For large a, projection of S_n on the (a, c) plane approximates the region $9c^2 - 4a^3 \leq 0$.*

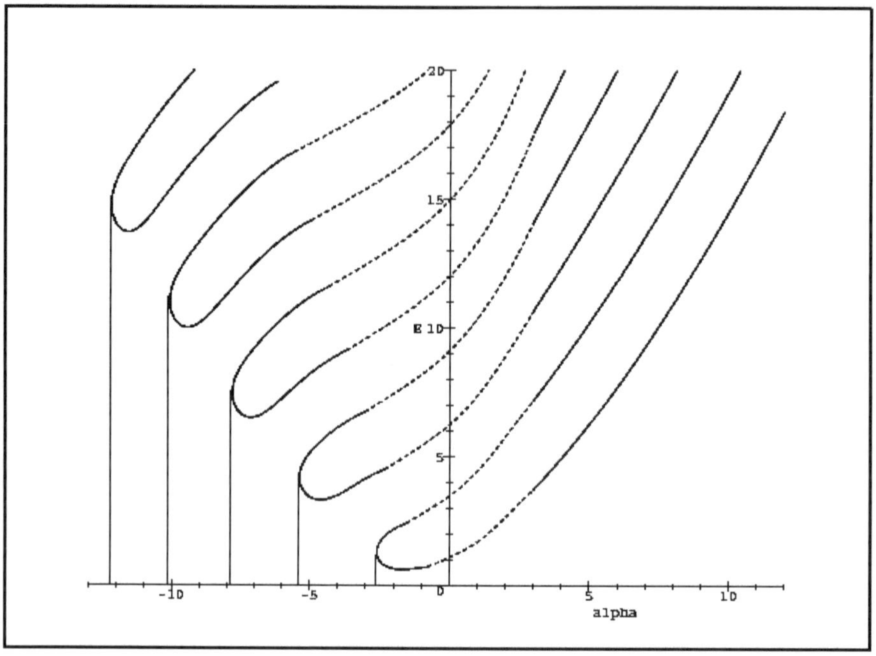

FIGURE 1. Real spectral locus for the PT-symmetric cubic.

Numerical computation suggests that the surfaces have the shape of infinite funnels with the sharp end stretching towards $a = -\infty$, $c = 0$, and that the section of S_n by every plane $a = a_0$ is a closed curve.

Theorem 5 implies that this section is compact for large a_0. The computer-generated plot in Figure 2 is taken from Trinh's thesis.

The PT-symmetric quartic family of the type II is more complicated, due to the presence of the QES spectrum. Let $Z_J^{QES}(\mathbf{R})$ be the real QES spectral locus of the operator L_J, defined in (2).

THEOREM 6 ([**12**]). *For $J = n+1 > 0$, $Z_{n+1}^{QES}(\mathbf{R})$ consists of $[n/2]+1$ disjoint analytic curves $\Gamma_{n,m}$, $0 \leq m \leq [n/2]$.*

For $(b, \lambda) \in \Gamma_{n,m}$, the eigenfunction has n zeros, $n - 2m$ of them real.

If n is odd, then $b \to +\infty$ on both ends of $\Gamma_{m,n}$. If n is even, the same holds for $m < n/2$, but on the ends of $\Gamma_{n,n/2}$ we have $b \to \pm\infty$.

If $(b, \lambda) \in \Gamma_{n,m}$ and $(b, \mu) \in \Gamma_{n,m+1}$ and b is sufficiently large, then $\mu > \lambda$.

It follows from these theorems that in each family, there are infinitely many parameter values where pairs of real eigenvalues collide and escape from the real line to the complex plane.

In the quartic family of the type II, another interesting feature of the real spectral locus is present: for some parameter values the QES spectral locus crosses the rest of the spectral locus. This is called "level crossing".

THEOREM 7 ([**11**]). *The points $(b, \lambda) \in Z_J^{QES}$ where the level crossing occurs are the intersection points of Z_J^{QES} with Z_{-J}. For each $J \geq 1$ there are infinitely many such points, in general, complex. When J is odd, there are infinitely many*

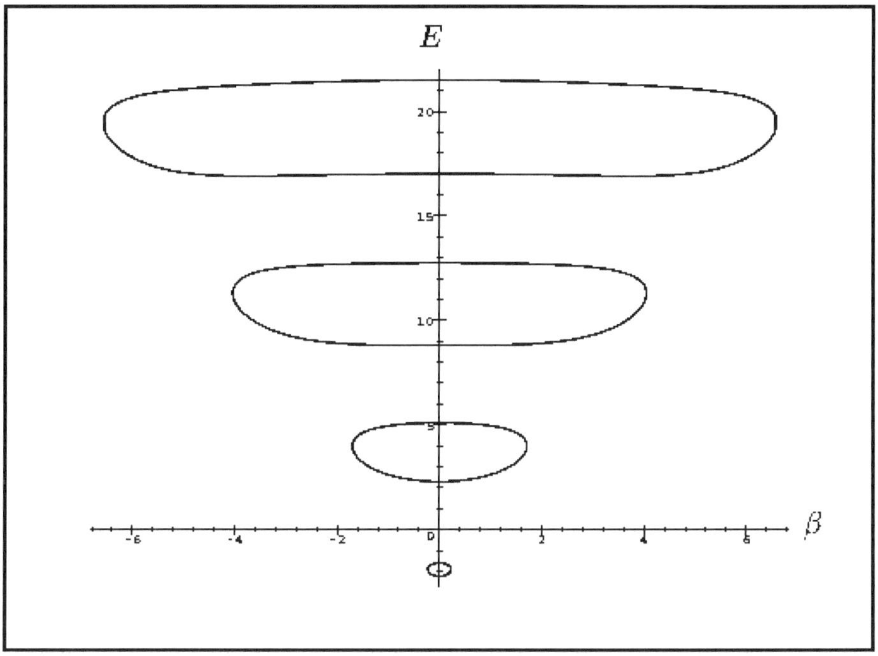

FIGURE 2. Section of the surfaces $S_0, \dots, S_3$ by the plane $a = -9$.

level crossing points with $b_k < 0$ and real λ_k. We have

$$b_k \sim -((3/4)\pi k)^{2/3}, \; k \to \infty.$$

The only known general result of reality of eigenvalues of PT-symmetric operators is a theorem of K. Shin [**19**], which for our quartic of type II implies that all eigenvalues are real if $J \leq 0$.

We have the following extensions of this result.

THEOREM 8 ([**13**]). *For every positive integer J, all non-QES eigenvalues of L_J are real.*

THEOREM 9 ([**13**]). *All eigenvalues of L_J are real for every real $J \leq 1$ (not necessarily integer).*

The QES part of the spectral loci is shown with dotted lines in Figures 4, 5, 6.

We briefly describe the methods of proofs. There are three main ingredients:

a) The Nevanlinna parametrization of the spectral locus.

b) Asymptotics at infinity (singular perturbation theory).

c) The Darboux transform of the QES quartic.

The Nevanlinna parametrization is the following. Let Z be the spectral locus of the problem

$$-y'' + P(z, a)y = \lambda y, \quad y(z) \to 0, \quad z \in S_j \cup S_k,$$

where S_j and S_j are non-adjacent Stokes sectors. Let $(a, \lambda) \in Z$, and y_0 an eigenfunction. Let y_1 be a second linearly independent solution. Then $f = y_0/y_1$ satisfies

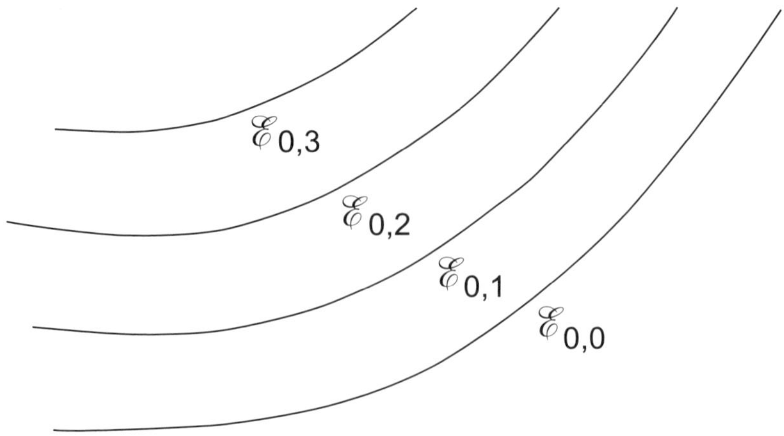

FIGURE 3. $Z_0(\mathbf{R})$.

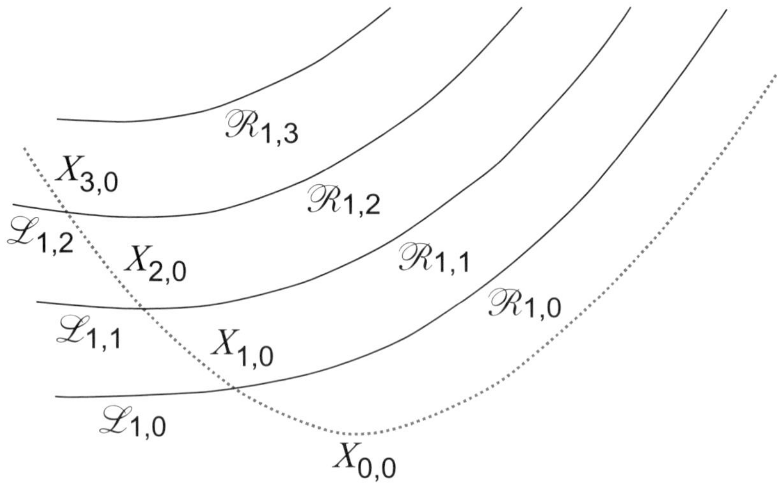

FIGURE 4. $Z_1(\mathbf{R})$.

the Schwarz differential equation

$$\frac{f'''}{f'} - \frac{3}{2}\left(\frac{f''}{f'}\right)^2 = -2(P - \lambda).$$

This function f is meromorphic in $\mathbf{C}$, has no critical points and has $d+2$ asymptotic values (radial limits along rays), one asymptotic value in each Stokes sector. The asymptotic values in S_j and S_k are 0. The asymptotic values in adjacent sectors are distinct.

In the opposite direction: If we have a meromorphic function in $\mathbf{C}$ without critical points and with finitely many asymptotic values, then it satisfies a Schwarz equation whose RHS is a polynomial [**18**]. The degree of this polynomial is the number of asymptotic tracts minus 2. (The asymptotic tracts are in one-to-one correspondence with the logarithmic singularities of the inverse function.)

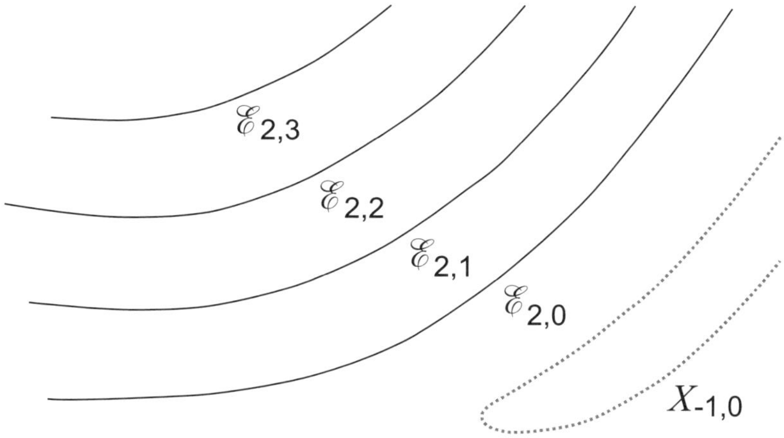

FIGURE 5. $Z_2(\mathbf{R})$.

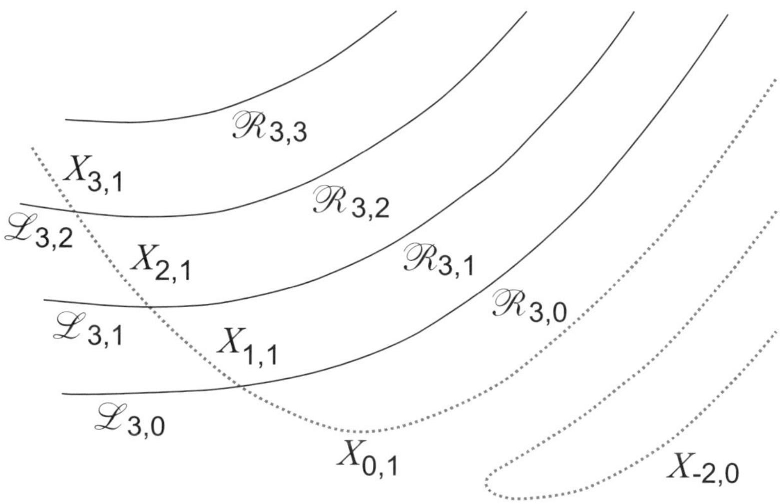

FIGURE 6. $Z_3(\mathbf{R})$.

The asymptotic values are meromorphic functions on Z which serve as local parameters. These are called the Nevanlinna parameters; they are simply related to the Stokes multipliers of the linear ODE [**20, 15, 16**].

Functions f of the above type with a given set of asymptotic values $A = \{a_0, \ldots, a_{d+1}\}$ have the property that

$$f : \mathbf{C}\backslash f^{-1}(A) \to \overline{\mathbf{C}}\backslash A$$

is a covering map. For a fixed A such a covering map can be completely described by certain combinatorial information, a cell decomposition of the plane. These cell decompositions label the charts of our description of the spectral locus.

It is important that we know exactly which cell decompositions can occur and how the cell decomposition changes when the point $(a_0 \ldots, a_{d+1})$ describes a closed loop in $\mathbf{C}^{d+2}$. This induces an action of the braid group on the set of special cell decompositions of the plane which can be explicitly computed.

This reduces the problem of the parametrization of a spectral locus to combinatorial topology.

For QES operators we use the Darboux transform. Let $-D^2+V$ be a second order linear differential operator with potential V. Let $\phi_0, \ldots \phi_n$ be some eigenfunctions with eigenvalues $\lambda_0, \ldots, \lambda_n$. The transformed operator is

$$-D^2+V-2\frac{d^2}{dz^2}\log W(\phi_0,\ldots,\phi_n),$$

where W is the Wronski determinant. The eigenvalues of the transformed operator are exactly those eigenvalues of $-D^2+V$ which are *distinct* from $\lambda_0, \ldots, \lambda_n$.

We use the Darboux transform to kill the QES part of the spectrum of L_J, and it turns out that the transformed operator is L_{-J}! Our study of the QES locus of the quartic family gives the following interesting identities with elementary intergals.

To describe these identities, we address the question studied by Heine and Stieltjes in the 19th century. Let h and p be polynomials. When does $y = pe^h$ satisfy a linear differential equation $y'' + Py = 0$ with a polynomial P?

THEOREM 10. *The following conditions are equivalent:*
a) $p'' + 2p'h'$ *is divisible by* p,
b) $p^{-2}e^h$ *has no residues,*
c) zeros of p *satisfy the system of equations*

$$\sum_{j:j\neq k}\frac{1}{z_k-z_j} = -h'(z_k),\ 1\leq k\leq \deg p.$$

Now take $h(z) = z^3/3 - bz$.

THEOREM 11. *Let* p *be a polynomial. All residues of* $y = p^{-2}e^{-2h}$ *vanish if and only if there exists a constant* C *and a polynomial* q *such that*

$$\left(p^2(-z)-\frac{C}{p^2(z)}\right)e^{-2h(z)} = \frac{d}{dz}\left(\frac{q(z)}{p(z)}e^{-2h(z)}\right).$$

Moreover, if this happens, then

$$C = (-1)^n 2^{-2n}\frac{\partial}{\partial\lambda}Q_{n+1},$$

where $\lambda = y''/y - z^4 + 2bz^2 - 2(n+1)z$, *and* $Q_{n+1}(b,\lambda) = 0$ *is the equation of the QES spectral locus of* L_{n+1}.

This was conjectured in [**11**], on the basis of calculations with the Darboux transform of L_{n+1}, and proved by E. Mukhin and V. Tarasov [**17**].

References

[1] P. Alexandersson, *On eigenvalues of the Schrödinger operator with an even complex-valued polynomial potential*, Comput. Methods Funct. Theory **12** (2012), no. 2, 465–481. MR3058517

[2] P. Alexandersson and A. Gabrielov, *On eigenvalues of the Schrödinger operator with a complex-valued polynomial potential*, Comput. Methods Funct. Theory **12** (2012), no. 1, 119–144. MR2977283

[3] C. Bender and S. Boettcher, *Quasi-exactly solvable quartic potential*, J. Phys. A **31** (1998), L273–L277. MR1621102 (99a:81023)

[4] C. Bender and T. -T. Wu, *Anharmonic oscillator*, Phys. Rev. **184** (1969), 1231–1260. MR0260323 (41:4951)

[5] M. Crum, *Associated Sturm–Liouville systems*, Quart. J. Math. **6** (1955), 121–127. MR0072332 (17:266g)

[6] G. Darboux, *Sur la représentation sphérique des surfaces*, C. R. Acad. Sci., XCIV, No. 20 (1882), 1343–1345.

[7] E. Delabaere and F. Pham, *Eigenvalues of complex Hamiltonians with PT-symmetry. I, II* Phys. Lett. A **250** (1998), no. 1-3, 25–28, 29–32. MR1742959 (2000k:81079)

[8] A. Eremenko and A. Gabrielov, *Analytic continuation of eigenvalues of a quartic oscillator*, Comm. Math. Phys. **287** (2009) 431–457. MR2481745 (2010b:34189)

[9] A. Eremenko and A. Gabrielov, *Irreducibility of some spectral determinants*, arXiv:0904.1714.

[10] A. Eremenko and A. Gabrielov, *Singular perturbation of polynomial potentials in the complex domain with applications to PT-symmetric families*, Moscow Math. J. **11** (2011), 473–503. MR2894426 (2012m:34160)

[11] A. Eremenko and A. Gabrielov, *Quasi-exactly solvable quartic: elementary integrals and asymptotics*, J. Phys. A: Math. Theor. **44** (2011), 312001. MR2817969 (2012m:81198)

[12] A. Eremenko and A. Gabrielov, *Quasi-exactly solvable quartic: real algebraic spectral locus*, J. Phys. A: Math. Theor. **45** (2012), 175205. MR2913475

[13] A. Eremenko and A. Gabrielov, *Two-parametric PT-symmetric quartic family*, J. Phys. A: Math. Theor. **45** (2012), 175206. MR2913476

[14] H. Habsch, *Die Theorie der Grundkurven und Äquivalenzproblem bei der Darstellung Riemannscher Flachen*, Mitt. Math. Seminar Giessen, 42, 1952. MR0054728 (14:969h)

[15] D. Masoero, *Poles of Integrale Tritronquee and Anharmonic Oscillators. A WKB Approach*, J. Phys. A: Math. Theor. **43** (2010), 095201. MR2592326 (2011b:81086)

[16] D. Masoero, *Y-System and Deformed Thermodynamic Bethe Ansatz*, Lett. Math. Phys. **94** (2010), 151–164. MR2733559 (2011j:34293)

[17] E. Mukhin and V. Tarasov, *On conjectures of A. Eremenko and A. Gabrielov*, Lett. Math. Phys. **103** (2013), no. 6, 653–663. MR3054650

[18] R. Nevanlinna, *Über Riemannsche Flächen mit endlich vielen Windungspunkten*, Acta Math. **58** (1932), 295–373. MR1555350

[19] K. Shin, *On the reality of eigenvalues for a class of PT-symmetric oscillators*, Comm. Math. Phys. **229** (2002), 543–564. MR1924367 (2003g:34188)

[20] Y. Sibuya, *Global theory of a second order linear ordinary differential equation with a polynomial coefficient*, North-Holland, Amsterdam, 1975. MR0486867 (58:6561)

[21] D. T. Trinh, *Asymptotique et analyse spectrale de l'oscillateur cubique*, These, 2002.

[22] A. Turbiner, *Quasi–exactly–solvable problems and* $sl(2)$ *algebra*, Comm. Math. Phys. **118** (1988), 467–474. MR958807 (89j:58051)

[23] A. Ushveridze, *Quasi-exactly solvable models*, Inst. of Physics, Bristol and Philadelphia, 1994. MR1329549 (96i:81297)

[24] J. Zinn-Justin and U. Jentschura, *Imaginary cubic perturbation: numerical and analytic study*, J. Phys. A **43** (2010), 425301. MR2726718 (2011m:81103)

PURDUE UNIVERSITY
E-mail address: eremenko@math.purdue.edu
URL: http://www.math.purdue.edu/~eremenko/

PURDUE UNIVERSITY
E-mail address: agabriel@math.purdue.edu
URL: http://www.math.purdue.edu/~agabriel/

Amer. Math. Soc. Transl.
Volume **233**, 2014

The Central Limit Theorem for Linear Eigenvalue Statistics of the Sum of Independent Random Matrices of Rank One

Olivier Guédon, Anna Lytova, Alain Pajor, and Leonid Pastur

Dedicated to Professor V. A. Marchenko on the occasion of his 90th birthday

Abstract. We consider $n \times n$ random matrices $M_n = \sum_{\alpha=1}^{m} \tau_\alpha \mathbf{y}_\alpha \otimes \mathbf{y}_\alpha$, where $\tau_\alpha \in \mathbb{R}$, $\{\mathbf{y}_\alpha\}_{\alpha=1}^{m}$ are i.i.d. isotropic random vectors of $\mathbb{R}^n$ (see Definition 1.1), whose components are not necessarily independent. It was shown in [**26**] that if $m, n \to \infty$, $m/n \to c \in [0, \infty)$, the Normalized Counting Measures of $\{\tau_\alpha\}_{\alpha=1}^{m}$ converge weakly and $\{\mathbf{y}_\alpha\}_{\alpha=1}^{m}$ are *good* (see Definition 1.2), then the Normalized Counting Measures of eigenvalues of M_n converge weakly in probability to a non-random limit found in [**24**]. In this paper we indicate a subclass of good vectors, which we call *very good* (see Definition 1.6) and for which the linear eigenvalue statistics of the corresponding matrices converge in distribution to the Gaussian law, i.e., the Central Limit Theorem is valid (see Theorem 1.8). An important example of good vectors, studied in [**26**], are the vectors with log-concave distribution (see Definition 1.1). We discuss the conditions for them, guaranteeing the validity of the Central Limit Theorem for linear eigenvalue statistics of corresponding matrices.

1. Introduction: Problem and Main Result

Let $\{\mathbf{y}_\alpha\}_{\alpha=1}^{m}$ be i.i.d. random vectors of $\mathbb{R}^n$, and $\{\tau_\alpha\}_{\alpha=1}^{m}$ be a collection of real numbers. Consider the $n \times n$ real symmetric random matrix

$$M_n = \sum_{\alpha=1}^{m} \tau_\alpha L_{\mathbf{y}_\alpha}, \tag{1.1}$$

where $L_{\mathbf{y}} = \mathbf{y} \otimes \mathbf{y}$ is an $n \times n$ rank-one matrix defined as $L_{\mathbf{y}}\mathbf{x} = (\mathbf{y}, \mathbf{x})\mathbf{y}$, $\forall \mathbf{x} \in \mathbb{R}^n$ and $(\ ,\ )$ is the standard Euclidean scalar product in $\mathbb{R}^n$.

We denote $\{\lambda_l^{(n)}\}_{l=1}^{n}$ the eigenvalues of M_n counting their multiplicity and introduce their Normalized Counting Measure (NCM) N_n, setting for any $\Delta \subset \mathbb{R}$

$$N_n(\Delta) = \mathrm{Card}\{l \in [1, n] : \lambda_l^{(n)} \in \Delta\}/n. \tag{1.2}$$

2010 *Mathematics Subject Classification.* Primary 60B20, 47A75; Secondary 60F05.

The authors' research was partly supported by the Network of Mathematical Research (France-Ukraine) 2013–2015.

Leonid Pastur is grateful to the Bezout Research Foundation (Labex Bezout) for the invitation to the University Marne la Vallée (France), where the final version of this paper was prepared.

Likewise, we define the NCM σ_m of $\{\tau_\alpha\}_{\alpha=1}^m$,

$$\sigma_m(\Delta) = \text{Card}\{\alpha \in [1,m] : \tau_\alpha \in \Delta\}/m, \tag{1.3}$$

and assume that the sequence $\{\sigma_m\}_{m=1}^\infty$ converges weakly:

$$\lim_{m\to\infty} \sigma_m = \sigma, \ \sigma(\mathbb{R}) = 1. \tag{1.4}$$

It follows from the results of [**24**] that if (1.4) holds, the mixed moments up to the 4th order of the components of $\{\mathbf{y}_\alpha\}_{\alpha=1}^m$ satisfy certain conditions as $n \to \infty$, valid, in particular, for the vectors uniformly distributed over the unit sphere of $\mathbb{R}^n$ (or $\mathbb{C}^n$) and for vectors with independent i.i.d. components, and

$$n \to \infty, \ m \to \infty, \ m/n \to c \in [0, \infty), \tag{1.5}$$

then there exists a non-random measure N of total mass 1 such that for any interval $\Delta \subset \mathbb{R}$ we have the convergence in probability

$$\lim_{n\to\infty,\ m\to\infty,\ m/n\to c} N_n(\Delta) = N(\Delta). \tag{1.6}$$

The measure N can be found as follows. We introduce its Stieltjes transform (see e.g. [**1**])

$$f(z) = \int \frac{N(d\lambda)}{\lambda - z}, \ \Im z \neq 0. \tag{1.7}$$

Here and below the integrals without limits denote the integrals over $\mathbb{R}$. Then f is uniquely determined by the functional equation

$$zf(z) = c - 1 - c\int (1 + \tau f(z))^{-1}\sigma(d\tau) \tag{1.8}$$

considered in the class of functions analytic in $\mathbb{C} \setminus \mathbb{R}$ and such that $\Im f(z)\Im z \geq 0, \ \Im z \neq 0$. Since the Stieltjes transform determines N uniquely by the formula

$$\lim_{\varepsilon\to 0+} \frac{1}{\pi}\int \varphi(\lambda)\Im f(\lambda + i\varepsilon)d\lambda = \int \varphi(\lambda)N(d\lambda), \tag{1.9}$$

valid for any continuous function with compact support, (1.8) determines N uniquely.

Note that if the components $\{y_{\alpha j}\}_{j=1}^n$ of $\mathbf{y}_\alpha, \ \alpha = 1, ..., m$ are i.i.d. random variables of zero mean and variance $1/n$, the matrix M_n is usually written as

$$M_n = Y^T TY, \quad Y = \{y_{\alpha k}\}_{\alpha,k=1}^{m,n}, \quad T = \{\tau_\alpha \delta_{\alpha\beta}\}_{\alpha,\beta=1}^m, \tag{1.10}$$

and is closely related to the sample covariance matrix of statistics. A particular case of this for $T = I_m$ and Gaussian $\{y_{\alpha j}\}_{\alpha=j=1}^{m,n}$ has been known since the 1930s as the Wishart matrix (see e.g. [**25**]).

The random matrices (1.1) also appear in the local theory of Banach spaces and asymptotic convex geometry (see e.g. [**10, 32**]). A particular important case in this framework is related to the study of geometric parameters associated to i.i.d. random points uniformly distributed over a convex body in $\mathbb{R}^n$ and of the asymptotic geometry of random convex polytopes generated by these points (see e.g. [**3, 9, 13, 14, 22**]). This motivated us to consider random vectors known as isotropic with a log-concave distribution. We recall the corresponding definitions.

DEFINITION 1.1. (i) A random vector $\mathbf{y} = \{y_j\}_{j=1}^n \in \mathbb{R}^n$ is called *isotropic* if

$$\mathbf{E}\{y_j\} = 0, \quad \mathbf{E}\{y_j y_k\} = n^{-1}\delta_{jk}, \ j,k = 1,...,n. \tag{1.11}$$

(ii) A measure μ on $\mathbb{C}^n$ is *log-concave* if for any measurable subsets A, B of $\mathbb{C}^n$ and any $\theta \in [0,1]$,

$$\mu(\theta A + (1-\theta)B) \geq \mu(A)^\theta \mu(B)^{(1-\theta)}$$

whenever $\theta A + (1-\theta)B = \{\theta y_1 + (1-\theta)y_2 \, : \, y_1 \in A, \ y_2 \in B\}$ is measurable.

It was proved in [**26**] that (1.6) and (1.8) remain valid in the case where the probability law of the i.i.d. vectors $\{\mathbf{y}_\alpha\}_{\alpha=1}^m$ is isotropic and log-concave.

In fact, a more general result was established in [**26**]. We introduce

DEFINITION 1.2. A random isotropic vector $\mathbf{y} \in \mathbb{R}^n$ is called *good* if for any $n \times n$ complex matrix A_n which does not depend on $\mathbf{y}$, we have

$$\mathbf{Var}\{(A_n\mathbf{y}, \mathbf{y})\} \leq ||A_n||^2\delta_n, \quad \delta_n = o(1), \ n \to \infty, \tag{1.12}$$

where $||A_n||$ is the operator norm of A_n.

We then have [**26**]:

THEOREM 1.3. *Let n and m be positive integers satisfying (1.5), $\{\mathbf{y}_\alpha\}_{\alpha=1}^m$ be i.i.d. good vectors of $\mathbb{R}^n$, and $\{\tau_\alpha\}_{\alpha=1}^m$ be real numbers satisfying (1.4). Consider the random matrix M_n (1.1) and the Normalized Counting Measure of its eigenvalues N_n (1.2). Then for any interval $\Delta \subset \mathbb{R}$ we have in probability*

$$\lim_{n\to\infty, m\to\infty, m/n\to c\in[0,\infty)} N_n(\Delta) = N(\Delta), \tag{1.13}$$

where the limiting non-random measure N is given by (1.6)–(1.9).

It follows from Theorem 1.3 that if M_n is given by (1.1), where $\{\mathbf{y}_\alpha\}_{\alpha=1}^{m_n}$ are i.i.d. good vectors and

$$\mathcal{N}_n[\varphi] = \sum_{j=1}^n \varphi(\lambda_j^{(n)}) \tag{1.14}$$

is the *linear eigenvalue statistic* corresponding to any continuous and bounded *test-function* $\varphi : \mathbb{R} \to \mathbb{C}$, then we have in probability

$$\lim_{n\to\infty, m\to\infty, m/n\to c\in[0,\infty)} n^{-1}\mathcal{N}_n[\varphi] = \int \varphi(\lambda)dN(\lambda). \tag{1.15}$$

This can be viewed as an analog of the law of large numbers of probability theory for (1.14). In this paper we deal with the fluctuations of $\mathcal{N}_n$ around its limit (1.14), i.e. with an analog of the Central Limit Theorem (CLT) of probability theory. Our goal is to find a class of i.i.d. good vectors and a class of test functions such that the centered and appropriately normalized linear eigenvalue statistics

$$\mathcal{N}_n^\circ[\varphi]/\nu_n, \quad \mathcal{N}_n^\circ[\varphi] = \mathcal{N}_n[\varphi] - \mathbf{E}\{\mathcal{N}_n[\varphi]\} \tag{1.16}$$

converge in distribution to a Gaussian random variable.

There is a number of papers on the CLT for linear eigenvalue statistics of matrices (1.1) where $\{y_{\alpha j}\}_{\alpha,j=1}^{m,n}$ are independent, i.e. for sample covariance matrices (1.10) (see [**4, 15, 23, 27, 30**] and references therein). Unfortunately, much less is known in the case where the components of $\mathbf{y}_\alpha$'s are dependent (see e.g. [**29**], Chapter 17 and references therein).

An important step in proving the CLT is the asymptotic analysis of the variance of the corresponding linear statistic

$$\mathbf{Var}\{\mathcal{N}_n[\varphi]\} := \mathbf{E}\{(\mathcal{N}_n^\circ[\varphi])^2\}, \tag{1.17}$$

in particular, the proof of a bound

$$\mathbf{Var}\{\mathcal{N}_n[\varphi]\} \le C_n||\varphi||_H^2, \tag{1.18}$$

where $||...||_H$ is a functional norm and C_n depends only on n, or even an asymptotic of the variance. This determines the normalization factor ν_n in (1.16) and the class H of test-functions for which the CLT is valid.

It appears that for many random matrices normalized so that there exists a limit of their NCM, in particular for sample covariance matrices (1.10), the variance of linear eigenvalue statistic with $\varphi \in C^1$ admits the bound

$$\mathbf{Var}\{\mathcal{N}_n[\varphi]\} = O(1), \quad n \to \infty, \tag{1.19}$$

or even a limit as $n \to \infty$. Thus the CLT has to be valid for (1.16) without any n-dependent normalization factor ν_n **[29]**. This has to be compared with the generic situation in probability theory, where the variance of a linear statistic of i.i.d. random variables are proportional to n for any bounded φ; hence the CLT is valid for an analog of (1.16) with $\nu_n = n^{-1/2}$.

To formulate the version of (1.19), which we will prove and use in this paper, we introduce

DEFINITION 1.4. The distribution of random vector $\mathbf{y} \in \mathbb{R}^n$ is called *unconditional* if its components $\{y_j\}_{j=1}^n$ have the same joint distribution as $\{\pm y_j\}_{j=1}^n$ for any choice of signs.

LEMMA 1.5. *Let* $\mathbf{y} = \{y_i\}_{j=1}^n$ *be an isotropic random vector having an unconditional distribution and satisfying*

$$a_{2,2} := \mathbf{E}\{y_j^2 y_k^2\} = n^{-2} + O(n^{-3}),\ j \neq k, \quad \kappa_4 := \mathbf{E}\{y_j^4\} - 3a_{2,2} = O(n^{-2}). \tag{1.20}$$

We consider the random matrix M_n *of (1.1) in which* m *and* n *satisfy (1.5),* $\{y_\alpha\}_{\alpha=1}^m$ *are i.i.d. random vectors satisfying (1.20), and* $\{\tau_\alpha\}_{\alpha=1}^m$ *are non-negative real numbers with the limiting counting distribution* σ *(1.4) having a finite fourth moment:*

$$m_4 := \int_0^\infty \tau^4 d\sigma(\tau) < \infty. \tag{1.21}$$

Then we have for all sufficiently large m *and* n

$$\mathbf{Var}\{\mathcal{N}_n[\varphi]\} \le C||\varphi||_{2+\delta}^2, \tag{1.22}$$

where C *is an absolute constant and*

$$||\varphi||_{2+\delta}^2 = \int (1+2|k|)^{2(2+\delta)}|\widehat{\varphi}(k)|^2 dk, \quad \widehat{\varphi}(k) = \int e^{ikx}\varphi(x)dx. \tag{1.23}$$

The proof of the lemma is given in Section 3. However, it turns out that the validity of the CLT, more exactly, its proof in this paper, requires more conditions on the components of random vectors $\{y_\alpha\}_{\alpha=1}^m$ in (1.1). Namely, we introduce

DEFINITION 1.6. A random isotropic vector $\mathbf{y} = \{y_j\}_{j=1}^n \in \mathbb{R}^n$ is called *very good* if its distribution is unconditional, there exist n-independent $a, b \in \mathbb{R}$ such that (cf. (1.20))

$$a_{2,2} := \mathbf{E}\{y_j^2 y_k^2\} = n^{-2} + an^{-3} + o(n^{-3}), \quad j \neq k, n \to \infty, \tag{1.24}$$

$$\kappa_4 := \mathbf{E}\{y_j^4\} - 3a_{2,2} = bn^{-2} + o(n^{-2}), \quad n \to \infty, \tag{1.25}$$

and

$$\mathbf{E}\{|(A_n\mathbf{y}, \mathbf{y})^\circ|^4\} \leq ||A_n||^4 \widetilde{\delta}_n, \quad \widetilde{\delta}_n = O(n^{-2}),\ n \to \infty, \tag{1.26}$$

for any $n \times n$ complex matrix A_n which does not depend on $\mathbf{y}$.

It is easy to check that the vectors $\mathbf{y} = \mathbf{x}/n^{1/2}$, where $\mathbf{x}$ has i.i.d. components with even distribution such that $\mathbf{E}\{x_j^8\} < \infty$, are very good as well as vectors uniformly distributed over the unit ball or unit sphere of $\mathbb{R}^n$. For other examples of very good random vectors of geometric origin, see Section 2.

REMARK 1.7. Here, instead of unconditionality, one can assume that $\mathbf{y}$ satisfies condition (2.8) below (as it was assumed in [**24**]).

Now we are ready to formulate our main result:

THEOREM 1.8. *Let m and n be positive integers satisfying (1.5), $\{\mathbf{y}_\alpha\}_{\alpha=1}^m$ be i.i.d. very good vectors in the sense of Definition 1.6, and $\{\tau_\alpha\}_{\alpha=1}^m$ be non-negative numbers satisfying (1.4) and (1.21). Consider the corresponding random matrix M_n of (1.1) and a linear statistic $\mathcal{N}_n[\varphi]$ (1.14) of its eigenvalues. Assume that $\varphi : \mathbb{R} \to \mathbb{R}$, $\varphi \in H_{2+\delta}$, $\delta > 0$ (see (1.23)). Then $\mathcal{N}_n^\circ[\varphi]$ of (1.16) converges in distribution to the Gaussian random variable with zero mean and variance $V[\varphi] = \lim_{y\downarrow 0} V_y[\varphi]$, where*

$$V_\eta[\varphi] = \frac{1}{2\pi^2} \int\int \Re\big[C(z_1, \overline{z_2}) - C(z_1, z_2)\big]\varphi(\lambda_1)\varphi(\lambda_2)d\lambda_1 d\lambda_2 \ , \tag{1.27}$$

$$C(z_1, z_2) = \frac{\partial^2}{\partial z_1 \partial z_2}\Big(2\log\frac{\Delta f}{\Delta z} - (a+b)f(z_1)f(z_2)\frac{\Delta z}{\Delta f}\Big), \tag{1.28}$$

with $z_{1,2} = \lambda_{1,2} + i\eta$, $\Delta f = f(z_1) - f(z_2)$, $\Delta z = z_1 - z_2$, and f is given by (1.8).

REMARK 1.9. Note that in fact

$$C(z_1, z_2) = \lim_{n\to\infty} \mathbf{E}\{\gamma_n^\circ(z_1)\gamma_n(z_2)\},$$

where $\gamma_n(z) = \mathrm{Tr}(M_n - zI_n)^{-1}$ and $\gamma_n^\circ = \gamma_n - \mathbf{E}\{\gamma_n\}$.

REMARK 1.10. The condition $\tau_\alpha \geq 0$, $\alpha = 1, ..., m$, is a purely technical one, and it can be shown that the results remain valid for $\tau_\alpha \in \mathbb{R}$.

REMARK 1.11. One can also rewrite $V_\eta[\varphi]$ in the form

$$\begin{aligned} V_\eta[\varphi] =& \frac{1}{2\pi^2} \int\int \Re\big[L(z_1, z_2) - L(z_1\overline{z_2})\big](\varphi(\lambda_1) - \varphi(\lambda_2))^2 d\lambda_1 d\lambda_2 \\ &+ \frac{(a+b)c}{\pi^2} \int_0^\infty \tau^2 d\sigma(\tau)\Big(\Im\int \frac{f'(z_1)}{(1+\tau f(z_1))^2}\varphi(\lambda_1)d\lambda_1\Big)^2, \end{aligned} \tag{1.29}$$

$$L(z_1, z_2) = \frac{\partial^2}{\partial z_1 \partial z_2} \log \frac{\Delta f}{\Delta z}.$$

In particular, if $\tau_\alpha = 1$, $\alpha = 1, ..., m$, then

$$V[\varphi]|_{\tau=1} = \frac{1}{2\pi^2} \int_{a_-}^{a_+} \int_{a_-}^{a_+} \left(\frac{\Delta\varphi}{\Delta\lambda}\right)^2 \frac{(4c - (\lambda_1 - a_m)(\lambda_2 - a_m))d\lambda_1 d\lambda_2}{\sqrt{(a_+ - \lambda_1)(\lambda_1 - a_-)}\sqrt{(a_+ - \lambda_2)(\lambda_2 - a_-)}}$$

$$+ \frac{a+b}{4c\pi^2} \left(\int_{a_-}^{a_+} \varphi(\mu) \frac{\mu - a_m}{\sqrt{(a_+ - \mu)(\mu - a_-)}} d\mu \right)^2, \tag{1.30}$$

where $\Delta\varphi = \varphi(\lambda_1) - \varphi(\lambda_2)$, $a_\pm = (1 \pm \sqrt{c})^2$, $a_m = 1 + c$. This expression coincides with the expression for the limiting variance of linear eigenvalue statistics of sample covariance matrices (see [**23**]), in which the fourth cumulant of matrix entries is replaced with $a + b$.

The paper is organized as follows. In Section 2 we present some facts on the isotropic random vectors with a log-concave unconditional symmetric distribution. In Section 3 we prove Lemma 1.5. In Section 4 we present the proof of our main result, Theorem 1.8. Section 5 contains auxiliary results.

2. Isotropic random vectors with log-concave distribution

Let $\mathbf{y} \in \mathbb{R}^n$ be a random isotropic vector with a log-concave density (see Definition 1.1). A typical example from convex geometry is a vector uniformly distributed over a convex body in $\mathbb{R}^n$. The study of the concentration of the Euclidean norm of $\mathbf{y}$ around its average is a part of an important branch of high dimensional convex geometry related to a famous conjecture of Kannan, Lovász and Simonovits [**18**]. (See also the surveys [**16, 33**] on the validity of the Poincaré type inequality.)

In particular, the so-called thin shell conjecture claims that

$$\mathbf{P}\{| \,||\mathbf{y}|| - 1| > t\} \leq 2\exp(-ct\sqrt{n}),$$

where $c > 0$ is a universal constant. A weaker conjecture, known as the variance conjecture, claims that

$$\mathbf{Var}\{||\mathbf{y}||^2\} \leq C/n, \tag{2.1}$$

where C is a universal constant. The conjecture in full generality is still open.

The first breakthrough was obtained by [**12, 19**] where the bound

$$\mathbf{Var}\{||\mathbf{y}||^2\} = o(1), \; n \to \infty \tag{2.2}$$

was proved. The bound is the basic tool to prove Berry-Esséen type inequalities for one-dimensional marginals of $\mathbf{y}$ [**2, 19, 7**] and is sometimes called CLT for convex bodies. The best-known improvement of (2.2) is [**17**]:

$$\mathbf{Var}\{||\mathbf{y}||^2\} = O(n^{-1/3}), \; n \to \infty.$$

The variance conjecture (2.1) has been proved in certain special cases. Anttila, et al. [**2**] considered random isotropic vectors uniformly distributed over the unit ball B_p^n of the ℓ_p^n norm in $\mathbb{R}^n$, and Wojtaszczyk [**34**] considered the same setting for a generalized Orlicz unit ball. Klartag [**20**] studied vectors with the log-concave unconditional isotropic distribution (see Definitions 1.1 and 1.4).

While in high dimensional convex geometry one focuses mainly on *quantitative* estimates as above; in this paper, we will also need precise asymptotics for mixed moments of the components of $\mathbf{y}$. This raises new questions in high dimensional convex geometry, related to general quadratic forms rather than for norms. More precisely, let A_n be a $n \times n$ complex matrix such that $\|A_n\| \leq 1$. It was proved in [**26**] that

$$\mathbf{Var}\{(A_n\mathbf{y}, \mathbf{y})\} = o(1), \ n \to \infty.$$

According to Lemma 1.5, we need the best possible bound, i.e., an analog of (2.1) for quadratic forms. It is for instance known when $\mathbf{y}$ is uniformly distributed over the unit ball B_p^n and follows from the corresponding Poincaré type estimates [**21, 31**]. We prove below in Lemma 2.1 that the analog (2.10) of (2.1) is valid for any random vector with a log-concave unconditional symmetric isotropic distribution.

To prove the CLT (see Theorem 1.8), we will need precise asymptotics for the mixed moments of components of $\mathbf{y}$ (see (1.20) and (1.24)–(1.25)), as well as more bounds for quadratic forms (see (1.26)). Given the parameters $a_{2,2}$ and κ_4, we are considering in fact a sequence of n-dimensional log-concave isotropic distributions satisfying (1.20) or (1.24)–(1.25). From a geometric point of view, one may consider a sequence of n-dimensional convex bodies, such as unit balls of norms, and their uniform distributions (normalized to be isotropic). A natural example is given by the sequence of the unit balls $\{B_p^n\}_{n\in\mathbb{N}}$ for which we check that (1.20) and (1.24)–(1.25)) are valid with a and b depending only on p (see (2.12)). As for the general case, we have

LEMMA 2.1. *If $\mathbf{y} = (y_1, ..., y_n) \in \mathbb{R}^n$ is an isotropic vector with a log-concave unconditional symmetric distribution, then it satisfies* (1.20), (1.12) *with $\delta_n = O(n^{-1})$, $n \to \infty$, and* (1.26).

PROOF. We will use the dimension free Khinchine-Kahane-type inequality by Bourgain [**8**] (see also [**6**]): if P_d is a polynomial of degree d, and $\mathbf{y} \in \mathbb{R}^n$ has a log-concave distribution, then

$$\mathbf{E}\{|P_d(\mathbf{y})|^q\} \leq C(d, q)\mathbf{E}\{|P_d(\mathbf{y})|\}^q, \tag{2.3}$$

where $C(d, q)$ depends only on d and q and does not depend on n. By using (2.3) and (1.11), we obtain for the fourth moments of coordinates $\{y_j\}_{j=1}^n$ of $\mathbf{y}$:

$$\mathbf{E}\{|y_j|^4\} \leq C\mathbf{E}\{y_j^2\}^2 = Cn^{-2}. \tag{2.4}$$

If the distribution of $\mathbf{y}$ is symmetric, then

$$a_{2,2} = \frac{1}{n-1}\mathbf{E}\Big\{\sum_{k=1}^n y_j^2 y_k^2 - y_j^4\Big\} = \frac{1}{n(n-1)}\mathbf{E}\{||\mathbf{y}||^4\} - \frac{1}{n-1}\mathbf{E}\{y_j^4\}. \tag{2.5}$$

It follows from (2.1) and (1.11) that

$$\mathbf{E}\{||\mathbf{y}||^4\} = 1 + \mathbf{Var}\{||\mathbf{y}||^2\} \leq 1 + C/n. \tag{2.6}$$

This and (2.5) yield

$$a_{2,2} \leq n^{-2} + C/n^3. \tag{2.7}$$

On the other hand, $\mathbf{E}\{||\mathbf{y}||^4\} \geq \mathbf{E}\{||\mathbf{y}||^2\}^2 = 1$, which together with (2.4) and (2.5) lead to $a_{2,2} \geq n^{-2} + C'/n^3$, and we get the first part of (1.20). The second part follows from the first one and (2.4).

Since for any random $\mathbf{y} = \{y_j\}_{j=1}^n$ with unconditional distribution

$$\mathbf{E}\{y_j y_k y_p y_q\} = a_{2,2}(\delta_{jk}\delta_{pq} + \delta_{jp}\delta_{kq} + \delta_{jq}\delta_{kp}) + \kappa_4\delta_{jk}\delta_{jp}\delta_{jq}, \tag{2.8}$$

we have for a symmetric matrix A_n

$$\mathbf{Var}\{(A_n\mathbf{y},\mathbf{y})\} = (a_{2,2}-n^{-2})|\operatorname{Tr} A_n|^2 + 2a_{2,2}\operatorname{Tr}|A_n|^2 + \kappa_4\sum_{j=1}^{n}|A_{njj}|^2, \tag{2.9}$$

where $|A_n|^2 = A_nA_n^*$. This and (1.20) lead to

$$\mathbf{Var}\{(A_n\mathbf{y},\mathbf{y})\} = O(n^{-1}). \tag{2.10}$$

In addition, it follows from (2.3)

$$\mathbf{E}\{|(A_n\mathbf{y},\mathbf{y})^\circ|^4\} \le C\mathbf{Var}\{(A_n\mathbf{y},\mathbf{y})\}^2, \tag{2.11}$$

which together with (2.10) yield (1.26). □

Note, however, that not too much is known about isotropic vectors with a log-concave distribution, which satisfy (1.24)–(1.25), i.e., are very good in the sense of Definition 1.6. Thus, some of them might not satisfy (1.24) and/or (1.25), for instance the coefficient in front of n^{-3} and/or the coefficient in front of n^{-2} would "oscillate" in n. This would mean that different subsequences of vectors can have different coefficients a and b in (1.24)–(1.25). Correspondingly, different subsequences of characteristic functions of $\mathcal{N}_n^\circ[\varphi]$ would have Gaussian limits with different variances (1.27)–(1.28). If this is the case, this situation could be compared with that of [**28**], where it was shown that the limiting forms of the variance and the probability law of fluctuations of linear eigenvalue statistics for certain unitary invariant matrix ensembles depend on a sequence $n_j\to\infty$.

Here are the examples of very good vectors. It was mentioned in the previous section that the vectors $\mathbf{y} = n^{-1/2}\mathbf{x}$, where components $\{x_j\}_{j=1}^n$ of $\mathbf{x}$ are i.i.d. random variables with an even distribution such that $\mathbf{E}\{x_j^2\} = 1$, $\mathbf{E}\{x_j^4\} = m_4$, $\mathbf{E}\{x_j^8\} < \infty$ are very good with $a_{2,2} = n^{-2}$ and $\kappa_4 = n^{-2}(m_4-3)$; thus $a = 0$ and $b = m_4 - 3$ in (1.24)–(1.25). Note that in this case Lemma 2.1 is valid without the assumption of log-concave distribution of $\mathbf{y}$.

It can also be shown that the vectors uniformly distributed over the Euclidean unit ball in $\mathbb{R}^n$ are also very good. Let us consider a more general case, where $\mathbf{x}$ is uniformly distributed over the unit ball

$$B_p^n = \{\mathbf{x}\in\mathbb{R}^n : ||\mathbf{x}||_p = \big(\sum_{j=1}^n |x_j|^p\big)^{1/p} \le 1\}$$

of the space l_p^n. According to [**5**], we have

$$\begin{aligned}\mathbf{E}_{B_p^n}\{f\} :&= \frac{1}{|B_p^n|}\int_{B_p^n} f(\mathbf{x})d\mathbf{x}\\ &= \frac{1}{(2\Gamma(1+1/p))^n}\int_0^\infty dte^{-t}\int_{\mathbb{R}^n} d\mathbf{x}e^{-||\mathbf{x}||_p^p} f(\mathbf{x}(||\mathbf{x}||_p^p+t)^{-1/p}).\end{aligned}$$

This allows us to calculate the moments of $\mathbf{x}$ and to show that the vector

$$\mathbf{y} = \left(\frac{1}{n}\frac{B(1/p,2/p)}{B(n/p+1,2/p)}\right)^{1/2}\mathbf{x},$$

where B is the β-function, is isotropic and satisfies (1.24)–(1.25) with

$$a = -\frac{8}{p},\quad b = \frac{\Gamma(1/p)\Gamma(5/p)}{\Gamma(3/p)^2} - 3. \tag{2.12}$$

3. Proof of Lemma 1.5

We will essentially follow the scheme proposed in [**30**], which is based on two main ingredients. The first is an inequality that allows us to transform bounds for the variances of the trace of resolvent of a random matrix into bounds for the variances of linear eigenvalue statistics with a sufficiently smooth test function:

PROPOSITION 3.1. *Let M_n be an $n \times n$ random matrix and $\mathcal{N}_n[\varphi]$ be a linear statistic of its eigenvalues (see (1.14)). Then we have for any $s > 0$*

$$\mathbf{Var}\{\mathcal{N}_n[\varphi]\} \le C_s ||\varphi||_s^2 \int_0^\infty d\eta e^{-\eta}\eta^{2s-1} \int_{-\infty}^\infty \mathbf{Var}\{\gamma_n(\lambda + i\eta)\} d\lambda, \tag{3.1}$$

where C_s depends only on s, $||\varphi||_s$ is defined in (1.23), $G(z) = (M_n - z)^{-1}$ is the resolvent of M_n, and

$$\gamma_n(z) = \operatorname{Tr} G(z). \tag{3.2}$$

The second ingredient of the scheme of [**30**] is an improved version of the martingale approach, providing the bound $\mathbf{Var}\{\gamma_n(z)\} \le C(z)$ instead of the bound $\mathbf{Var}\{\gamma_n(z)\} \le C(z)n$ (see e.g. [**29**], Theorem 19.1.6). Using this version, we prove

LEMMA 3.2. *Consider matrix M_n of (1.1), where $\{\mathbf{y}_\alpha\}_{\alpha=1}^{m_n}$ are i.i.d. random isotropic vectors having an unconditional distribution and satisfying (1.20), and $\{\tau_\alpha\}_{\alpha=1}^{m_n}$ are non-negative numbers satisfying (1.4) and (1.21). Then we have*

$$\mathbf{Var}\{\gamma_n(z)\} \le C|\Im z|^{-6}, \tag{3.3}$$

and also $\forall \varepsilon \in (0, 1/2]$

$$\mathbf{Var}\{\gamma_n(z)\} \le \frac{C}{n|\Im z|^4} \sum_{\alpha=1}^{m} (\tau_\alpha(|\Im z| + \tau_\alpha))^{3/2+\varepsilon} \mathbf{E}\{n^{-1} \operatorname{Tr} |G^\alpha(z)|^{2(1/2+\varepsilon)}\}, \tag{3.4}$$

where C does not depend on n and z, $|G^\alpha|^2 = G^\alpha G^{\alpha *}$, and

$$G^\alpha(z) = G(z)|_{\tau_\alpha = 0}. \tag{3.5}$$

If $\{\mathbf{y}_\alpha\}_{\alpha=1}^{m_n}$ are i.i.d. very good vectors in the sense of Definition 1.6, then additionally

$$\mathbf{E}\{|\gamma_n(z)^\circ|^4\} \le C|z|^4 |\Im z|^{-12}. \tag{3.6}$$

PROOF. Given an integer $\alpha \in [1, m]$, denote $\mathbf{E}_{\le\alpha}$ and $\mathbf{E}_\alpha$ the expectation with respect to $\{\mathbf{y}_1, ..., \mathbf{y}_\alpha\}$ and $\mathbf{y}_\alpha$, so that for any random variable ξ, depending on $\{\mathbf{y}_\alpha\}_{\alpha=1}^m$ we have $\mathbf{E}_{\le 0} = \xi$, $\mathbf{E}_{\le m} = \mathbf{E}\{\xi\}$ and

$$\xi - \mathbf{E}\{\xi\} = \sum_{\alpha=1}^m (\mathbf{E}_{\le\alpha-1}\{\xi\} - \mathbf{E}_{\le\alpha}\{\xi\}).$$

By using the definition of $\mathbf{E}_{\le\alpha}$ and the above identity it is easy to find that

$$\mathbf{Var}\{\xi\} = \sum_{\alpha=1}^m \mathbf{E}\{|\mathbf{E}_{\le\alpha-1}\{\xi\} - \mathbf{E}_{\le\alpha}\{\xi\}|^2\}. \tag{3.7}$$

Denote also

$$M_n^\alpha = M_n|_{\tau_\alpha=0}, \quad \gamma_n^\alpha(z) = \operatorname{Tr} G^\alpha(z), \quad \xi_\alpha^\circ = \xi - \mathbf{E}_\alpha\{\xi\}, \tag{3.8}$$

where $G^\alpha(z)$ is defined in (3.5). Applying (3.7) to $\xi = \gamma_n$ (see (3.2)) and using the Schwarz inequality, we get

$$\mathbf{Var}\{\gamma_n\} \le \sum_{\alpha=1}^{m} \mathbf{E}\{|(\gamma_n)^\circ_\alpha|^2\}. \tag{3.9}$$

Furthermore, since $M_n - M_n^\alpha = \tau_\alpha L_\alpha$ is the rank one matrix (see (1.1) and (3.8)), we can write the formula

$$G - G^\alpha = -\frac{\tau_\alpha G^\alpha L_\alpha G^\alpha}{1+\tau_\alpha(G^\alpha \mathbf{y}_\alpha, \mathbf{y}_\alpha)}, \tag{3.10}$$

implying for γ_n and γ_n^α of (3.5) and (3.8)

$$\gamma_n - \gamma_n^\alpha = -\frac{\tau_\alpha((G^\alpha)^2 \mathbf{y}_\alpha, \mathbf{y}_\alpha)}{1+\tau_\alpha(G^\alpha \mathbf{y}_\alpha, \mathbf{y}_\alpha)} = -\frac{B_{\alpha n}}{A_{\alpha n}}, \tag{3.11}$$

where

$$A_{\alpha n} = 1 + \tau_\alpha(G^\alpha \mathbf{y}_\alpha, \mathbf{y}_\alpha), \tag{3.12}$$

$$B_{\alpha n} = \frac{d}{dz} A_{\alpha n} = \tau_\alpha((G^\alpha)^2 \mathbf{y}_\alpha, \mathbf{y}_\alpha). \tag{3.13}$$

It follows from the spectral theorem for real symmetric matrices that there exists a non-negative measure m^α such that

$$(G^\alpha \mathbf{y}_\alpha, \mathbf{y}_\alpha) = \int_0^\infty \frac{m^\alpha(d\lambda)}{\lambda - z}. \tag{3.14}$$

This and (3.12)–(3.13) yield

$$|A_\alpha| \ge |\Im A_\alpha| = \tau_\alpha |\Im (G^\alpha \mathbf{y}_\alpha, \mathbf{y}_\alpha))| = \tau_\alpha |\Im z| \int_0^\infty \frac{m^\alpha(d\lambda)}{|\lambda - z|^2}, \tag{3.15}$$

and

$$|B_\alpha| \le |\tau_\alpha| \int_0^\infty \frac{m^\alpha(d\lambda)}{|\lambda - z|^2}, \tag{3.16}$$

implying

$$|B_{\alpha n}/A_{\alpha n}| \le 1/|\Im z|. \tag{3.17}$$

This and the identity

$$\frac{1}{A} = \frac{1}{\mathbf{E}\{A\}} - \frac{A^\circ}{A\mathbf{E}\{A\}}, \quad A^\circ = A - \mathbf{E}\{A\}, \tag{3.18}$$

allows us to write

$$\begin{aligned}
\mathbf{E}\{|(\gamma_n)^\circ_\alpha|^2\} &= \mathbf{E}\{|\gamma_n - \gamma_n^\alpha - \mathbf{E}_\alpha\{\gamma_n - \gamma_n^\alpha\}|^2\} \\
&\le \mathbf{E}\Big\{\Big|\frac{B_{\alpha n}}{A_{\alpha n}} - \frac{\mathbf{E}_\alpha\{B_{\alpha n}\}}{\mathbf{E}_\alpha\{A_{\alpha n}\}}\Big|^2\Big\} = \mathbf{E}\Big\{\Big|\frac{(B_{\alpha n})^\circ_\alpha}{\mathbf{E}_\alpha\{A_{\alpha n}\}} - \frac{B_{\alpha n}}{A_{\alpha n}} \cdot \frac{(A_{\alpha n})^\circ_\alpha}{\mathbf{E}_\alpha\{A_{\alpha n}\}}\Big|^2\Big\} \\
&\le 2\mathbf{E}\Big\{\Big|\frac{(B_{\alpha n})^\circ_\alpha}{\mathbf{E}_\alpha\{A_{\alpha n}\}}\Big|^2\Big\} + \frac{2}{|\Im z|^2}\mathbf{E}\Big\{\Big|\frac{(A_{\alpha n})^\circ_\alpha}{\mathbf{E}_\alpha\{A_{\alpha n}\}}\Big|^2\Big\}.
\end{aligned} \tag{3.19}$$

Let us estimate the second term on the right-hand side of (3.19). Since $(A_{\alpha n})^\circ_\alpha = \tau_\alpha (G^\alpha \mathbf{y}_\alpha, \mathbf{y}_\alpha)^\circ_\alpha$, then in view of definitions of (1.20)

$$\frac{1}{\tau_\alpha^2}\mathbf{E}_\alpha\{|(A_{\alpha n})^\circ_\alpha|^2\} = \big(a_{2,2} - \frac{1}{n^2}\big)|\operatorname{Tr} G^\alpha|^2 + 2a_{2,2}\operatorname{Tr}|G^\alpha|^2 + \kappa_4 \sum_{j=1}^{n} |G^\alpha_{jj}|^2. \tag{3.20}$$

This, (1.20), and the bound $||G^\alpha|| \le |\Im z|^{-1}$ yield

$$\mathbf{E}\{|(A_{\alpha n})^\circ_\alpha|^2\} \le \frac{C\tau_\alpha^2}{n^2} \operatorname{Tr}|G^\alpha|^2, \tag{3.21}$$

$$\mathbf{E}\{|(B_{\alpha n})^\circ_\alpha|^2\} \le \frac{C\tau_\alpha^2}{n^2} \operatorname{Tr}|G^\alpha|^4 \le \frac{C\tau_\alpha^2}{n^2|\Im z|^2} \operatorname{Tr}|G^\alpha|^2. \tag{3.22}$$

It follows then from (3.9), (3.19), and (3.21)–(3.22) that

$$\mathbf{Var}\{\gamma_n(z)\} \le \frac{C}{n|\Im z|^2} \sum_{\alpha=1}^m \tau_\alpha^2 \mathbf{E}\Big\{\frac{n^{-1}\operatorname{Tr}|G^\alpha(z)|^2}{|\mathbf{E}_\alpha\{A_{\alpha n}\}|^2}\Big\}. \tag{3.23}$$

Let N_n^α be the normalized counting measure of M_n^α. Then we have in view of (3.8), (cf. (3.14))

$$\gamma_n^\alpha(z) = \int_0^\infty \frac{N_n^\alpha(d\lambda)}{\lambda - z}. \tag{3.24}$$

In addition, (1.11) and (3.8) imply

$$\mathbf{E}_\alpha\{A_{\alpha n}\} = 1 + \tau_\alpha n^{-1}\gamma_n^\alpha. \tag{3.25}$$

This and an argument similar to that leading to (3.17) yield

$$\frac{n^{-1}\operatorname{Tr}|G^\alpha|^2}{|\mathbf{E}_\alpha\{A_{\alpha n}\}|} \le \frac{1}{\tau_\alpha|\Im z|}. \tag{3.26}$$

It also follows from (3.10) that

$$A_{\alpha n}^{-1} = 1 + \tau_\alpha(G\mathbf{y}_\alpha, \mathbf{y}_\alpha) \tag{3.27}$$

implying, together with (1.11) and the Jensen inequality

$$|\mathbf{E}_\alpha\{A_{\alpha n}\}|^{-1} \le \mathbf{E}_\alpha\{|A_{\alpha n}|^{-1}\} \le 1 + \tau_\alpha|\Im z|^{-1}. \tag{3.28}$$

Now (3.23), (3.28), and (1.21) yield (3.3).

Applying (3.26), (3.28) and Jensen inequality again, we get

$$\begin{aligned}\frac{n^{-1}\operatorname{Tr}|G^\alpha(z)|^2}{|\mathbf{E}_\alpha\{A_{\alpha n}\}|^2} &= \frac{1}{|\mathbf{E}_\alpha\{A_{\alpha n}\}|^{3/2+\varepsilon}}\Big(\frac{\operatorname{Tr}|G^\alpha(z)|^2}{n|\mathbf{E}_\alpha\{A_{\alpha n}\}|}\Big)^{1/2-\varepsilon}(n^{-1}\operatorname{Tr}|G^\alpha(z)|^2)^{1/2+\varepsilon} \\ &\le (1+\tau_\alpha|\Im z|^{-1})^{3/2+\varepsilon}(\tau_\alpha|\Im z|)^{-1/2+\varepsilon}\cdot n^{-1}\operatorname{Tr}|G^\alpha(z)|^{2(1/2+\varepsilon)}.\end{aligned}$$

This and (3.23) lead to (3.4).

To prove (3.6) we will use an analog of (3.7) for the 4th moment of $\gamma_n^\circ = \gamma_n - \mathbf{E}\{\gamma_n\}$ (see e.g. [**11**] and [**29**], Section 18.1.2), which together with an argument analogous to that leading to (3.19), yields

$$\begin{aligned}\mathbf{E}\{|\gamma_n^\circ|^4\} &\le m\sum_{\alpha=1}^m \mathbf{E}\{|(\gamma_n)^\circ_\alpha|^4\} \\ &\le Cm\sum_{\alpha=1}^m \mathbf{E}\Big\{\Big|\frac{(B_{\alpha n})^\circ_\alpha}{\mathbf{E}_\alpha\{A_{\alpha n}\}}\Big|^4\Big\} + \frac{1}{|\Im z|^4}\mathbf{E}\Big\{\Big|\frac{(A_{\alpha n})^\circ_\alpha}{\mathbf{E}_\alpha\{A_{\alpha n}\}}\Big|^4\Big\},\end{aligned} \tag{3.29}$$

where (see (1.26))

$$\mathbf{E}_\alpha\{|(A_{\alpha n})^\circ_\alpha|^4\} \le \tau_\alpha^4|/|\Im z|^4 n^2, \quad \mathbf{E}_\alpha\{|(B_{\alpha n})^\circ_\alpha|^4\} \le \tau_\alpha^4|/|\Im z|^8 n^2. \tag{3.30}$$

Since the matrix M_n with non-negative τ_α, $\alpha = 1, ..., m$ is positive definite, it follows from (3.14) that

$$\Im(z(G^\alpha \mathbf{y}_\alpha, \mathbf{y}_\alpha)) = \Im \int_0^\infty \frac{\lambda m^\alpha(d\lambda)}{\lambda - z} = \Im z \cdot \int_0^\infty \frac{\lambda m^\alpha(d\lambda)}{|\lambda - z|^2}.$$

This yields the inequality $\Im z \cdot \Im(z(G^\alpha \mathbf{y}_\alpha, \mathbf{y}_\alpha)) \geq 0$, so that

$$|A_{\alpha n}|^{-1} \leq \left| \frac{z}{\Im z + \tau_\alpha \Im(z(G^\alpha \mathbf{y}_\alpha, \mathbf{y}_\alpha))} \right| \leq |z||\Im z|^{-1}, \tag{3.31}$$

and by the Jensen inequality

$$|\mathbf{E}_\alpha\{A_{\alpha n}\}|^{-1} \leq |z||\Im z|^{-1}. \tag{3.32}$$

Now (3.6) follows from (3.29)–(3.30) and (3.32). □

Proof of Lemma 1.5. It follows from (3.1) and (3.4) that

$$\begin{aligned} \mathbf{Var}\{\mathcal{N}_n[\varphi]\} \leq C_s' ||\varphi||_s^2 \int_0^\infty d\eta e^{-\eta} \eta^{2s-5} \frac{1}{n} \sum_{\alpha=1}^m (\tau_\alpha(\eta + \tau_\alpha))^{3/2+\varepsilon} \\ \times \int_{-\infty}^\infty \mathbf{E}\{n^{-1} \operatorname{Tr} |(G^\alpha(z)|^{1+2\varepsilon}\} d\lambda, \end{aligned} \tag{3.33}$$

and we have for $z = \lambda + i\eta$ (cf. (3.24))

$$\int_{-\infty}^\infty \mathbf{E}\{n^{-1}\mathrm{Tr}|G^\alpha(z)|^{1+2\varepsilon)}\} d\lambda = \int_0^\infty \mathbf{E}\{N_n^\alpha(d\lambda)\} \int_{-\infty}^\infty \frac{d\mu}{((\lambda - \mu)^2 + \eta^2)^{1/2+\varepsilon}} \leq \frac{C}{\eta^{2\varepsilon}}.$$

Thus, for any $s = 2 + \delta$, $\delta > \varepsilon$, the integral over η in (3.33) converges. Lemma 1.5 is proved.

4. Proof of Theorem 1.8

It suffices to show that if

$$Z_n(x) = \mathbf{E}\{e_n(x)\}, \quad e_n(x) = e^{ix\mathcal{N}_n^\circ[\varphi]}, \tag{4.1}$$

then we have uniformly in $|x| \leq C$

$$\lim_{n \to \infty} Z_n(x) = \exp\{-x^2 V[\varphi]/2\} \tag{4.2}$$

with $V[\varphi]$ of (1.27). Define for any test-functions $\varphi \in H_{2+\delta}$

$$\varphi_y = P_y * \varphi, \tag{4.3}$$

where P_y is the Poisson kernel

$$P_y(x) = \frac{y}{\pi(x^2 + y^2)}, \tag{4.4}$$

and "$*$" denotes the convolution. We have

$$\lim_{y \downarrow 0} ||\varphi - \varphi_y||_{2+\delta} = 0. \tag{4.5}$$

Denote for the moment the characteristic function (4.1) by $Z_n[\varphi]$, to make explicit its dependence on the test function. We then have for any converging subsequence $\{Z_{n_j}[\varphi]\}_{j=1}^\infty$

$$\lim_{n_j \to \infty} Z_{n_j}[\varphi] = \lim_{y \downarrow 0} \lim_{n_j \to \infty} (Z_{n_j}[\varphi] - Z_{n_j}[\varphi_y]) + \lim_{y \downarrow 0} \lim_{n_j \to \infty} Z_{n_j}[\varphi_y].$$

Since by (1.22) and (4.5)

$$|Z_{n_j}[\varphi] - Z_{n_j}[\varphi_y]| \le |x|\mathbf{Var}\{\mathcal{N}_{n_j}[\varphi] - \mathcal{N}_{n_j}[\varphi_y]\}^{1/2} \le C|x|||\varphi - \varphi_y||_{2+\delta}\to 0,$$

as $y \downarrow 0$, then

$$\lim_{n_j\to\infty} Z_{n_j}[\varphi] = \lim_{y\downarrow 0}\lim_{n_j\to\infty} Z_{n_j}[\varphi_y]. \tag{4.6}$$

Hence it suffices to find the limit of $Z_{yn} := Z_n[\varphi_y] = \mathbf{E}\{e_{yn}(x)\}$ with $e_{yn}(x) = e^{ix\mathcal{N}_n^\circ[\varphi_y]}$, as $n \to \infty$.

It follows from (4.3)–(4.4) that

$$\mathcal{N}_n[\varphi_y] = \frac{1}{\pi}\int \varphi(\mu)\Im\gamma_n(z)d\mu, \quad z = \mu + iy. \tag{4.7}$$

This allows us to write

$$\frac{d}{dx}Z_{yn}(x) = \frac{1}{2\pi}\int \varphi(\mu)(Y_n(z,x) - Y_n(\overline{z},x))d\mu, \tag{4.8}$$

where $Y_n(z,x) = \mathbf{E}\{\gamma_n(z)e_{yn}^\circ(x)\}$. Now (3.3) yields

$$|Y_n(z,x)| \le 2\mathbf{Var}\{\gamma_n(z)\}^{1/2} \le C|\Im z|^{-3}.$$

This and the dominated convergence theorem imply that if $\varphi \in L^1$, then the limit of the integral in (4.8) as $n \to \infty$ can be obtained from that of Y_n for any fixed non-real z.

We have from the resolvent identity and (3.2)

$$\gamma_n(z) = -\frac{n}{z} + \frac{1}{z}\operatorname{Tr} M_nG(z) = \frac{-n+m}{z} - \frac{1}{z}\sum_{\alpha=1}^m A_{\alpha n}^{-1}(z),$$

where $A_{\alpha n}$ is defined in (3.12). This implies

$$\begin{aligned} Y_n(z,x) &= -\frac{1}{z}\sum_{\alpha=1}^m \mathbf{E}\{A_{\alpha n}^{-1}(z)(e_{yn}^\alpha(x))^\circ\} - \frac{1}{z}\sum_{\alpha=1}^m \mathbf{E}\{A_{\alpha n}^{-1}(z)(e_{yn}(x) - e_{yn}^\alpha(x))^\circ\} \\ &=: T_1^{(n)} + T_2^{(n)}. \end{aligned} \tag{4.9}$$

Iterating (3.18) three times, we get for $T_1^{(n)}$ of (4.9):

$$\begin{aligned} T_1^{(n)} &= \frac{1}{z}\sum_{\alpha=1}^m \frac{\mathbf{E}\{A_{\alpha n}^\circ(z)(e_{yn}^\alpha(x))^\circ\}}{\mathbf{E}\{A_{\alpha n}(z)\}^2} - \frac{1}{z}\sum_{\alpha=1}^m \frac{\mathbf{E}\{(A_{\alpha n}^\circ(z))^2(e_{yn}^\alpha(x))^\circ\}}{\mathbf{E}\{A_{\alpha n}(z)\}^3} \\ &\quad + \frac{1}{z}\sum_{\alpha=1}^m \frac{\mathbf{E}\{(A_{\alpha n}^\circ(z))^3A_{\alpha n}^{-1}(z)(e_{yn}^\alpha(x))^\circ\}}{\mathbf{E}\{A_{\alpha n}(z)\}^3} =: T_{11}^{(n)} - T_{12}^{(n)} + T_{13}^{(n)}. \end{aligned} \tag{4.10}$$

It follows from (5.3) and (3.31)–(3.32) that

$$T_{13}^{(n)} = O(n^{-1/2}), \quad n \to \infty. \tag{4.11}$$

Consider now $T_{11}^{(n)}$. Since e_{yn}^α does not depend on $\mathbf{y}_\alpha$, (3.25) implies

$$\begin{aligned} \mathbf{E}\{A_{\alpha n}(z)(e_{yn}^\alpha(x))^\circ\} &= \mathbf{E}\{\mathbf{E}_\alpha\{A_{\alpha n}(z)\}(e_{yn}^\alpha(x))^\circ\} \\ &= \tau_\alpha n^{-1}\mathbf{E}\{\gamma_n^\alpha(z)(e_{yn}^\alpha(x))^\circ\} = \tau_\alpha n^{-1}Y_n(z,x) + R_n, \end{aligned} \tag{4.12}$$

where $R_n = \tau_\alpha n^{-1}\mathbf{E}\{\gamma_n^\alpha (e_{yn}^\alpha)^\circ - \gamma_n e_{yn}^\circ\}$. Consequently, applying (4.7), the Schwarz inequality and then (3.3), (5.2) and (4.3), we get

$$\tau_\alpha^{-1} n|R_n| \le 2\mathbf{E}\{|(\gamma_n - \gamma_n^\alpha)^\circ(z)|\} + \frac{|x|}{\pi}\int |\varphi(\lambda_2)|\mathbf{E}\{|\gamma_n^\circ(z)||(\gamma_n - \gamma_n^\alpha)^\circ(z_2)|\}d\lambda_2$$

$$\le C_z \tau_\alpha n^{-1/2}. \tag{4.13}$$

Here and below we denote by C_z any positive quantity depending only on $|\Im z|$ and $|z|$. It follows then from (4.12)–(4.13) that

$$\mathbf{E}\{A_{\alpha n}(z)(e_{yn}^\alpha(x))^\circ\} = \tau_\alpha n^{-1} Y_n(z,x) + O(n^{-3/2}).$$

Plugging this and (5.4) in $T_{11}^{(n)}$ of (4.10), we get

$$T_{11}^{(n)} = Y_n(z,x)\frac{1}{nz}\sum_{\alpha=1}^m \frac{\tau_\alpha}{(1+\tau_\alpha f(z))^2} + o(1), \quad n \to \infty,$$

and by (1.4)–(1.5)

$$T_{11}^{(n)} = Y_n(z,x)\frac{c}{z}\int_0^\infty \frac{\tau d\sigma(\tau)}{(1+\tau f(z))^2} + o(1), \quad n\to\infty. \tag{4.14}$$

Next, it follows from (5.7)–(5.8), the Schwarz inequality and (5.5) that

$$|\mathbf{E}\{(A_{\alpha n}^\circ)^2(e_{yn}^\alpha)^\circ\}| = |\mathbf{E}\{\mathbf{E}_\alpha\{((A_{\alpha n})_\alpha^\circ)^2\}(e_{yn}^\alpha)^\circ\} + \tau_\alpha^2 n^{-2}\mathbf{E}\{((\gamma_n^\alpha)^\circ)^2(e_{yn}^\alpha)^\circ\}|$$

$$= o(n^{-1}). \tag{4.15}$$

Hence,

$$T_{12}^{(n)} = o(1), \quad n\to\infty. \tag{4.16}$$

Now (4.11), (4.14) and (4.16) yield for $T_1^{(n)}$ of (4.10):

$$T_1^{(n)} = Y_n(z,x)\frac{c}{z}\int_0^\infty \frac{\tau d\sigma(\tau)}{(1+\tau f(z))^2} + o(1), \quad n\to\infty. \tag{4.17}$$

Consider $T_2^{(n)}$ of (4.9). Since by (4.7)

$$e_{yn} - e_{yn}^\alpha = \frac{ixe_{yn}^\alpha}{\pi}\int \varphi(\lambda_1)\Im(\gamma_n - \gamma_n^\alpha)^\circ(z_1)d\lambda_1$$
$$+ O\Big(\Big|\int \varphi(\lambda)\Im(\gamma_n - \gamma_n^\alpha)^\circ(z_1)d\lambda_1\Big|^2\Big),$$

then in view of (3.11)

$$\mathbf{E}\{A_{\alpha n}^{-1}(z)(e_{yn} - e_{yn}^\alpha)^\circ(x)\} = \frac{ix}{\pi}\int \varphi(\lambda_1)\mathbf{E}\{e_{yn}^\alpha(x)(A_{\alpha n}^{-1})^\circ(z)\Im(B_{\alpha n}A_{\alpha n}^{-1})^\circ(z_1)\}d\lambda_1$$

$$+ \int\int O(R_n^{(1)})\varphi(\lambda_1)\varphi(\lambda_2)d\lambda_1 d\lambda_2, \tag{4.18}$$

where

$$R_n^{(1)} = \mathbf{E}\{(A_{\alpha n}^{-1})^\circ(z)\Im(B_{\alpha n}A_{\alpha n}^{-1})^\circ(z_1)\Im(B_{\alpha n}A_{\alpha n}^{-1})^\circ(z_2)\},$$

$z_{1,2} = \lambda_{1,2} + iy$. Applying (3.18) to $A_{\alpha n}^{-1}(z)$, $A_{\alpha n}^{-1}(z_1)$, $A_{\alpha n}^{-1}(z_2)$, we get

$$\mathbf{E}\{(A_{\alpha n}^{-1})^\circ(z)(B_{\alpha n}A_{\alpha n}^{-1})^\circ(z_1)(B_{\alpha n}A_{\alpha n}^{-1})^\circ(z_2)\}$$
$$= \frac{\mathbf{E}\{(A_{\alpha n}^\circ A_{\alpha n}^{-1})^\circ(z)(B_{\alpha n} - A_{\alpha n}^\circ A_{\alpha n}^{-1}B_{\alpha n})^\circ(z_1)(B_{\alpha n} - A_{\alpha n}^\circ A_{\alpha n}^{-1}B_{\alpha n})^\circ(z_2)\}}{\mathbf{E}\{A_{\alpha n}(z)\}\mathbf{E}\{A_{\alpha n}(z_1)\}\mathbf{E}\{A_{\alpha n}(z_2)\}}.$$

This and (5.3) yield

$$R_n^{(1)} = O(n^{-1/2}), \quad n \to \infty. \tag{4.19}$$

Similarly, iterating twice (3.18), applying (5.3) and (5.5) and taking into account that only linear terms in $A^\circ_{\alpha n}$ and $B^\circ_{\alpha n}$ give a non-vanishing contribution, we get

$$\begin{aligned}
&\mathbf{E}\{e^\alpha_{yn}(x)(A^{-1}_{\alpha n})^\circ(z)(B_{\alpha n}A^{-1}_{\alpha n})^\circ(z_1)\} \\
&\quad = -\frac{\mathbf{E}\{e^\alpha_{yn}(x)A^\circ_{\alpha n}(z)B^\circ_{\alpha n}(z_1)\}}{\mathbf{E}\{A_{\alpha n}(z)\}^2\mathbf{E}\{A_{\alpha n}(z_1)\}} + \frac{\mathbf{E}\{e^\alpha_{yn}(x)A^\circ_{\alpha n}(z)A^\circ_{\alpha n}(z_1)B_{\alpha n}(z_1)\}}{\mathbf{E}\{A_{\alpha n}(z)\}^2\mathbf{E}\{A_{\alpha n}(z_1)\}^2} + O(n^{-3/2}) \\
&\quad = \mathbf{E}\{e^\alpha_{yn}(x)\}\Big[-\frac{\mathbf{E}\{A^\circ_{\alpha n}(z)B^\circ_{\alpha n}(z_1)\}}{\mathbf{E}\{A_{\alpha n}(z)\}^2\mathbf{E}\{A_{\alpha n}(z_1)\}} \\
&\qquad\qquad + \frac{\mathbf{E}\{A^\circ_{\alpha n}(z)A^\circ_{\alpha n}(z_1)\}\mathbf{E}\{B_{\alpha n}(z_1)\}}{\mathbf{E}\{A_{\alpha n}(z)\}^2\mathbf{E}\{A_{\alpha n}(z_1)\}^2}\Big] + O(n^{-3/2}).
\end{aligned}$$

This, the bound $|Z_{yn} - \mathbf{E}\{e^\alpha_{yn}\}| = O(n^{-1/2})$ (cf. (4.13)), and (3.13) imply

$$\begin{aligned}
&\mathbf{E}\{e^\alpha_{yn}(x)(A^{-1}_{\alpha n})^\circ(z)(B_{\alpha n}A^{-1}_{\alpha n})^\circ(z_1)\} \\
&\quad = -Z_{yn}(x)\frac{1}{\mathbf{E}\{A_{\alpha n}(z)\}^2}\frac{\partial}{\partial z_1}\frac{\mathbf{E}\{A^\circ_{\alpha n}(z)A^\circ_{\alpha n}(z_1)\}}{\mathbf{E}\{A_{\alpha n}(z_1)\}} + O(n^{-3/2}) \\
&\quad = -Z_{yn}(x)\frac{\tau_\alpha^2 D_{\tau_\alpha}(z,z_1)}{(1+\tau_\alpha f(z))^2} + O(n^{-3/2}),
\end{aligned} \tag{4.20}$$

where

$$D_{\tau_\alpha}(z,z_1) = \frac{\partial}{\partial z_1}\frac{2\Delta f/\Delta z + (a+b)f(z)f(z_1)}{1+\tau_\alpha f(z_1)}, \tag{4.21}$$

and we used (5.4) and (5.6). Plugging (4.19)–(4.21) in (4.18) and applying (1.4)–(1.5), we get for $T_2^{(n)}$ of (4.9):

$$T_2^{(n)} = -\frac{xZ_{yn}(x)}{2\pi z}\int d\lambda_1\varphi(\lambda_1)\, c\int_0^\infty \frac{\tau^2 d\sigma(\tau)}{(1+\tau f(z))^2}[D_\tau(z,z_1) - D_\tau(z,\overline{z_1})] + o(1).$$

This, (4.9), and (4.17) yield

$$\begin{aligned}
Y_n(z,x) =& \Big(c\int_0^\infty \frac{\tau d\sigma(\tau)}{(1+\tau f(z))^2} - z\Big)^{-1} \\
&\times \frac{xZ_{yn}(x)}{2\pi}\int d\lambda_1\varphi(\lambda_1)\, c\int_0^\infty \frac{\tau^2 d\sigma(\tau)}{(1+\tau f(z))^2}[D_\tau(z,z_1) - D_\tau(z,\overline{z_1})] + o(1),
\end{aligned}$$

and after some calculations based on (1.8) we finally get

$$Y_n(z,x) = \frac{xZ_{yn}(x)}{2\pi}\int d\lambda_1\varphi(\lambda_1)\,[C(z,z_1) - C(z,\overline{z_1})] + o(1), \tag{4.22}$$

where $C(z,z_1)$ is defined in (1.28). Now it follows from (4.8) and (4.22) that

$$\frac{d}{dx}Z_{yn}(x) = -xV_y[\varphi]Z_{yn}(x) + o(1), \tag{4.23}$$

where $V_y[\varphi]$ is given in (1.27). If we consider $\widetilde{Z}_{yn}(x) = e^{x^2 V_y[\varphi]}Z_{yn}(x)$, then (4.23) yields

$$d\widetilde{Z}_{yn}(x)/dx = o(1), \quad n \to \infty$$

for any $|x| \le C$, and since $\widetilde{Z}_{yn}(0) = Z_{yn}(0) = 1$, we obtain $\widetilde{Z}_{yn}(x) = 1 + o(1)$ uniformly in $|x| \le C$. Hence,

$$\lim_{n\to\infty} Z_{yn}(x) = \exp\{-x^2 V_y[\varphi]\}.$$

Now we take into account (4.6), allowing us to pass to the limit $y \downarrow 0$, and obtain (1.27). The theorem is proved.

5. Auxiliary results

LEMMA 5.1. *Under the conditions of Theorem* 1.8 *we have:*

$$(5.1)\qquad (i)\ \lim_{n\to\infty} \mathbf{E}\{n^{-1}\gamma_n^\alpha\} = f,\ |\gamma_n - \gamma_n^\alpha| = O(1), \quad n\to\infty,$$

$$(5.2)\qquad (ii)\ \mathbf{Var}\{\gamma_n - \gamma_n^\alpha\} = O(n^{-1}), \quad n\to\infty,$$

$$(5.3)\qquad (iii)\ \mathbf{E}\{|A_{\alpha n}^\circ|^p\} \le \frac{C\tau_\alpha^p}{n^{p/2}|\Im z|^p},\ \mathbf{E}\{|B_{\alpha n}^\circ|^p\} \le \frac{C\tau_\alpha^p}{n^{p/2}|\Im z|^{2p}}, \quad p = 1,2,3,4,$$

$$(5.4)\qquad (iv)\ \mathbf{E}\{A_{\alpha n}\}^{-1} = (1+\tau_\alpha f)^{-1} + o(1), \quad n\to\infty,$$

$$(5.5)\qquad (v)\ \mathbf{Var}\{n\mathbf{E}_\alpha\{A_{\alpha n}^\circ(z_1)A_{\alpha n}(z_2)\}\} = o(1), \quad n\to\infty,$$

$$(5.6)\qquad (vi)\ \lim_{n\to\infty} \tau_\alpha^{-2} n\mathbf{E}\{A_{\alpha n}^\circ(z_1)A_{\alpha n}(z_2)\} = (a+b)f(z_1)f(z_2) + 2\Delta f/\Delta z,$$

where γ_n^α, $A_{\alpha n}$, $B_{\alpha n}$ *are defined in* (3.8) *and* (3.12)*–*(3.13)*, and* f *is a unique solution of* (1.8)*.*

PROOF. *(i)* It follows from (3.11) and (3.17) that $|\gamma_n - \gamma_n^\alpha| \le |\Im z|^{-1}$. On the other hand, Theorem 1.3 implies $\lim_{n\to\infty} \mathbf{E}\{n^{-1}\gamma_n\} = f$. This leads to (5.1).

(ii) Consider $V = \mathbf{Var}\{\Delta\gamma_n\}$, $\Delta\gamma_n = \gamma_n - \gamma_n^\alpha$. By (3.11) and (3.18), we have

$$\begin{aligned} V = \mathbf{E}\{\Delta\gamma_n \Delta\overline{\gamma}_n^\circ\} &= -\mathbf{E}\{(B_{\alpha n}/A_{\alpha n})\Delta\overline{\gamma}_n^\circ\} \\ &= -\mathbf{E}\{B_{\alpha n}\Delta\overline{\gamma}_n^\circ\}/\mathbf{E}\{A_{\alpha n}\} - \mathbf{E}\{A_{\alpha n}^\circ(B_{\alpha n}/A_{\alpha n})\Delta\overline{\gamma}_n^\circ\}/\mathbf{E}\{A_{\alpha n}\}, \end{aligned}$$

and by (3.17), (3.28), the Schwarz inequality, and (3.21)–(3.22)

$$V \le CV^{1/2}(\mathbf{Var}\{B_{\alpha n}\}^{1/2} + \mathbf{Var}\{A_{\alpha n}\}^{1/2}/|\Im z|) \le CV^{1/2}n^{-1/2}.$$

This yields (5.2).

(iii) Note that

$$(5.7)\qquad A_{\alpha n}^\circ = (A_{\alpha n})_\alpha^\circ + \tau_\alpha n^{-1}(\gamma_n^\alpha)^\circ,$$

and that (3.3) and (3.6) imply

$$(5.8)\qquad \mathbf{Var}\{\gamma_n^\alpha\},\ \mathbf{E}\{|(\gamma_n^\alpha)^\circ|^4\} = O(1), \quad n\to\infty.$$

This allows us to replace $A_{\alpha n}^\circ$ by $(A_{\alpha n})_\alpha^\circ$ as $n\to\infty$ (see e.g. (4.15)). In view of (3.21) and (5.7)–(5.8) we have

$$(5.9)\qquad \mathbf{E}\{|A_{\alpha n}^\circ|^2\} = \mathbf{E}\{|(A_{\alpha n})_\alpha^\circ|^2\} + \tau_\alpha^2 n^{-2}\mathbf{Var}\{\gamma_n^\alpha\} = O(n^{-1}).$$

This yields (5.3) for $p = 2$. Similarly, (5.3) for $p = 3,4$ follows from (1.26) and (5.7)–(5.8).

(iv) We have $\mathbf{E}\{A_{\alpha n}\}^{-1} = (1+\tau_\alpha f)^{-1} + r_n$, where

$$r_n = \tau_\alpha(f - \mathbf{E}\{n^{-1}\gamma_n^\alpha\})\mathbf{E}\{A_{\alpha n}\}^{-1}(1+\tau_\alpha f)^{-1}.$$

The bound $|(1+\tau_\alpha f)^{-1}| \le |z/\Im z|$, (3.32) and (5.1) imply $r_n = o(1)$, hence (5.4).

(v) It follows from (2.8)–(2.9) and (1.24)–(1.25) that

(5.10)
$$n\tau_\alpha^{-2}\mathbf{E}_\alpha\{(A_{\alpha n})^\circ_\alpha(z_1)A_{\alpha n}(z_2)\} = n\Big[(a_{2,2}-n^{-2})\operatorname{Tr}G^\alpha(z_1)\operatorname{Tr}G^\alpha(z_2) + 2a_{2,2}\operatorname{Tr}G^\alpha(z_1)G^\alpha(z_2) + \kappa_4\sum_{j=1}^n G^\alpha_{jj}(z_1)G^\alpha_{jj}(z_2)\Big]$$
$$= an^{-2}\gamma_n^\alpha(z_1)\gamma_n^\alpha(z_2) + 2n^{-1}\frac{\gamma_n^\alpha(z_1)-\gamma_n^\alpha(z_2)}{z_1-z_2} + bg_n^\alpha(z_1,z_2) + O(n^{-1}),$$

where g_n^α is defined in (5.19) (see Lemma 5.2). Now (5.5) follows from (5.8)–(5.9) and (5.13).

(vi) The relation (5.6) follows from (5.1), (5.10) and (5.12). □

LEMMA 5.2. *Consider*

$$g_n(z_1,z_2) = n^{-1}\sum_{j=1}^n G_{jj}(z_1)G_{jj}(z_2). \tag{5.11}$$

Then under the conditions of Theorem 1.8 *we have as* $n\to\infty$

$$\mathbf{E}\{g_n(z_1,z_2)\} = f(z_1)f(z_2) + o(1), \tag{5.12}$$
$$\mathbf{Var}\{g_n(z_1,z_2)\} = o(1), \tag{5.13}$$

where f *is given by* (1.8).

PROOF. Since by the resolvent identity

$$G_{jj}(z_1) = -\frac{1}{z_1} + \frac{1}{z_1}(MG)_{jj}(z_1) = -\frac{1}{z_1} + \frac{1}{z_1}\sum_{\alpha=1}^m \tau_\alpha y_{\alpha j}(G\mathbf{y}_\alpha)_j(z_1),$$

then

$$\mathbf{E}\{g_n(z_1,z_2)\} = -\frac{\mathbf{E}\{n^{-1}\gamma_n(z_2)\}}{z_1} + \sum_{\alpha,j=1}^{m,n}\frac{\tau_\alpha\mathbf{E}\{y_{\alpha j}(G\mathbf{y}_\alpha)_j(z_1)G_{jj}(z_2)\}}{nz_1}. \tag{5.14}$$

It follows from (3.10) that $(G\mathbf{y}_\alpha)_j = A_{\alpha n}^{-1}(G^\alpha\mathbf{y}_\alpha)_j$, $G_{jj} = G^\alpha_{jj} - \tau_\alpha A^{-1}_{\alpha n}(G^\alpha\mathbf{y}_\alpha)^2_j$, where $A_{\alpha n}$ is defined in (3.12). Applying (3.18), we obtain

(5.15)
$$\mathbf{E}\{y_{\alpha j}(G\mathbf{y}_\alpha)_j(z_1)G_{jj}(z_2)\} = \mathbf{E}\{y_{\alpha j}(A^{-1}_{\alpha n}(G^\alpha\mathbf{y}_\alpha)_j)(z_1)G^\alpha_{jj}(z_2)\} - \tau_\alpha\mathbf{E}\{y_{\alpha j}(A^{-1}_{\alpha n}(G^\alpha\mathbf{y}_\alpha)_j)(z_1)(A^{-1}_{\alpha n}(G^\alpha\mathbf{y}_\alpha)^2_j)(z_2)\}$$
$$= \mathbf{E}\{A_{\alpha n}(z_1)\}^{-1}\mathbf{E}\{\mathbf{E}_\alpha\{y_{\alpha j}(G^\alpha\mathbf{y}_\alpha)_j(z_1)\}G^\alpha_{jj}(z_2)\} + r_n,$$

where

(5.16)
$$r_n = \frac{\mathbf{E}\{y_{\alpha j}(A^\circ_{\alpha n}A^{-1}_{\alpha n}(G^\alpha\mathbf{y}_\alpha)_j)(z_1)G^\alpha_{jj}(z_2)\}}{\mathbf{E}\{A_{\alpha n}(z_1)\}} - \frac{\mathbf{E}\{y_{\alpha j}((1+A^\circ_{\alpha n}A^{-1}_{\alpha n})(G^\alpha\mathbf{y}_\alpha)_j)(z_1)((1+A^\circ_{\alpha n}A^{-1}_{\alpha n})^{-1}(G^\alpha\mathbf{y}_\alpha)^2_j)(z_2)\}}{\mathbf{E}\{A_{\alpha n}(z_1)\}\mathbf{E}\{A_{\alpha n}(z_2)\}}.$$

Since $|(G^\alpha \mathbf{y}_\alpha)_j|^2 \le |\Im z|^{-2}||\mathbf{y}_\alpha||^2$, then by the Schwarz inequality, (5.3), and (1.24) the first term in the right-hand side of (5.16) is less than

$$C\mathbf{Var}\{A_{\alpha n}\}^{1/2}\mathbf{E}\{y_{\alpha j}^2|(G^\alpha \mathbf{y}_\alpha)_j|^2\} = O(n^{-3/2}).$$

By the same reason, all the terms in the numerator of the second term on the right-hand side of (5.16), which contain $A^\circ_{\alpha n}$, are of the order $O(n^{-3/2})$. For the only term that does not contain $A^\circ_{\alpha n}$ we have

$$\begin{aligned}\mathbf{E}\{y_{\alpha j}(G^\alpha \mathbf{y}_\alpha)_j(z_1)(G^\alpha \mathbf{y}_\alpha)_j^2(z_2)\} &= \sum_{p,q,s=1}^n G^\alpha_{jp}(z_1)G^\alpha_{jq}(z_2)G^\alpha_{js}(z_2)\mathbf{E}\{y_{\alpha j}y_{\alpha p}y_{\alpha q}y_{\alpha s}\} \\ &= 3a_{2,2}G^\alpha_{jj}(G^\alpha G^\alpha)_{jj} + \kappa_4 G^\alpha_{jj}G^\alpha_{jj}G^\alpha_{jj} = O(n^{-2}),\end{aligned}$$

where we used (2.8) and (1.20). Hence

$$r_n = O(n^{-3/2}), \quad n \to \infty. \tag{5.17}$$

Besides, it follows from (1.11) that

$$\mathbf{E}_\alpha\{y_{\alpha j}(G^\alpha(z_1)\mathbf{y}_\alpha)_j\} = n^{-1}G^\alpha_{jj}(z_1). \tag{5.18}$$

Plugging (5.17)–(5.18) in (5.15) and then in (5.14), we get

$$\mathbf{E}\{g_n(z_1,z_2)\} = -\frac{1}{z_1}f(z_2) + \frac{1}{n^2 z_1}\sum_{\alpha,j=1}^{m,n}\frac{\tau_\alpha \mathbf{E}\{g_n^\alpha(z_1,z_2)\}}{1+\tau_\alpha f(z_1)} + o(1), \quad n \to \infty,$$

where

$$g_n^\alpha(z_1,z_2) = n^{-1}\sum_{j=1}^n G^\alpha_{jj}(z_1)G^\alpha_{jj}(z_2), \tag{5.19}$$

and we took into account (5.1), (5.4). Using the argument similar to that leading to (5.17), we obtain $\mathbf{E}\{g_n - g_n^\alpha\} = o(1)$, $n \to \infty$. This and (1.4) show that

$$\lim_{n\to\infty}\mathbf{E}\{g_n(z_1,z_2)\} = \left(c\int\frac{\tau d\sigma(\tau)}{1+\tau f(z_1)} - z_1\right)^{-1} f(z_2) = f(z_1)f(z_2),$$

and we get (5.12). The proof of (5.13) follows the scheme of the proof of Lemma 3.2. □

References

[1] N. I. Akhiezer and I. M. Glazman, *Theory of Linear Operators in Hilbert Space*, Dover, New York, 1993. MR1255973 (94i:47001)

[2] M. Anttila, K. Ball, and T. Perissinaki, *The central limit problem for convex bodies*, Trans. Amer. Math. Soc. **355** (2003), 4723–4735. MR1997580 (2005b:52010)

[3] G. Aubrun, *Sampling convex bodies: a random matrix approach*, Proc. Amer. Math. Soc. **135** (2007), 1293–1303. MR2276637 (2007k:52005)

[4] Z. D. Bai and J. W. Silverstein, *CLT for linear spectral statistics of large dimensional sample covariance matrices*, Ann. Prob. **32** (2004), 553–605. MR2040792 (2005b:60046)

[5] F. Barthe, O. Guédon, S. Mendelson, and A. Naor, *A probabilistic approach to the geometry of the l^p-ball*, Ann. Prob. **33** (2005) 2, 480–513. MR2123199 (2006g:46014)

[6] S. G. Bobkov, *Remarks on the growth of L^p-norms of polynomials*, Geometric Aspects of Functional Analysis, 27–35, Lecture Notes in Math., **1745**, Springer, Berlin, 2000. MR1796711 (2002b:46016)

[7] S. Bobkov, *Concentration of normalized sums and a central limit theorem for noncorrelated random variables*, Ann. Prob. **32** (2004), 2884–2907. MR2094433 (2005i:60041)

[8] J. Bourgain, *On the distribution of polynomials on high dimensional convex sets*, Lecture Notes in Math. **1469** (1991), 127–137. MR1122617 (92j:52007)

[9] J. Bourgain, *Random points in isotropic convex sets*, in Convex Geometric analysis (Berkeley, CA, 1996), Math. Sci. Res. Inst. Publ., vol. 34, Cambridge Univ. Press, Cambridge, 1999, pp. 53–58. MR1665576 (99m:60021)

[10] K. R. Davidson and S. Szarek, *Local operator theory, random matrices and Banach spaces*, in Handbook on the Geometry of Banach spaces, Volume 1 (W. B. Johnson and J. Lindenstrauss, eds.), Elsevier Science, 2001, pp. 317–366. MR1863696 (2004f:47002a)

[11] S. W. Dharmadhikari, V. Fabian, and K. Jogdeo, *Bounds on the moments of martingales*, Ann. Math. Statist. **39** (1968), 1719–1723. MR0230363 (37:5925)

[12] B. Fleury, O. Guédon, and G. Paouris, *A stability result for mean width of L^p-centroid bodies*, Advances in Mathematics **214.2** (2007), 865–877. MR2349721 (2008g:52012)

[13] A. Giannopoulos and V. Milman, *Concentration property on probability spaces*, Advances in Mathematics **156** (2000), 77–106. MR1800254 (2001m:28001)

[14] A. Giannopoulos and A. Tsolomitis, *Volume radius of a random polytope in a convex body*, Math. Proc. Cambridge Philos. Soc. **134** (2003), 13–21. MR1937788 (2003g:52002)

[15] V. L. Girko, *Theory of Stochastic Canonical Equations*, vols. I, II. Kluwer, Dordrecht, 2001. MR1887675 (2003f:60048)

[16] O. Guédon, *Concentration phenomena in high dimensional geometry*, to appear in Proceedings of ESAIM.

[17] O. Guédon, E. Milman, *Interpolating thin-shell and sharp large-deviation estimates for isotropic log-concave measures*, Geom. Funct. Anal. **21** (2011), 1043–1068. MR2846382

[18] R. Kannan, L. Lovász, and M. Simonovits, *Isoperimetric problems for convex bodies and a localization lemma*, Discrete Comput. Geom. **13** (1995), 541–559. MR1318794 (96e:52018)

[19] B. Klartag, *A central limit theorem for convex sets*, Invent. Math. **168** (2007), 91–131. MR2285748 (2008d:60034)

[20] B. Klartag, *A Berry-Esseen type inequality for convex bodies with an unconditional basis*, Prob. Theory Related Fields **145** (2009), no. 1–2, 1–33. MR2520120 (2010h:60072)

[21] R. Latałla and J. O. Wojtaszczyk, *On the infimum convolution inequality*, Studia Math. **189** (2008), 147–187. MR2449135 (2009k:52011)

[22] A. Litvak, A. Pajor, M. Rudelson, and N. Tomczak-Jaegermann, *Smallest singular value of random matrices and geometry of random polytopes*, Advances in Mathematics **195** (2005), 491–523. MR2146352 (2006g:52009)

[23] A. Lytova and L. Pastur, *Central limit theorem for linear eigenvalue statistics of random matrices with independent entries*, Ann. Prob. **37** (2009), 5, 1778–1840. MR2561434 (2011c:60024)

[24] V. Marchenko and L. Pastur, *The eigenvalue distribution in some ensembles of random matrices*, Math. USSR Sbornik **1** (1967), 457–483.

[25] R. J. Muirhead, *Aspects of Multivariate Statistical Theory*, Wiley, New York, 1982. MR652932 (84c:62073)

[26] A. Pajor and L. Pastur, *On the limiting empirical measure of eigenvalues of the sum of rank one matrices with log-concave distribution*, Studia Math. **195** (2009), 11–29. MR2539559 (2010h:15022)

[27] G. M. Pan and W. Zhou, *Central limit theorem for signal-to-interference ratio of reduced rank linear receiver*, Ann. Appl. Probab. **18** (2008), 1232–1270. MR2418244 (2009k:62120)

[28] L. Pastur, *Limiting laws of linear eigenvalue statistics for Hermitian matrix models*, J. Math. Phys. **47** (2006), 103–303. MR2268864 (2008a:82040)

[29] L. Pastur and M. Shcherbina. *Eigenvalue Distribution of Large Random Matrices*, Mathematical Surveys and Monographs, **171**, Amer. Math. Soc., Providence, RI, 2011, p. 634. MR2808038 (2012d:60015)

[30] M. Shcherbina, *Central limit theorem for linear eigenvalue statistics of the Wigner and sample covariance random matrices*, Math. Physics, Analysis, Geometry **7** (2011), 2, 176–192. MR2829615 (2012m:60064)

[31] S. Sodin, *An isoperimetric inequality on the l_p balls*, Ann. Inst. Henri Poincaré, Probab. Stat. **44** (2008), 362–373. MR2446328 (2009j:60038)

[32] S. Szarek, *Convexity, complexity, and high dimensions*, in International Congress of Mathematicians, Vol. II, Eur. Math. Soc., Zürich, 2006, pp. 1599–1621. MR2275661 (2008f:46003)

[33] S. Vempala, *Geometric Random Walks: A Survey*, in Combinatorial and Computational Geometry, Math. Sci. Res. Inst. Publ., vol. 52, Cambridge Univ. Press, Cambridge, 2005, pp. 577–616. MR2178341

[34] J. O. Wojtaszczyk, *The square negative correlation property for generalized Orlicz balls*, Geometric Aspects of Functional Analysis, Lecture Notes in Math., **1910**, Springer, Berlin, 2007, pp. 305–313. MR2349615 (2008h:52006)

Department of Mathematics, University Paris-Est, 5, Bd Descartes, Champs sur Marne, 77454, Marne la Vallée, Cédex 2, France

E-mail address: olivier.guedon@univ-mlv.fr

Department of Mathematics, B. Verkin Institute for Low Temperature Physics and Engineering of the NASU, 47 Lenin Ave., Kharkiv 61103, Ukraine

E-mail address: anna.lytova@gmail.com;lytova@ilt.kharkov.ua

Department of Mathematics, University Paris-Est, 5, Bd Descartes, Champs sur Marne, 77454, Marne la Vallée, Cédex 2, France

E-mail address: Alain.Pajor@univ-mlv.fr

Department of Mathematics, B. Verkin Institute for Low Temperature Physics and Engineering of the NASU, 47 Lenin Ave., Kharkiv 61103, Ukraine

E-mail address: pastur@ilt.kharkov.ua

Amer. Math. Soc. Transl.
Volume **233**, 2014

On the Spectrum of Nonself-Adjoint Zakharov-Shabat Operators on $\mathbb{R}$

T. Kappeler, P. Lohrmann, and P. Topalov

Dedicated to Vladimir Marchenko on the occasion of his 90th birthday

ABSTRACT. We describe the spectrum of nonself-adjoint Zakharov-Shabat operators on the line in the case where the potential is periodic.

1. Introduction

The *focusing* nonlinear Schrödinger equation (fNLS) with periodic boundary conditions is an integrable PDE. Unlike for the *defocusing* one (dNLS), its phase space is *not* exclusively foliated by regular tori, but is also known to have as invariant sets critical tori with hyperbolic structure; see [**8**]. This principal difference between fNLS and dNLS is reflected by the fact that the Zakharov-Shabat operator $L(\varphi)$ appearing in the Lax pair formulation of these equations is self-adjoint in the case of dNLS, but nonself-adjoint for fNLS. The Zakharov-Shabat operators (ZS) are differential operators of first order of the form ($x \in \mathbb{R}$, $\partial_x = \partial/\partial x$)

$$L(\varphi) := i \begin{pmatrix} 1 & 0 \\ 0 & -1 \end{pmatrix} \partial_x + \begin{pmatrix} 0 & \varphi_1 \\ \varphi_2 & 0 \end{pmatrix}, \tag{1.1}$$

where the potential $\varphi = (\varphi_1, \varphi_2)$ is in L^2_r in the case of dNLS, respectively in iL^2_r for fNLS. Here L^2_r is the real Hilbert space $\{\varphi = (\varphi_1, \varphi_2) \,|\, \varphi_1 \in L^2(\mathbb{T}, \mathbb{C});\ \varphi_2 = \bar{\varphi}_1\}$ and $\mathbb{T} = \mathbb{R}/\mathbb{Z}$. Paramount for the study of the geometry of the phase space of fNLS are the spectral properties of $L(\varphi)$. Such an analysis was initiated in [**1**] and later in more detail, taken up in [**8, 10, 11**] and, more recently, in [**5**]. However, for $\varphi \in iL^2_r$, a full understanding of the periodic spectrum, the Dirichlet spectrum or the spectrum of $L(\varphi)$ when considered on the whole real line constitutes one of the major open problems in the area of integrable PDEs in one space dimension; see [**2**] for more details. The aim of this paper is to study the spectrum of $L(\varphi)$ when considered on the whole line. To state our main results we first have to introduce some more notation. Consider the ZS operator $L(\varphi)$ for $\varphi = (\varphi_1, \varphi_2)$ in

2010 *Mathematics Subject Classification.* Primary 35P25.

The first author was supported in part by the Swiss National Science Foundation.

The second author was supported in part by the Swiss National Science Foundation and the European Research Council under FP7 "New connections between dynamical systems and Hamiltonian PDE with small divisor phenomena".

The third author was supported in part by NSF Grant DMS-0901443.

$L^2_c := L^2(\mathbb{T}, \mathbb{C}) \times L^2(\mathbb{T}, \mathbb{C})$. For any $\lambda \in \mathbb{C}$, let $M = M(x, \lambda, \varphi)$ be the fundamental 2×2 matrix of the equation

$$L(\varphi)M = \lambda M, \quad M = \begin{pmatrix} m_1 & m_2 \\ m_3 & m_4 \end{pmatrix},$$

satisfying the initial condition $M(0, \lambda, \varphi) = \mathrm{Id}_{2\times 2}$. The fundamental solution $M(x, \lambda, \varphi)$ is a continuous function on $\mathbb{R} \times \mathbb{C} \times L^2_c$ and for any given $x \in \mathbb{R}$, analytic in λ, φ on $\mathbb{C} \times L^2_c$; see e.g. Theorem 1.1 in [**4**]. Moreover, the proof of Theorem 1.4 in [**4**] shows that the following stronger statement holds.

LEMMA 1.1. *The fundamental matrix M defines an analytic map*

$$M : \mathbb{C} \times L^2_c \to C([0, 2], \mathbb{C}^{2\times 2}), \quad (\lambda, \varphi) \mapsto M(\cdot, \lambda, \varphi),$$

where $C([0, 2], \mathbb{C}^{2\times 2})$ denotes the Banach space of 2×2 matrix valued continuous functions on the interval $[0, 2]$, endowed with the supremum norm.

We denote by $\mathrm{spec}_p L(\varphi)$ the spectrum of the operator $L = L(\varphi)$ with domain

$$\mathrm{dom_p}(L) := \{F \in H^1_{\mathrm{loc}} \times H^1_{\mathrm{loc}} | \, F(2) = F(0)\}.$$

As $L(\varphi)$ has a compact resolvent this spectrum is discrete and each of its eigenvalues has finite algebraic multiplicity. It is referred to as the periodic spectrum of $L(\varphi)$ and its eigenvalues as periodic eigenvalues of $L(\varphi)$ – or by a slight abuse of terminology as periodic eigenvalues of φ. Recall that the algebraic multiplicity of a periodic eigenvalue λ of $L(\varphi)$ equals the dimension of the root space $R_\lambda(\varphi)$, defined as the following $\mathbb{C}$-subspace of $\mathrm{dom}_p L(\varphi)$:

$$R_\lambda(\varphi) := \{f \in \mathrm{dom}_p(L(\varphi)) \,|\, \exists n \in \mathbb{Z}_{\geq 1} \text{ with } (\lambda - L(\varphi))^n f = 0\}.$$

By the definition of the fundamental solution M, any solution f of the equation $L(\varphi)f = \lambda f$ is given by $f(x) = M(x, \lambda)f(0)$. Hence a complex number λ is a periodic eigenvalue of $L(\varphi)$ iff there exists a nonzero solution of $L(\varphi)f = \lambda f$ with $f(1) = M(1, \lambda)f(0) = \pm f(0)$. It means that 1 or -1 is an eigenvalue of the Floquet matrix $M(1, \lambda)$. Denote by $\Delta(\lambda) \equiv \Delta(\lambda, \varphi)$ the discriminant of $L(\varphi)$, $\Delta(\lambda, \varphi) := \mathrm{Tr}\, M(1, \lambda, \varphi) = m_1(1, \lambda, \varphi) + m_4(1, \lambda, \varphi)$. Note that by Lemma 1.1 it is analytic on $\mathbb{C} \times L^2_c$. In view of the Wronskian identity, $\det M(1, \lambda) = 1$, it then follows that λ is a periodic eigenvalue of $L(\varphi)$ iff $\Delta(\lambda) = \pm 2$. For a proof of the following result see e.g. [**5**].

LEMMA 1.2. *For any $\varphi \in L^2_c$, the periodic spectrum of $L(\varphi)$ coincides as a set with the zero set of the function $\Delta^2(\cdot, \varphi) - 4$. More precisely, for any periodic eigenvalue λ of $L(\varphi)$, the algebraic multiplicity of λ coincides with the multiplicity of λ as a root of $\Delta^2(\cdot, \varphi) - 4$.*

By Floquet theory, the spectrum $\mathrm{spec}_{\mathbb{R}} L(\varphi)$ of the operator $L(\varphi)$, $\varphi \in L^2_c$, viewed as an unbounded operator on $L^2(\mathbb{R}; \mathbb{C}) \times L^2(\mathbb{R}; \mathbb{C})$ with domain $\mathrm{dom}_{\mathbb{R}} L(\varphi) := H^1(\mathbb{R}; \mathbb{C}) \times H^1(\mathbb{R}; \mathbb{C})$, is absolutely continuous and given by (cf. e.g. [**8**])

$$\mathrm{spec}_{\mathbb{R}} L(\varphi) = \{\lambda \in \mathbb{C} \,|\, \Delta(\lambda) \text{ is real and } |\Delta(\lambda)| \leq 2\}.$$

Note that $\mathrm{spec}_{\mathbb{R}} L(\varphi)$ coincides with the set of $\lambda \in \mathbb{C}$ such that the Floquet multipliers of the equation $L(\varphi)F = \lambda F$ lie on the unit circle in $\mathbb{C}$. To state our main results on $\mathrm{spec}_{\mathbb{R}} L(\varphi)$ we need to describe some features of the periodic spectrum $\mathrm{spec}_p L(\varphi)$. First let us recall the so-called *Counting Lemma*, providing a rough estimate of the periodic eigenvalues of $L(\varphi)$ for φ in L^2_c; for a proof see e.g. [**5**]. For any $\rho \in \mathbb{Z}$ define

$$\text{Box}_\rho := \{\lambda = u + iv \in \mathbb{C} \mid |u|, |v| < \rho\pi + \pi/4\}.$$

LEMMA 1.3. *For any potential in L^2_c there exist a neighborhood $\mathcal{W}$ in L^2_c and an integer $\rho \geq 0$ so that when counted with their algebraic multiplicities, for any φ in $\mathcal{W}$, $L(\varphi)$ has exactly two periodic eigenvalues inside any box $D_n := \{\lambda = u + iv \in \mathbb{C} \mid |u - n\pi|, |v| < \pi/4\}$ with $|n| > \rho$ and exactly $4\rho + 2$ eigenvalues in Box_ρ. There are no other eigenvalues.*

For a potential $\varphi \in L^2_r$, the operator $L(\varphi)$, when considered with domain $\text{dom}_{\text{p}} L(\varphi)$ or with domain $\text{dom}_{\mathbb{R}} L(\varphi)$, is self-adjoint, and hence its spectrum real. The following lemma is well known; see e.g. [**4**]. We recall that for two sequences $(a_n)_{n\in\mathbb{Z}}$, $(b_n)_{n\in\mathbb{Z}}$ in $\mathbb{C}$, $a_n = b_n + \ell^2_n$ means that $(a_n - b_n)_{n\in\mathbb{Z}}$ is a ℓ^2-sequence, i.e. $\sum_{n\in\mathbb{Z}} |a_n - b_n|^2 < \infty$.

THEOREM 1.4. *For any potential $\varphi \in L^2_r$, all periodic eigenvalues of $L(\varphi)$ are real. When counted with their multiplicities they are given by two doubly infinite sequences, $(\lambda^+_n)_{n\in\mathbb{Z}}$ and $(\lambda^-_n)_{n\in\mathbb{Z}}$, listed in such a way that $\lambda^\pm_n = n\pi + \ell^2_n$ and $\cdots < \lambda^-_n \leq \lambda^+_n < \lambda^-_{n+1} \leq \lambda^+_{n+1} < \cdots$. Furthermore, $\text{spec}_{\mathbb{R}} L(\varphi) = \mathbb{R} \setminus \bigcup_{n\in\mathbb{Z}} (\lambda^-_n, \lambda^+_n)$. The open intervals $(\lambda^-_n, \lambda^+_n) \subseteq \mathbb{R}, n \in \mathbb{Z}$, are pairwise disjoint and referred to as the spectral gaps of $L(\varphi)$.*

As an example, consider the zero potential. Note that the fundamental solution $M(x, \lambda, 0)$ is given by the diagonal matrix $\text{diag}(e^{-i\lambda x}, e^{i\lambda x})$, and hence $\Delta(\lambda, 0) = 2\cos(\lambda)$. A straightforward computation then yields $\text{spec}_{\mathbb{R}} L(0) = \mathbb{R}$ and

$$\text{spec}_p L(0) = \{\lambda^+_n = \lambda^-_n = n\pi | n \in \mathbb{Z}\}.$$

For $\varphi \in iL^2_r$, the spectrum $\text{spec}_{\mathbb{R}} L(\varphi)$ is more complicated. Note that if $\varphi \neq 0$, then $L(\varphi)$, when considered with domain $\text{dom}_{\text{p}} L(\varphi)$ or $\text{dom}_{\mathbb{R}} L(\varphi)$, is nonself-adjoint and hence its spectrum not necessarily real. On the other hand, $L(\varphi)$ has various symmetries; see [**3**] for a detailed account. In this paper, the following one is used frequently. For any function $f : \mathbb{R} \to \mathbb{C}^2$ with components f_1, f_2, introduce the functions $\breve{f}, \hat{f} : \mathbb{R} \to \mathbb{C}^2$, given by $\breve{f} = (-\bar{f}_2, \bar{f}_1)$ and $\hat{f} = -(\bar{f}_2, \bar{f}_1)$. Note that for any $\varphi \in L^2_c$, one has $\varphi = \hat{\varphi}$ iff $\varphi \in iL^2_r$ and that $\mathfrak{J} f := \breve{f}$ is an anti-involution, $\mathfrak{J}^2 f = -f$. The following result has been proved in [**5**]. As usual, for any complex number λ, we denote by $\bar{\lambda}$ its complex conjugate.

LEMMA 1.5. *Assume that $\varphi \in L^2_c$, $\lambda \in \mathbb{C}$, and f in $H^1_{loc}(\mathbb{R}, \mathbb{C}^2)$ solves $(L(\varphi) - \lambda)^n f = 0$ for some $n \in \mathbb{Z}_{\geq 1}$. Then $(L(\hat{\varphi}) - \bar{\lambda})^n \breve{f} = 0$.*

As a corollary we get

COROLLARY 1.6. *One has the following relation on the fundamental matrix $M(x, \lambda, \varphi)$:*

$$\begin{pmatrix} m_1 & m_2 \\ m_3 & m_4 \end{pmatrix}\Bigg|_{x,\lambda,\varphi} = \begin{pmatrix} \overline{m}_4 & -\overline{m}_3 \\ -\overline{m}_2 & \overline{m}_1 \end{pmatrix}\Bigg|_{x,\bar{\lambda},\hat{\varphi}}. \tag{1.2}$$

In particular, for any $x \in \mathbb{R}$, $\lambda \in \mathbb{R}$, and $\varphi \in iL^2_r$, the fundamental matrix $M(x, \lambda, \varphi)$ belongs to the group $SU(2)$. It follows from Lemma 1.5 that for any φ in iL^2_r, the complex conjugate of any periodic eigenvalue $\lambda \in \text{spec}_p L(\varphi)$ is again a periodic eigenvalue with the same algebraic and geometric multiplicity as λ. The following lemma summarizes the features of $\mathrm{spec}_p L(\varphi)$ needed in this paper; see [**5**] for a proof.

LEMMA 1.7. *Let $\varphi \in iL^2_r$. Then any real periodic eigenvalue of $L(\varphi)$ has geometric multiplicity two and even algebraic multiplicity. For any periodic eigenvalue λ in $\mathbb{C}\backslash\mathbb{R}$, its complex conjugate $\bar{\lambda}$ is also a periodic eigenvalue and has the same algebraic and geometric multiplicity as λ. It then follows that the periodic eigenvalues of $L(\varphi)$, when counted with their algebraic multiplicities, are given by two infinite sequences $(\lambda_n^+)_{n\in\mathbb{Z}}$ and $(\lambda_n^-)_{n\in\mathbb{Z}}$ with $\lambda_n^- = \overline{\lambda_n^+}$ and $\mathrm{Im}\,\lambda_n^+ \geq 0$ for any $n \in \mathbb{Z}$ so that $\lambda_n^+ = n\pi + \ell_n^2$ and $(\lambda_n^+)_{n\in\mathbb{Z}}$ is lexicographically ordered.*

Finally, denote by $\dot{\Delta}(\lambda,\varphi)$ the partial derivative of the discriminant $\Delta(\lambda,\varphi)$ with respect to λ. Note that by Lemma 1.1, $\dot{\Delta}$ is analytic on $\mathbb{C}\times L^2_c$. The following properties of Δ and $\dot{\Delta}$ follow from Lemma 1.2 and are well known; see e.g. [**4**, Proposition 3.4 and 3.5]. To state them, introduce $\pi_n := n\pi$ for $n \in \mathbb{Z}\setminus\{0\}$ and $\pi_0 := 1$.

LEMMA 1.8. *(i) For any φ in L^2_c, $\Delta^2(\lambda,\varphi) - 4$ and $\dot{\Delta}(\lambda,\varphi)$ admit product representations,*

$$\Delta^2(\lambda,\varphi) - 4 = -4\prod_{n\in\mathbb{Z}} \frac{(\lambda_n^+(\varphi)-\lambda)\,(\lambda_n^-(\varphi)-\lambda)}{\pi_n^2} \quad \text{and} \quad \dot{\Delta}(\lambda,\varphi) = 2\prod_{n\in\mathbb{Z}} \frac{\dot{\lambda}_n - \lambda}{\pi_n},$$

where the roots $\dot{\lambda}_n$ of $\dot{\Delta}(\lambda,\varphi)$ are counted with their multiplicities and listed in such a way that they are lexicographically ordered and satisfy the asymptotic estimates $\dot{\lambda}_n = n\pi + \ell_n^2$. These asymptotics hold locally uniformly in φ.

(ii) For any $\varphi \in iL^2_r$ and $\lambda \in \mathbb{C}$, $\Delta(\bar{\lambda},\varphi) = \bar{\Delta}(\lambda,\varphi)$ and $\bar{\dot{\Delta}}(\lambda,\varphi) = \dot{\Delta}(\bar{\lambda},\varphi)$. In particular, the zero set of $\dot{\Delta}(\cdot,\varphi)$ is invariant under complex conjugation. In view of the asymptotics stated in (i), for n sufficiently large, $\dot{\lambda}_n$ is real and simple.

Our main result is the following one where the word "smooth" means C^∞–differentiable.

THEOREM 1.9. *For any $\varphi \in iL^2_r$, there exists $R \in \mathbb{Z}_{\geq -1}$ such that $\lambda_n^\pm \in D_n$ for any $|n| > R$, $\lambda_n^\pm \in \mathrm{Box}_R$ for any $|n| \leq R$,*

$$spec_{\mathbb{R}} L(\varphi) \subseteq \mathbb{R} \cup \mathrm{Box}_R \cup \bigcup_{|n|>R} D_n, \qquad \overline{spec_{\mathbb{R}} L(\varphi)} = spec_{\mathbb{R}} L(\varphi),$$

and the following properties hold: (i) for any $|n| > R$, $spec_{\mathbb{R}} L(\varphi) \cap D_n$ consists of the interval $D_n \cap \mathbb{R}$ and an arc $\mathcal{B}_n$ which is the graph of a smooth function above the interval $i[\mathrm{Im}\,\lambda_n^-, \mathrm{Im}\,\lambda_n^+] \subseteq i\mathbb{R}$, intersecting the real line in $\dot{\lambda}_n$ at a right angle. In addition, $\mathcal{B}_n$ is invariant under complex conjugation and connects λ_n^- with λ_n^+. (ii) The intersection $spec_{\mathbb{R}} L(\varphi) \cap \mathrm{Box}_R$ consists of the interval $\mathrm{Box}_R \cap \mathbb{R}$ and $2R+1$ closed, smooth, nonempty arcs $(\mathcal{B}_n)_{|n|\leq R}$ without self-intersection. (These $2R+1$ arcs are listed in an arbitrary way.) The endpoints of each of these arcs are periodic eigenvalues, and in their interiors, $|\Delta| < 2$. In case $\mathcal{B}_n$ consists of a single point it is a periodic eigenvalue of algebraic multiplicity ≥ 2. Two arcs can meet at most once. If they touch each other at their endpoints, this endpoint is a periodic eigenvalue of algebraic multiplicity ≥ 2. In case some of the $(\mathcal{B}_n)_{|n|\leq R}$ intersect in a point in $\mathbb{C}\setminus\mathbb{R}$, they do this at equal angles. If some of them intersect in a point *on $\mathbb{R}$, then together with the real axis they meet at equal angles. The points where some of the $\mathcal{B}_n$'s intersect are zeros of $\dot{\Delta}(\lambda)$, and the number of arcs intersecting at such a point equals the order of the zero of $\dot{\Delta}(\lambda)$ plus one. The integer R can be chosen locally uniformly with respect to φ.*

REMARK 1.10. Note that the periodic eigenvalues of $L(0)$ are $n\pi$, $n \in \mathbb{Z}$, each having algebraic and geometric multiplicity equal to two. For any $\varphi \in iL^2_r$ in a sufficiently small neighborhood of zero, the number R in Theorem 1.9 can be chosen to be -1. Note that by definition Box_{-1} is empty.

Organisation: In section 2, we study the locus of $\operatorname{Im}\Delta = 0$ where $\operatorname{Im}\Delta(\lambda)$ denotes the imaginary part of $\Delta(\lambda)$ and prove Theorem 1.9. In section 3 we discuss some examples.

Acknowledgment: We would like to thank the referee for pointing out to us some additional references and making valuable suggestions for improving the exposition of the paper.

2. Locus of $\operatorname{Im}\Delta = 0$

In this section we study for an arbitrary $\varphi \in iL^2_r$ the subset

$$\Lambda \equiv \Lambda_\varphi := \{\lambda \in \mathbb{C} \mid \operatorname{Im}\Delta(\lambda,\varphi) = 0\},$$

which by Lemma 1.8 is invariant under complex conjugation. The main results are summarized in Theorem 2.7 below and will be the key ingredients for the proof of Theorem 1.9. The starting point is the observation that for the zero potential, $\Delta(\lambda, 0) = 2\cos\lambda$ and

$$\Lambda \equiv \Lambda_0 = \{\lambda = u + iv \in \mathbb{C} \mid -2\sin u \cdot \sinh v = 0\} = \mathbb{R} \cup \bigcup_{n\in\mathbb{Z}} \{\lambda \in \mathbb{C} \mid \operatorname{Re}\lambda = n\pi\}.$$

Note that for $\lambda \in \mathbb{R} \setminus \pi\mathbb{Z}$, one has $-2 < \Delta(\lambda, 0) < 2$, whereas for $\lambda = n\pi \pm iv$ with $v > 0$, $(-1)^n\Delta(n\pi \pm iv, 0) > 2$ is strictly increasing and satisfies

$$\lim_{v\to\infty} (-1)^n \Delta(n\pi \pm iv, 0) = \infty\,.$$

For $\varphi \in L^2_r$ one can see that Λ_φ consists of the real axis $\mathbb{R} \subseteq \mathbb{C}$ and countably many nonintersecting smooth lines $\mathcal{L}_n \subseteq \mathbb{C}$, $n \in \mathbb{Z}$, where for any $n \in \mathbb{Z}$, $\mathcal{L}_n$ intersects the real axis only in the point $\dot{\lambda}_n$ and approaches asymptotically the line $n\pi + i\mathbb{R}$. In addition, $\dot{\lambda}_n$ is the *only* critical point of $\Delta(\cdot, \varphi)\big|_{\mathcal{L}_n}$ and $(-1)^n\Delta(\lambda, \varphi)$ is strictly increasing along the half lines $\mathcal{L}_n^{\pm} := \{\lambda \in \mathbb{C} \mid \pm \operatorname{Im}\lambda > 0\}$

It turns out that for potentials $\varphi \in iL^2_r$, the set Λ is more complicated. Without further reference we will use the notation introduced in section 1. First note that for $\varphi \in iL^2_r$, the real axis is always contained in Λ.

LEMMA 2.1. *$\mathbb{R} \subseteq \mathrm{spec}_{\mathbb{R}} L(\varphi)$ for any $\varphi \in iL^2_r$.*

PROOF. It follows from Lemma 1.8 (*ii*) that $\Delta(\lambda, \varphi) = \overline{\Delta(\bar{\lambda}, \varphi)}$. Hence for $\lambda \in \mathbb{R}$, $\Delta(\lambda)$ is real valued. Furthermore, using that $\hat{\varphi} = \varphi$ for $\varphi \in iL^2_r$, the Wronskian identity, and Corollary 1.6, one has for $\lambda \in \mathbb{R}$, $|m_1(x, \lambda, \varphi)| = |m_4(x, \lambda, \varphi)|$ and

$$1 = m_1 m_4 - m_2 m_3\big|_{x,\lambda,\varphi} = m_1\overline{m}_1 - m_2(-\overline{m}_2)\big|_{x,\lambda,\varphi} = |m_1|^2 + |m_2|^2\big|_{x,\lambda,\varphi}\,.$$

Hence $|m_1| \le 1$ and thus $|\Delta(\lambda)| \le |m_1| + |m_4| \le 2$. □

To study the zero set Λ write $\lambda = u + iv$ with $u, v \in \mathbb{R}$ and let $\Delta = \Delta_1 + i\Delta_2$, where $\Delta_1(u, v) := \mathrm{Re}(\Delta(u + iv))$ and $\Delta_2(u, v) := \mathrm{Im}(\Delta(u + iv))$.

LEMMA 2.2. *For any $\varphi \in iL^2_r$ and $\mu \in \Lambda$, the following statements hold: (i) If $\dot\Delta(\mu) \neq 0$, then there exists a neighborhood U of μ in $\mathbb{C}$ such that $\Lambda \cap U$ is a C^∞ arc. (ii) If for some $n \geq 2$, $\partial^m_\lambda \Delta(\mu) = 0$ $\forall 1 \leq m < n$, but $\partial^n_\lambda \Delta(\mu) \neq 0$, then there exists a neighborhood U of μ in $\mathbb{C}$ and n smooth arcs $\mathcal{C}_1, \cdots, \mathcal{C}_n$ of finite length so that $\Lambda \cap U = \bigcup_{k=1}^n \mathcal{C}_k$, $\bigcap_{k=1}^n \mathcal{C}_k = \{\mu\}$, and $\mathcal{C}_1, \cdots, \mathcal{C}_n$ meet at equal angles in μ.*

PROOF. *(i)* Let $\mu \in \Lambda$ with $\dot\Delta(\mu) \neq 0$. Writing $\mu = a+ib$ one has $\partial_u \Delta_2(a, b) \neq 0$ or $\partial_v \Delta_2(a, b) \neq 0$ by the Cauchy-Riemann equations. By the implicit function theorem there then exists a neighborhood U of $\mu \in \mathbb{C}$ such that $\Lambda \cap U$ is the graph of a C^∞ function defined on an interval of the imaginary axis in the case $\partial_u \Delta_2(a, b) \neq 0$, or on one of the real axis in the case $\partial_v \Delta_2(a, b) \neq 0$.

(ii) Let $\mu \in \Lambda$ be such that $\partial^m_\lambda \Delta(\mu) = 0$ for any $1 \leq m < n$, but $\partial^n_\lambda \Delta(\mu) \neq 0$. Then the Taylor expansion of Δ at μ is of the form $\Delta(\lambda) - \Delta(\mu) = (\lambda - \mu)^n \cdot \big(c + O(\lambda - \mu)\big)$ with $\Delta(\mu) \in \mathbb{R}$ and c a nonzero complex number. Hence near μ there exists a holomorphic change of variable $F : z \mapsto \lambda$ with $F(0) = \mu$ so that $\Delta(F(z)) - \Delta(\mu) = z^n$. As a consequence, near μ, the zero level set of $\mathrm{Im}\,\Delta(\lambda) = \mathrm{Im}(\Delta(\lambda) - \Delta(\mu))$ consists of n smooth arcs, $\mathcal{C}_1, \cdots, \mathcal{C}_n$, satisfying the claimed properties. In particular, they intersect at equal angles in μ. □

LEMMA 2.3. *For any φ in iL^2_r there exist $P \in \mathbb{Z}_{\geq -1}$, $\delta > 0$, and $0 < \eta < 1/4$, so that for any $|n| > P$, the set $\Lambda \cap \mathrm{Box}^{\delta,\eta}_{n\pi}$ is the union of the open interval $(n\pi - \eta, n\pi + \eta) \subseteq \mathbb{R}$ and the graph of a smooth function above the interval $(-i\delta, i\delta) \subseteq i\mathbb{R}$, crossing the real axis in $\dot\lambda_n$ at a right angle. Here $\mathrm{Box}^{\delta,\eta}_{n\pi}$ denotes the box*

$$\mathrm{Box}^{\delta,\eta}_{n\pi} := \{\lambda \in \mathbb{C} \,|\, |\mathrm{Re}\,\lambda - n\pi| < \eta \text{ and } |\mathrm{Im}\,\lambda| < \delta\}\,.$$

PROOF. Fix $\varphi \in iL^2_r$. By Lemma 2.1, $\Delta_2(u, 0) = 0$ for any u in $\mathbb{R}$. Hence the function (introduced in [**6**])

$$F(u, v) := \Delta_2(u, v)/v \tag{2.1}$$

is well defined on $\mathbb{R} \times \mathbb{R}$ and satisfies

$$F(u, v) = \int_0^1 (\partial_v \Delta_2)(u, vt)dt. \tag{2.2}$$

In particular, F is a real analytic function of the variables u and v. As $\Delta_2(u, 0)$ vanishes on $\mathbb{R}$, so does $\partial_u \Delta_2(u, 0)$. By the Cauchy-Riemann equation it then follows that for $u \in \mathbb{R}$, $\dot\Delta(u, 0) = 0$ iff $\partial_v \Delta_2(u, 0) = 0$; i.e., the real roots of $\dot\Delta$ are roots of $\partial_v \Delta_2(\cdot, 0)$ and vise versa. As F is real analytic on $\mathbb{R} \times \mathbb{R}$ it has an analytic extension to a neighborhood of $\mathbb{R} \times \mathbb{R}$ in $\mathbb{C} \times \mathbb{C}$ which we again denote by F. Note that the functions $F(\cdot, \cdot)$ and $\Delta_2(\cdot, \cdot)$ have the same zeros in $\mathbb{R} \times (\mathbb{R} \setminus \{0\})$; hence it suffices to study the zero set of F. By Lemma 1.8 (ii), there exists $P \in \mathbb{Z}_{\geq -1}$ such that for any $|n| > P$, $\dot\lambda_n$ is a real simple root of $\dot\Delta$. In order to study the zero level set of F, consider the map

$$\mathcal{F} = (\mathcal{F}_n)_{n \in \mathbb{Z} \setminus [-P,P]} : B^\infty_1 \times (-1, 1) \to \ell^\infty \equiv \ell^\infty(\mathbb{Z} \setminus [-P, P], \mathbb{R}),$$

where for any $|n| > P$, $\mathcal{F}_n(\mathfrak{u}, v) = F(\dot\lambda_n + u_n, v)$ with $\mathfrak{u} := (u_n)_{n \in \mathbb{Z} \setminus [-P,P]}$, and for any $r > 0$, B^∞_r is the ball $\{\mathfrak{u} \in \ell^\infty \,|\, \|\mathfrak{u}\|_\infty < r\}$ of radius r in $\ell^\infty \equiv \ell^\infty(\mathbb{Z} \setminus [-P, P], \mathbb{R})$, centered at the origin. Note that $\mathcal{F}_n(\mathfrak{u}, v)$ only depends on u_n and

v, but not on u_k, $k \neq n$. It follows from (2.2) and [**4**, Proposition 2.3] that $\mathcal{F} : B_1^\infty \times (-1,1) \to \ell^\infty$ extends to a locally bounded map $\mathcal{F}_{\mathbb{C}} : \mathcal{V}_{\mathbb{C}} \to \ell^\infty_{\mathbb{C}}$, where $\mathcal{V}_{\mathbb{C}}$ is an open neighborhood of $B_1^\infty \times (-1,1)$ in $(B_1^\infty \times (-1,1)) \otimes \mathbb{C}$. As by (2.2) each component $\mathcal{F}_n$, $|n| > P$, is analytic on $\mathcal{V}_{\mathbb{C}}$, we conclude from Theorem A.3 in [**7**] that $\mathcal{F}$ is real analytic on $B_1^\infty \times (-1,1)$. As for any $|n| > P$, $\dot\lambda_n$ is a real simple root of $\dot\Delta(\cdot)$ one gets from (2.2) with $v = 0$,

(2.3)
$$\mathcal{F}|_{u=0,v=0} = \partial_v \Delta_2(\dot\lambda_n, 0) = 0 \quad \text{and} \quad \partial_{u_n}\mathcal{F}_n|_{u_n=0,v=0} = \partial_u[\partial_v\Delta_2(u,0)]|_{u=\dot\lambda_n}.$$

By the asymptotics $\dot\lambda_n = n\pi + \ell^2_n$ and [**4**, Proposition 2.3] one has

$$\partial_u\partial_v\Delta_2(u,v,\varphi)|_{u=\dot\lambda_n,v=0} = \partial_u\partial_v \operatorname{Im}\Delta(u+iv,0)|_{u=n\pi,v=0} + \ell^2_n. \tag{2.4}$$

On the other hand, for the zero potential,

$$\Delta(\lambda,0) = 2\cos\lambda \quad \text{and} \quad \operatorname{Im}\Delta(u+iv,0) = -2\sin u \sinh v. \tag{2.5}$$

From (2.3)-(2.5) we then conclude that

$$\partial_{\mathrm{u}}\mathcal{F}|_{\mathrm{u}=0,v=0} = 2\ \mathrm{diag}\left(((-1)^{n+1})_{n\in\mathbb{Z}\setminus[-P,P]}\right) + \ell^2_n. \tag{2.6}$$

By choosing P even larger if necessary, one can ensure that $\partial_{\mathrm{u}}\mathcal{F}|_{\mathrm{u}=0,v=0} : \ell^\infty \to \ell^\infty$ is invertible. Hence in view of (2.3) and the implicit function theorem there exist $0 < \delta < 1$ and a real analytic function $\mathfrak{h} : (-\delta,\delta) \to \ell^\infty$, $v \mapsto \mathfrak{h}(v) = (h_n(v))_{n\in\mathbb{Z}\setminus[-P,P]}$, with $\mathfrak{h}(0) = 0$, so that for any $v \in (-\delta,\delta)$, $\mathcal{F}(\mathfrak{h}(v), v) = 0$. Moreover, there exists $0 < \eta < 1/4$ such that the map $(-\delta,\delta) \to \ell^\infty \times (-\delta,\delta)$, $v \mapsto (\mathfrak{h}(v), v)$ parametrizes the zero level set of $\mathcal{F}$ in $B_\eta^\infty \times (-\delta,\delta)$. In particular, for any $n \in \mathbb{Z}\setminus[-P,P]$, the intersection of the zero level set of F with

$$\mathrm{Box}^{\delta,\eta}_{\dot\lambda_n} := \left\{\lambda \in \mathbb{C} \,\middle|\, |\operatorname{Re}\lambda - \dot\lambda_n| < \eta, |\operatorname{Im}\lambda| < \delta\right\}$$

is parametrized by $z_n : (-\delta,\delta) \mapsto \mathrm{Box}^{\delta,\eta}_{\dot\lambda_n}$, $v \mapsto \dot\lambda_n + h_n(v) + iv$. Let $g_n := \mathrm{Image}(z_n) \subseteq \mathrm{Box}^{\delta,\eta}_{\dot\lambda_n}$. By the definition (2.1) of F, the set $g_n \setminus \mathbb{R}$ then coincides with the intersection of the zero level set of Δ_2 with $\mathrm{Box}^{\delta,\eta}_{\dot\lambda_n} \setminus \mathbb{R}$. As $\Delta(\lambda)$ is real for $\lambda \in \mathbb{R}$, we see that the intersection of the zero level set of Δ_2 with $\mathrm{Box}^{\delta,\eta}_{\dot\lambda_n}$ coincides with $g_n \cup (\dot\lambda_n - \eta, \dot\lambda_n + \eta) \subseteq \mathbb{C}$. Furthermore, by Lemma 2.2, g_n crosses the real axis at a right angle. In view of the asymptotics $\dot\lambda_n = n\pi + \ell^2_n$, and by choosing $0 < \eta < 1/4$ smaller and P larger, if necessary, one finally obtains the assertion of Lemma 2.3. □

Recall that $\mathrm{Box}_\rho \subseteq \mathbb{C}$ denotes the box $\mathrm{Box}_\rho := \left\{\lambda = u+iv \,\middle|\, |u|, |v| < \rho\pi + \pi/4\right\}$.

LEMMA 2.4. *For any given $\varphi \in iL^2_r$, let P, δ, η and $\mathrm{Box}^{\delta,\eta}_{n\pi}$ be defined as in Lemma 2.3. Then for any $0 < \varepsilon < 1/2$ there exists $Q \in \mathbb{Z}_{\geq -1}$ with $Q \geq P$, depending on $\varphi, \varepsilon, \delta, \eta$, so that for any $\lambda \in \mathbb{C} \setminus \left(\mathrm{Box}_Q \cup \left(\bigcup_{n\in\mathbb{Z}} \mathrm{Box}^{\delta,\eta}_{n\pi}\right)\right)$,*

$$\frac{\Delta^2(\lambda,\varphi)-4}{\Delta^2(\lambda,0)-4},\ \frac{\Delta(\lambda,\varphi)-2}{\Delta(\lambda,0)-2},\ \frac{\Delta(\lambda,\varphi)+2}{\Delta(\lambda,0)+2},\ \text{and}\ \frac{\dot\Delta(\lambda,\varphi)}{\dot\Delta(\lambda,0)} \in D^\varepsilon(1),$$

where $D^\varepsilon(z)$ denotes the open disc in $\mathbb{C}$ with radius ε, centered at $z \in \mathbb{C}$.

PROOF. Let $0 < \varepsilon < 1/2$ be arbitrarily fixed. Note that for any $\lambda \in \mathbb{C} \setminus \mathrm{Box}_{n\pi}^{\delta,\eta}$ one has $|\lambda - n\pi| \geq \min(\delta, \eta)$. As $\lambda_n^\pm = n\pi + \ell_n^2$, it then follows that for any $\lambda \in \mathbb{C} \setminus \bigcup_{m\in\mathbb{Z}} B_{m\pi}^{\delta,\eta}$,

$$\frac{\lambda - \lambda_n^\pm}{\lambda - n\pi} = 1 - \frac{\lambda_n^\pm - n\pi}{\lambda - n\pi} = 1 + \ell_n^2.$$

Hence there exists an integer $P' \geq P$ such that $\left|\frac{\lambda-\lambda_n^\pm}{\lambda-n\pi} - 1\right| \leq 1/2$ for any $\lambda \in \mathbb{C} \setminus \bigcup_{m\in\mathbb{Z}} B_{m\pi}^{\delta,\eta}$, and $|n| > P'$. It then follows that for any $\lambda \in \mathbb{C} \setminus \bigcup_{m\in\mathbb{Z}} B_{m\pi}^{\delta,\eta}$,

$$\left|\log\left(\prod_{|n|>P'} \frac{\lambda - \lambda_n^\pm}{\lambda - n\pi}\right)\right| \leq \sum_{|n|>P'} \left|\log\left(1 + \frac{n\pi - \lambda_n^\pm}{\lambda - n\pi}\right)\right| \leq \sum_{|n|>P'} \left|\frac{n\pi - \lambda_n^\pm}{\lambda - n\pi}\right| < \infty.$$

In view of the Counting Lemma 1.3 there then exists $Q' \in \mathbb{Z}_{\geq -1}$ with $Q' \geq P'$ such that for any $\lambda \in \mathbb{C} \setminus \bigcup_{m\in\mathbb{Z}} B_{m\pi}^{\delta,\eta}$,

$$\prod_{|n|>Q'} \frac{(\lambda - \lambda_n^+)(\lambda - \lambda_n^-)}{(\lambda - n\pi)^2} \in D^{\varepsilon/4}(1), \tag{2.7}$$

and all periodic eigenvalues $\lambda_n^\pm$ with $|n| \leq Q'$ lie inside $\mathrm{Box}_{Q'}$. Now choose $Q \in \mathbb{Z}_{-1}$ with $Q \geq Q'$ so that for any $|\lambda| > \pi Q + \pi/4$ and $|n| \leq Q'$,

$$\frac{\lambda - \lambda_n^\pm}{\lambda - n\pi} = 1 + \frac{n\pi - \lambda_n^\pm}{\lambda - n\pi} \in D^a(1),$$

where $a > 0$ is chosen so that

$$\prod_{|n|\leq Q'} \frac{(\lambda - \lambda_n^+)(\lambda - \lambda_n^-)}{(\lambda - n\pi)^2} \in D^{\varepsilon/4}(1).$$

When combined with (2.7) and in view of the product representation of $\Delta^2 - 4$ in Lemma 1.8 (i) one gets

$$\frac{\Delta^2(\lambda, \varphi) - 4}{\Delta^2(\lambda, 0) - 4} = \prod_{n\in\mathbb{Z}} \frac{(\lambda - \lambda_n^+)(\lambda - \lambda_n^-)}{(\lambda - n\pi)^2} \in D^\varepsilon(1)$$

for any $\lambda \in \mathbb{C} \setminus \left(\mathrm{Box}_Q \cup \left(\bigcup_{n\in\mathbb{Z}} \mathrm{Box}_{n\pi}^{\delta,\eta}\right)\right)$, proving the first claim of the lemma. The remaining statements are shown in a similar way. □

For $0 < \eta < 1/4$, $\delta > 0$, introduce for any $n \in \mathbb{Z}$

$$\mathcal{I}_n := \{\lambda = u + iv \in \mathbb{C} \mid (n-1)\pi + \eta \leq u \leq n\pi - \eta,\ |v| > 0\},$$

$$\mathcal{J}_n := \{\lambda = u + iv \in \mathbb{C} \mid n\pi - \eta \leq u \leq n\pi + \eta,\ |v| > 0\}$$

and, for $R \in \mathbb{Z}_{\geq -1}$, the domain $\mathcal{E}_{R,\delta} := \big\{\lambda = u + iv \in \mathbb{C} \,\big|\, |v| \geq \delta\big\} \setminus \mathrm{Box}_R$.

LEMMA 2.5. *Given $\varphi \in iL_r^2$ let $0 < \eta < 1/4$ and $\delta > 0$ be as in Lemma 2.3 and let Q be as in Lemma 2.4. Then there exists $R \in \mathbb{Z}_{\geq -1}$ with $R \geq Q$ so that the following statements hold for any $n \in \mathbb{Z}$: (i) $\Lambda \cap \mathcal{E}_{R,\delta} \cap \mathcal{J}_n$ consists of a graph whose points are represented in the form $n\pi + h_n(v) + iv$, where h_n is a smooth function with* $\lim_{v\to\pm\infty} h_n(v) = 0$. *(ii) $\Lambda \cap \mathcal{I}_n \setminus \mathrm{Box}_R = \emptyset$.*

PROOF. By (2.5)

$$\operatorname{Im}\Delta(u+iv,0) = -2\sin u \sinh v, \qquad \operatorname{Re}\Delta(u+iv,0) = 2\cos u\cosh v, \tag{2.8}$$

and thus $|\Delta(u+iv,0)| = 2\sqrt{\sinh^2 v + \cos^2 u}$. Then for any $u+iv \in \mathcal{I}_n$ with $n \in \mathbb{Z}$ and $|v| \geq \delta$,

$$\frac{(-1)^n}{\sinh v}\operatorname{Im}\Delta(u+iv,\varphi) \geq \frac{(-1)^n \operatorname{Im}\Delta(u+iv,0)}{\sinh v} - \left|\frac{\Delta(u+iv,\varphi)-\Delta(u+iv,0)}{\sinh v}\right|.$$

As by assumption $(n-1)\pi + \eta \leq u \leq n\pi - \eta$ it follows that

$$\frac{(-1)^n \operatorname{Im}\Delta(u+iv,0)}{\sinh v} = (-1)^{n-1} 2\sin u \geq 2\sin\eta.$$

Choose $0 < \varepsilon' < 1/2$ so small that $2\varepsilon'\sqrt{1+\frac{1}{\sinh^2\delta}} \leq \sin\eta$. For this choice of ε', let Q be as in Lemma 2.4. Then for any $u+iv \in \mathcal{I}_n \cap \mathcal{E}_{Q,\delta}$ one has

$$\left|\frac{\Delta(u+iv,\varphi)-\Delta(u+iv,0)}{\sinh v}\right| \leq \left|\frac{\Delta(u+iv,\varphi)}{\Delta(u+iv,0)} - 1\right| \cdot \left|\frac{\Delta(u+iv,0)}{\sinh v}\right|.$$

It follows by Lemma 2.4 and the above choice of ε' that

$$\left|\frac{\Delta(u+iv,\varphi)-\Delta(u+iv,0)}{\sinh v}\right| < \varepsilon' \left|\frac{\Delta(u+iv,0)}{\sinh v}\right| \leq 2\varepsilon'\sqrt{1+\frac{1}{\sinh^2 v}} \leq \sin\eta.$$

Combining the above estimates yields

$$\frac{(-1)^n}{\sinh v}\operatorname{Im}\Delta(u+iv,\varphi) \geq \sin\eta \qquad \forall\, u+iv \in \mathcal{I}_n \cap \mathcal{E}_{Q,\delta}, \tag{2.9}$$

implying that $\Lambda \cap \mathcal{I}_n \cap \mathcal{E}_{Q,\delta} = \emptyset$. Furthermore, note that

$$\begin{aligned} \operatorname{Im}\dot\Delta(u+iv,0) &= \partial_u \operatorname{Im}\Delta(u+iv,0) = -2\cos u\sinh v, \\ \operatorname{Re}\dot\Delta(u+iv,0) &= \partial_u \operatorname{Re}\Delta(u+iv,0) = -2\sin u\cosh v. \end{aligned} \tag{2.10}$$

Arguing as above one then concludes that for any $u+iv \in \mathcal{J}_n \cap \mathcal{E}_{Q,\delta}$,

$$\frac{(-1)^{n+1}}{\sinh v}\operatorname{Im}\dot\Delta(u+iv,\varphi) \geq \cos\eta > 0\,. \tag{2.11}$$

Let us now discuss the sign of $\operatorname{Im}\Delta(u+iv,0)$. For a given $n \in \mathbb{Z}$, let us start at $u+iv \in \mathcal{I}_n$. Now increase u while keeping v fixed. As long as $u+iv$ stays in $\mathcal{I}_n$, the sign of $\operatorname{Im}\Delta(u+iv,0)$ remains constant. When u is increased further, λ moves across $\mathcal{J}_n$ and $\operatorname{Im}\Delta(u+iv,0)$ is a strictly monotone function of u. It changes sign at $u = n\pi$. The sign of $\operatorname{Im}\Delta(u+iv,0)$ for $u+iv \in \mathcal{I}_{n+1}$ again remains constant, but opposite to the one in $\mathcal{I}_n$. We then conclude from (2.11) that $\Lambda \cap \mathcal{E}_{Q,\delta} \cap \mathcal{J}_n$ consists of a graph contained in $\mathcal{E}_{Q,\delta}$ near the line $\{u = n\pi\}$. The points of this graph can be represented in the form $n\pi + h_n(v) + iv$, where h_n is a smooth function. To see that $\lim_{v\to\pm\infty} h_n(v) = 0$, use $\frac{(-1)^n}{\sinh v}\operatorname{Im}\Delta(n\pi + h_n(v) + iv,\varphi) = 0$ together with the mean value theorem to conclude that

$$\Big(\int_0^1 \frac{(-1)^{n+1}}{\sinh v}\operatorname{Im}\dot\Delta(n\pi + th_n(v) + iv,\varphi)\,dt\Big)\cdot h_n(v) = (-1)^{n+1}\frac{\operatorname{Im}\Delta(n\pi+iv,\varphi)}{\sinh v}.$$

Hence (2.11) implies that

$$\cos\eta\cdot|h_n(v)| \leq \frac{|\operatorname{Im}\Delta(n\pi+iv,\varphi)|}{|\sinh v|}.$$

By Lemma 2.4 it follows that for any $0 < \varepsilon'' < \varepsilon'$ there exists $Q' \in \mathbb{Z}_{-1}$ with $Q' \geq Q$ so that for any $|v| \geq Q'\pi + \pi/2$,

$$|\Delta(n\pi + iv, \varphi) - \Delta(n\pi + iv, 0)| \leq \varepsilon''|\Delta(n\pi + iv, 0)|.$$

As $\operatorname{Im}\Delta(n\pi + iv, 0) = 0$ one has

$$|\operatorname{Im}\Delta(n\pi + iv, \varphi)| \leq \varepsilon''|\Delta(n\pi + iv, 0)|,$$

and using that $\frac{|\Delta(n\pi+iv,0)|}{|\sinh v|} = 2\sqrt{1 + \frac{1}{\sinh^2 v}}$ one then concludes by combining the inequalities derived above that

$$\cos\eta \cdot |h_n(v)| \leq 2\varepsilon''\sqrt{1 + \frac{1}{\sinh^2 v}} \qquad \forall |v| > Q'\pi + \pi/2.$$

As $0 < \varepsilon'' < \varepsilon'$ can be chosen arbitrarily small, one concludes that

$$\lim_{v \to \pm\infty} |h_n(v)| = 0,$$

proving item (i). To show (ii) note that we already have proved that $\Lambda \cap \mathcal{E}_{Q,\delta} \cap \mathcal{I}_n = \emptyset$. Hence it suffices to show that there exists $R \in \mathbb{Z}_{\geq -1}$ with $R \geq Q$ so that for any $n \in \mathbb{Z}$, $\Lambda \cap \mathcal{I}_n \cap \{0 < |v| < \delta\} \setminus \mathrm{Box}_R = \emptyset$. First note that by the Cauchy-Riemann equation $\partial_v \operatorname{Im}\Delta(u + iv, \varphi) = \operatorname{Re}\dot{\Delta}(u + iv, \varphi)$, and hence

$$(-1)^n\partial_v \operatorname{Im}\Delta(u + iv, \varphi) \geq (-1)^n\partial_v \operatorname{Im}\Delta(u + iv, 0) - |\dot{\Delta}(u + iv, \varphi) - \dot{\Delta}(u + iv, 0)|.$$

By (2.8),

$$\partial_v \operatorname{Im}\Delta(u + iv, 0) = -2\sin u \cosh v,$$

implying that $(-1)^n\partial_v \operatorname{Im}\Delta(u + iv, 0) \geq 2\sin\eta$ for $(n-1)\pi + \eta \leq u \leq n\pi - \eta$ and $|v| \leq \delta$. Choose $0 < \varepsilon < \varepsilon'$ so that $2\varepsilon\sqrt{\sinh^2\delta + \sin^2\eta} \leq \sin\eta$, and for this choice of ε let $R \geq Q$ be a number as in Lemma 2.4. One then concludes from (2.10) that

$$|\dot{\Delta}(u + iv, \varphi) - \dot{\Delta}(u + iv, 0)| \leq \varepsilon|\dot{\Delta}(u + iv, 0)| \leq 2\varepsilon\sqrt{\sinh^2\delta + \sin^2\eta} \leq \sin\eta$$

for any $|v| \leq \delta$ and $(n-1)\pi + \eta \leq u \leq n\pi - \eta$ with $u + iv \notin \mathrm{Box}_R$. Combining all these estimates leads to $(-1)^n\partial_v \operatorname{Im}\Delta(u + iv, \varphi) \geq \sin\eta$ for $u + iv$ in $\mathcal{I}_n \cap \{|v| \leq \delta\} \setminus \mathrm{Box}_R$ with $n \in \mathbb{Z}$ arbitrary. Hence, $\Lambda \cap \mathcal{I}_n \cap \{0 < |v| < \delta\} \setminus \mathrm{Box}_R = \emptyset$ for any $n \in \mathbb{Z}$. $\square$

LEMMA 2.6. *For an arbitrary element $\varphi \in iL^2_r$, the following statements hold. (i) $\mathbb{C} \setminus \Lambda$ admits no connected component with compact closure. (ii) Let $G \subseteq \mathbb{C}$ be compact with a piecewise smooth boundary. Then $G \cap \Lambda$ consists of finitely many arcs of finite length, each of which connects two points in $\partial G \cap \Lambda$.*

PROOF. (i) Assume that $\mathbb{C} \setminus \Lambda$ admits a connected component $\mathcal{D}$, with compact closure. As Λ is a closed subset, $\mathcal{D}$ is open and $\partial\mathcal{D} \subseteq \Lambda$. By Lemma 2.2, $\partial\mathcal{D} \subseteq \Lambda$ is a union of smooth curves. As $\Delta_2(u, v) = \operatorname{Im}\Delta(u + iv)$ is harmonic, it follows from the maximum principle that Δ_2 vanishes identically on $\mathcal{D}$, contradicting the assumption $\mathcal{D} \subseteq \mathbb{C} \setminus \Lambda$.

(ii) For any $\lambda \in G \cap \Lambda$ denote by $U_\lambda \subseteq \mathbb{C}$ the open neighborhood of λ in Lemma 2.2. As G is compact and Λ is closed, the open cover $(U_\lambda)_{\lambda \in G \cap \Lambda}$ of $G \cap \Lambda$ admits a finite subcover. It follows that $\Lambda \cap G$ consists of finitely many arcs of finite length and that these arcs are either closed or end in ∂G. Furthermore, by (i), $G \cap \Lambda$ does not contain any closed curve. Thus it follows that each of these arcs begins and ends at points in $\partial G \cap \Lambda$. $\square$

THEOREM 2.7. *For any $\varphi \in iL^2_r$ there exists $R \in \mathbb{Z}_{\geq -1}$ so that the following statements hold. (i) The set $\Lambda \cap \{|\operatorname{Re}\lambda| < R\pi + \pi/4\}$ is the union of the open interval $(-(R\pi+\pi/4), R\pi+\pi/4)$ and $2R+1$ smooth curves $(\mathcal{C}_k)_{|k|\leq R}$. (These curves are listed in an arbitrary way.) Two curves can meet at most once. The points where some of the $\mathcal{C}_k$'s intersect are zeros of $\dot{\Delta}(\lambda)$, and the number of arcs intersecting in such a point equals the order of the zero of $\dot{\Delta}(\lambda)$ plus one. If some of the $\mathcal{C}_k$'s or if some of the $\mathcal{C}_k$'s and the real axis intersect in a point, they meet at equal angles. Each of these curves enters and leaves Box_R precisely once. Outside Box_R these curves are graphs above $\{iv \,|\, |v| \geq R\pi + \pi/4\}$. For any $|k| \leq R$, precisely one of these curves is asymptotically close to $\mathcal{L}^+_k := \{k\pi + iv \,|\, v \in \mathbb{R},\, v \geq 0\}$ and a possibly different one is asymptotically close to $\mathcal{L}^-_k := \{k\pi + iv \,|\, v \in \mathbb{R},\, v \leq 0\}$. (ii) The set $\Lambda \cap \{|\operatorname{Re}\lambda| \geq R\pi + \pi/4\}$ is the union of the two half lines $(-\infty, -(R\pi + \pi/4)]$, $[R\pi + \pi/4, \infty)$ with a disjoint union of smooth curves $(\mathcal{C}_k)_{|k|>R}$. For any $|k| > R$, $\mathcal{C}_k$ is a graph above the vertical line $\{\operatorname{Re}\lambda = k\pi\} = \mathcal{L}^-_k \cup \mathcal{L}^+_k$ and asymptotically close to this line. It crosses the real axis at a right angle in $\dot{\lambda}_n$. The number R can be chosen locally uniformly in φ.*

PROOF. Let δ and η be as in Lemma 2.3 and let R be as in Lemma 2.5. As, by definition, the vertical strips $\mathcal{I}_k$ are at a distance $0 < \eta < 1/4$ from all vertical lines $\{\lambda = n\pi + iv |\, v \in \mathbb{R}\}$, $n \in \mathbb{Z}$, it follows that

$$\mathcal{G}_{\pm R} := \{\pm(R\pi + \pi/4) + iv \mid 0 < |v| \leq R\pi + \pi/4\} \cap \Lambda \subseteq \mathcal{I}_{\pm(R+1)}.$$

Hence by Lemma 2.5 (ii), $\mathcal{G}_{\pm R} = \emptyset$. Furthermore, by Lemma 2.5 (i), $\partial\mathrm{Box}_R \cap \Lambda$ consists of $4R+2$ points $(p^\pm_k)_{|k|\leq R}$ with $p^\pm_k \in \mathcal{J}_k \cap \{\pm v = R\}$, and for any $|k| \leq R$, the intersection of Λ with the strip $\mathcal{J}_k \setminus \mathrm{Box}_R$ is a graph over the line $\{k\pi + iv \mid \pm v \geq R\}$ which is asymptotically close to it. Furthermore, by Lemma 2.6 applied to the closure G of Box_R, there are $2R+1$ smooth curves $\mathcal{C}_\ell \subseteq \Lambda \cap \mathrm{Box}_R$, $|\ell| \leq R$, each of which connects precisely two points in $\{p^\pm_k \mid |k| \leq R\}$. By Lemma 2.6, two curves can meet at most once. It follows from Lemma 2.2 that the points where the arcs $\mathcal{C}_k$ intersect are zeros of $\dot{\Delta}(\lambda)$ and that the number of arcs intersecting at such a point equals the order of the zero of $\dot{\Delta}(\lambda)$ plus one. Combined with Lemma 2.2, item (i) then follows. Item (ii) is a consequence of Lemma 2.3 and Lemma 2.5. Going through the arguments of the above proof and the ones of the lemmas used in the proof, one sees that R can be chosen locally uniformly in $\varphi \in iL^2_r$. □

PROOF OF THEOREM 1.9. The stated result follows directly from Theorem 2.7. □

3. Examples

As an example we compute $\operatorname{spec}_{\mathbb{R}} L(\varphi_a)$ for the constant potential $\varphi_a = (a, -\bar{a})$, $a \in \mathbb{C}$. In a straightforward way one verifies the following

LEMMA 3.1. *For any $a \in \mathbb{C}$,*

$$M(x, \lambda, \varphi_a) = \begin{pmatrix} \cos(\kappa x) - i\lambda \frac{\sin(\kappa x)}{\kappa} & ia \frac{\sin(\kappa x)}{\kappa} \\ i\bar{a} \frac{\sin(\kappa x)}{\kappa} & \cos(\kappa x) + i\lambda \frac{\sin(\kappa x)}{\kappa} \end{pmatrix}, \tag{3.1}$$

where $\kappa \equiv \kappa(\lambda) = \sqrt{\lambda^2 + |a|^2}$.

REMARK 3.2. Note that κ only depends on $|a|$ and that the right hand side of (3.1) does not depend on the choice of the sign of the root $\sqrt{\lambda^2+|a|^2}$, as cosine is an even function whereas sine is odd. Furthermore, note that the right hand side is well defined even at $\kappa=0$, as $\frac{\sin(\kappa x)}{\kappa}=x+O(\kappa^2)$.

It follows from Lemma 3.1 that $\Delta(\lambda)=2\cos\kappa(\lambda)$. Write $\kappa(\lambda)=\kappa_1(\lambda)+i\kappa_2(\lambda)$, where $\kappa_1(\lambda)$ denotes the real and $\kappa_2(\lambda)$ the imaginary part of $\kappa(\lambda)$. Then

$$\cos\kappa(\lambda)=\cos\kappa_1(\lambda)\cdot\cosh\kappa_2(\lambda)-i\sin\kappa_1(\lambda)\cdot\sinh\kappa_2(\lambda). \tag{3.2}$$

PROPOSITION 3.3. *For any $a\in\mathbb{C}$, $spec_{\mathbb{R}}L(\varphi_a)=\mathbb{R}\cup i[-|a|,|a|]$ and $\Lambda(\varphi_a)=\mathbb{R}\cup\bigcup_{k\in\mathbb{Z}}\Lambda_k$ where $\Lambda_0=i\mathbb{R}$, and for $k\in\mathbb{Z}_{>0}$ with $k^2\pi^2\le|a|^2$,*

$$\Lambda_k:=\left\{\lambda=u+iv\,\middle|\,v\ge\sqrt[+]{|a|^2-k^2\pi^2},\ u=\pm k\pi\Big(1-\frac{|a|^2}{v^2+k^2\pi^2}\Big)^{1/2}\right\},$$
$$\Lambda_{-k}:=-\Lambda_k,$$

whereas for $k\in\mathbb{Z}_{>0}$ with $k^2\pi^2>|a|^2$, $\Lambda_{\pm k}$ are graphs above the imaginary axis,

$$\Lambda_k:=\left\{\lambda=u+iv\,\middle|\,v\in\mathbb{R},\ u=k\pi\Big(1-\frac{|a|^2}{v^2+k^2\pi^2}\Big)^{1/2}\right\},\qquad\Lambda_{-k}:=-\Lambda_k\,.$$

PROOF. In view of (3.2), $\Delta(\lambda)$ is real valued iff $\kappa_1(\lambda)\in\pi\mathbb{Z}$ or $\kappa_2(\lambda)=0$. In addition, $-2\le\Delta(\lambda)\le2$ iff $\kappa_2(\lambda)=0$. Note that $\kappa_2(\lambda)=0$ iff $\lambda^2+|a|^2\in\mathbb{R}_{\ge0}$ or $\lambda\in\mathbb{R}\cup i[-|a|,|a|]$, showing that $\operatorname{spec}_{\mathbb{R}}L(\varphi_a)=\mathbb{R}\cup i[-|a|,|a|]$. If $\kappa_1(\lambda)=\pm k\pi$, $k\in\mathbb{Z}_{\ge0}$, then $\Delta(\lambda)=2\cos\kappa_1(\lambda)\cosh\kappa_2(\lambda)=(-1)^k2\cosh\kappa_2(\lambda)$. Hence $(-1)^k\Delta(\lambda)\ge 2$. Taking real and imaginary parts of $(\pm k\pi+i\kappa_2(\lambda))^2=(\kappa_1(\lambda)+i\kappa_2(\lambda))^2=(u+iv)^2+|a|^2$ leads to the identities

$$k^2\pi^2-\kappa_2^2(\lambda)=u^2-v^2+|a|^2\quad\text{and}\ \pm k\pi\kappa_2(\lambda)=uv. \tag{3.3}$$

First let us consider the case where $k=0$. Then $\kappa(\lambda)=i\kappa_2(\lambda)$, which says that $\lambda^2+|a|^2\le0$ or, equivalently, $\lambda\in i\mathbb{R}\setminus i(-|a|,|a|)$. In the case $k>0$, (3.3) leads to $\kappa_2^2(\lambda)=(uv/k\pi)^2$, yielding $k^2\pi^2-\frac{u^2v^2}{k^2\pi^2}=u^2-v^2+|a|^2$. Solving for u^2 one gets $u^2=k^2\pi^2\left(1-\frac{|a|^2}{v^2+k^2\pi^2}\right)$ with the restriction that $v^2\ge|a|^2-k^2\pi^2$. In particular, if $k^2\pi^2>|a|^2$,

$$u=\pm k\pi\Big(1-\frac{|a|^2}{v^2+k^2\pi^2}\Big)^{1/2}.$$

The case $0<k^2\pi^2\le|a|^2$ is treated in a similar way. Altogether, these computations imply the claimed statements. □

Note that on the imaginary axis, $\operatorname{Im}\Delta$ vanishes for *any* constant potential. If $0<n\pi<|a|<(n+1)\pi$, then $i\mathbb{R}$ meets $2n$ arcs contained in the set $\{\operatorname{Im}\Delta|_{\varphi_a}=0\}$.

References

[1] M. Ablowitz and Y. Ma, *The periodic cubic Schrödinger equation*, Studies in Applied Mathematics **65** (1981), 113–158. MR628138 (84e:58059)

[2] P. Deift, *Some open problems in random matrix theory and the theory of integrable systems*, in: Integrable Systems and Random Matrices: in Honor of Percy Deift, J. Baik at al., Editors, Contemporary Mathematics **458** (2008), AMS, Providence, 419–420. MR2411922 (2010a:37146)

[3] B. Grébert and T. Kappeler, *Symmetries of the Nonlinear Schrödinger equation*, Bull. Soc. Math. France **130** (2002), no. 4, 603–618. MR1947455 (2003j:35297)

[4] B. Grébert, T. Kappeler, and J. Pöschel, *Normal form theory for the nonlinear Schrödinger equation*, arXiv.org/abs/0907.3938v1. To appear in EMS series of Lectures in Mathematics, EMS Publishing House.

[5] T. Kappeler, P. Lohrmann, and P. Topalov, *Generic nonself-adjoint Zakharov-Shabat operators*, Math. Ann. **359** (2014), no. 1-2, 427–470. MR3201903

[6] T. Kappeler, P. Lohrmann, P. Topalov, and N.T. Zung, *Birkhoff coordinates for the focusing NLS-equation*, Comm. Math. Phys. **285** (2009), 1087–1107. MR2470917 (2010b:35435)

[7] T. Kappeler and J. Pöschel, *KdV & KAM* **45**, Ergeb. der Math. und ihrer Grenzgeb., Springer Verlag, 2003. MR1997070 (2004g:37099)

[8] Y. Li and D. McLaughlin, *Morse and Melnikov functions for NLS PDEs*, Comm. Math. Phys. **162** (1994), 175–214. MR1272771 (95g:58170)

[9] L. Nirenberg, *An extended interpolation inequality*, Ann. Sc. Norm. Sup. di Pisa, Ser. 6 **20** (1966), 733–737. MR0208360 (34:8170)

[10] V. Tkachenko, *Non-selfadjoint periodic Dirac operators with finite band spectra*, Integral Equations Operator Theory **36** (2000), no. 3, 325–348. MR1753422 (2001f:47079)

[11] V. Tkachenko, *Non-selfadjoint periodic Dirac operators*, Operator theory, system theory and related topics (Beer-Sheva/Rehovot, 1997), Oper. Theory Adv. Appl. **123** (2001), 485–512. MR1821926 (2002h:34185)

[12] V.E. Zakharov and A.B. Shabat, *A scheme for integrating nonlinear equations of mathematical physics by the method of the inverse scattering problem I*, Functional Anal. Appl. **8** (1974), 226–235.

Institute of Mathematics, University of Zürich, Winterthurerstrasse 190, CH-8057 Zürich

E-mail address: `thomas.kappeler@math.uzh.ch`

Department of Mathematics and Applications, University of Naples Federico II, Via Cintia, Monte S. Angelo, I-80126 Naples

E-mail address: `philipp.lohrmann@yahoo.com`

Department of Mathematics, Northeastern University, 360 Huntington Avenue, Boston, MA 02115

E-mail address: `p.topalov@neu.edu`

Amer. Math. Soc. Transl.
Volume **233**, 2014

Homogenized Model of Oscillations of Elastic Medium with Small Caverns Filled with Viscous Incompressible Fluid

E. Ya. Khruslov, M. V. Goncharenko, and N. K. Radyakin

ABSTRACT. The initial boundary value problem of the nonstationary vibration of elastic medium with a great number of small caverns filled with viscous incompressible fluid is considered. The asymptotic behavior of the solution is studied as the diameters of the caverns tends to zero and the number of caverns tends to infinity. The fluid viscosity depends on the characteristic scale of the microstructure ε such that $\mu_\varepsilon \to \mu \geq 0$ as $\varepsilon \to 0$. The homogenized equation that describes the first term of asymptotics is obtained in two cases. The first one is $\mu_\varepsilon \to 0$, and the second is $\mu_\varepsilon \to \mu > 0$. This equation is a model of wave propagation in medium such as wet soil, rocks, and biological tissues.

1. Introduction

We consider the composite medium, which consists of elastic and liquid phases. Namely, let Ω be a fixed domain in $\mathbb{R}^3$ with boundary $\partial\Omega$ and G_ε^j be a disjoint subdomains in Ω with smooth boundaries ∂G_ε^j that located in the whole volume of Ω. The sizes and the number of subdomains depend on a small parameter ε such that the diameters of G_ε^j tends to zero and the number of subdomains tends to infinity as $\varepsilon \to 0$. Assume that the domain $\Omega_\varepsilon = \Omega \setminus \bigcup_{j=1}^{N(\varepsilon)} \bar{G}_\varepsilon^j$ be an elastic medium and subdomains G_ε^j (the caverns in domain Ω) are filled with a viscous incompressible liquid.

Nonstationary oscillations of such a medium are described by the following system of equations

$$(1.1)\qquad \rho_s \frac{\partial^2 u_\varepsilon}{\partial t^2} - \sum_{n,p,q,r=1}^{3} \frac{\partial}{\partial x_n}(a_{npqr}\gamma_{qr}[u_\varepsilon])e^p = 0, x \in \Omega_\varepsilon$$

$$(1.2)\qquad \rho_f \frac{\partial v_\varepsilon}{\partial t} - \mu_\varepsilon \Delta v_\varepsilon = \nabla p_\varepsilon, \quad \mathrm{div} v_\varepsilon = 0, x \in G_\varepsilon = \bigcup_j G_\varepsilon^j$$

$$(1.3)\qquad \frac{\partial u_\varepsilon}{\partial t} = v_\varepsilon, x \in \partial G_\varepsilon$$

$$(1.4)\qquad T_f[v_\varepsilon] + p_\varepsilon \nu = T_s[u_\varepsilon], x \in \partial G_\varepsilon.$$

2010 *Mathematics Subject Classification.* Primary 35Q74.

Here $u_\varepsilon = u_\varepsilon(x,t)$ is an elastic displacement vector, $v_\varepsilon = v_\varepsilon(x,t)$ and p_ε are a velocity vector and a pressure of the liquid, $\rho_s, \rho_f = \text{const}$ are the densities of elastic and liquid phases, respectively; μ_ε is a dynamic viscosity of a liquid; e^p is the unit vector directed along an x_p axis,

$$\gamma_{qr}[u] = \frac{1}{2}\left(\frac{\partial u_q}{\partial x_r} + \frac{\partial u_r}{\partial x_q}\right)$$

are components of the deformation tensor of the medium, a_{npqr} are components of the elasticity tensor, which has the following properties of symmetry $a_{iklm} = a_{kilm} = a_{lmik} = a_{ikml}$ and is positively defined [**1**]

$$\sum_{n,p,q,r=1}^{3} a_{npqr} t_{np} t_{qr} \geq C \sum_{n,p=1}^{3} |t_{np}|^2, \quad C > 0, \forall \{t_{np}\}_{n,p=1}^{3}.$$

We denote by $T_f[v]$ the viscous stress vector on the liquid boundary ∂G_ε^j :

$$T_f[v] = 2\mu_\varepsilon \sum_{i,k=1}^{3} \gamma_{ik}[v] \nu_i e^k,$$

and by $T_s[u]$ the stress vector in the elastic medium on ∂G_ε^j:

$$T_s[u] = \sum_{n,p,q,r=1}^{3} a_{npqr} \gamma_{qr}[u] \nu_n e^q, \tag{1.5}$$

ν is an outer normal vector on ∂G_ε^j.

Condition (1.3) means the equivalence of the velocities of liquid and elastic medium on the interface, and condition (1.4) means the coincidence of stress vectors.

Let us supplement the system by initial and boundary conditions

$$u_\varepsilon = 0, \quad x \in \partial\Omega, \tag{1.6}$$

$$u_\varepsilon(x,0) = U_\varepsilon^0, \quad \frac{\partial u_\varepsilon}{\partial t} = U_\varepsilon^1(x), \quad x \in \Omega_\varepsilon, \tag{1.7}$$

$$v_\varepsilon(x,0) = V_\varepsilon^1(x), \quad x \in G_\varepsilon. \tag{1.8}$$

We assume that the initial vector-functions satisfy the following conditions

$$V_\varepsilon^1 \in W_2^1(G_\varepsilon), \quad \|V_\varepsilon^1\|_{W_2^1(\Omega_\varepsilon)} < C, \quad \operatorname{div} V_\varepsilon^1 = 0;$$

$$U_\varepsilon^k \in W_2^1(\Omega_\varepsilon), \|U_\varepsilon^k\|_{W_2^1(\Omega)} < C, k = 0, 1; \quad U_\varepsilon^0 = 0, x \in \partial\Omega; U_\varepsilon^1 = V_\varepsilon^1, x \in \partial G_\varepsilon;$$

$$\int_{\partial G_\varepsilon^j} U_{\varepsilon n}^0 ds = 0, \tag{1.9}$$

where C does not depend on ε. Condition (1.9) appears due to the incompressibility of liquid in caverns.

This problem describes wave propagation in a wet medium. We consider the simplest one: elastic medium with caverns, filled with liquid. Wet soil, seabed, porous rocks, and some biological tissues can be considered as examples of such a medium.

The goal of the paper is to derive homogenized equations that describe the oscillations of such a "composite" medium. We obtain the "macroscopic" model of the wet elastic medium. These questions have been studied by mechanics and mathematicians; see, for example, [2]–[7]. In particular, the homogenized equations in [7] were obtained in the case where the initial displacement $U^0_\varepsilon \equiv 0$ and the viscosity of the fluid in the pores did not depend on the characteristic scale of the microstructure ε .

In this paper we will obtain the homogenized equations when $U^0_\varepsilon \neq 0$ and the fluid viscosity μ_ε depends on ε in such a way that $\mu_\varepsilon \to \mu \geq 0$ as $\varepsilon \to 0$.

We show that the homogenized equation has the following form:

If $\mu > 0$, then

$$\rho\frac{\partial^2 u}{\partial t^2} - \sum_{n,p,q,r=1}^{3} \frac{\partial^2}{\partial x_n \partial t}\left(\int_0^t A_{npqr}(x,t-\tau)\gamma_{qr}[u(x,\tau)]d\tau\right) e^p = 0;$$

if $\mu = 0$, then

$$\rho\frac{\partial^2 u}{\partial t^2} - \sum_{n,p,q,r=1}^{3} \frac{\partial}{\partial x_n}\left(A_{npqr}(x)\gamma_{qr}[u(x,t)]\right) e^p = 0.$$

2. Formulation of the problem and the main result

We assume that the location of the caverns is locally periodic. This means that the caverns G^j_ε are of the same shape and located in periodically arranged parallelepipeds. The diameters, orientations, and coordinates of the mass centers of the neighboring caverns differ by a small value. Namely, we assume the space $\mathbb{R}^3$ divided into parallelepipeds $\Pi^j_\varepsilon = \{x \in \mathbb{R}^3 : |x_i - x^j_i| \leq (\theta_i\varepsilon)/2, i = 1,2,3\}$ with centers x^j and side length $\theta_i\varepsilon$. The sides of parallelepipeds are oriented along the coordinate axes. Each parallelepiped in the domain Ω contains the cavern G^j_ε. The set G^j_ε is a homothetic contraction and a rotation of a fixed set $G \in \mathbb{R}^3$ with unit diameter, the mass center in origin and the smooth boundary ∂G. The diameters of the sets G^j_ε are $d^j_\varepsilon = d(x^j)\varepsilon$, the mass centers are located in the points $x^j + a(x^j)\varepsilon$, and orientations are defined by the rotation operators $P(x^j)$. So we have

$$G^j_\varepsilon = \left\{x \in \Pi^j_\varepsilon : P^{-1}(x^j)\frac{x - x^j - a(x^j)\varepsilon}{d(x^j)\varepsilon} \in G\right\}. \tag{2.1}$$

We assume that the function $d(y)$, vector-function $a(y)$ and matrix $P(y)$ are continuously differentiable and the inequalities hold

$$\max_{y\in\Omega}(|d(y)| + |a(y)|) < \min_i \frac{\theta_i(1-2\delta)}{2}, \quad \delta > 0 \tag{2.2}$$

Sets G^j_ε do not overlap, due to (2.1) and (2.2). Their diameters and the distance between the neighboring ones are of the order ε.

In this paper we assume that the viscosity of the liquid in the caverns $\mu_\varepsilon = \mu\varepsilon^\alpha$ $(\mu > 0,\ 0 \leq \alpha < 2)$.

Let us introduce the notation

$$\Pi = \left\{\xi \in \mathbb{R}^3 : |\xi_i| < \frac{\theta_i}{2}, i = 1,2,3\right\},$$

$$G^y = \left\{\xi \in \Pi : P^{-1}(y)\frac{\xi - a(y)}{d(y)} \in G, y \in \Omega\right\}.$$

In the case of $\alpha = 0$ ($\mu_\varepsilon = \mu$), we consider the following boundary value problem in Π (the cell problem)

$$\mu\Delta\tilde{v}^{np}(x,\lambda) = \nabla\tilde{p}, \quad \mathrm{div}\tilde{v}^{np}(x,\lambda) = 0, \quad x \in G^y, \tag{2.3}$$

$$\sum_{i,k,l,m=1}^{3} \frac{\partial}{\partial x_i}\left(a_{iklm}\gamma_{lm}[\tilde{w}^{np}]\right) e^k = 0, \quad x \in \Pi \setminus G^y, \tag{2.4}$$

$$\tilde{v}^{np} = \tilde{w}^{np}, \quad x \in \partial G^y, \tag{2.5}$$

$$\mu\sum_{i,k=1}^{3} \gamma_{ik}[\tilde{v}^{np}]\nu_i e^k - \tilde{p}\nu = \frac{1}{\lambda}\sum_{i,k,l,m=1}^{3} a_{iklm}\gamma_{lm}[\tilde{w}^{np}]\nu_i e^k, \quad x \in \partial G^y, \tag{2.6}$$

$$\tilde{u}^{np} - \phi^{np} \in H^1_{\mathrm{per}}(\Pi), \tag{2.7}$$

where $\tilde{u}^{np} = \tilde{v}^{np}\chi_{G^y} + \tilde{w}^{np}(1-\chi_{G^y})$, χ_{G^y} is a characteristic function of the domain G^y, vector-function $\phi^{np}(x)$ is defined by equality $\phi^{np} = \frac{1}{2}(x_n e^p + x_p e^n)$, λ is a complex parameter, $\mathrm{Re}\lambda > 0$. We define by $H^1_{\mathrm{per}}(\Pi)$ restriction on Π of the closure in $H^1(\mathbb{R}^3, \mathrm{loc})$ of the set of smooth Π-periodic vector-functions in $\mathbb{R}^3$.

There exists a unique solution to the problem (2.3)–(2.7) (up to an additive constant vector) [**7**]. We define the tensor $\{A_{npqr}(y,\lambda)\}$ as follows:

$$A_{npqr}(y,\lambda) = 2\mu\int_{G^y}\sum_{i,k=1}^{3}\gamma_{ik}[\tilde{v}^{np}]\gamma_{ik}[\tilde{v}^{qr}]dx + \frac{1}{\lambda}\int_{\Pi\setminus G^y}\sum_{i,k,l,m=1}^{3} a_{iklm}\gamma_{ik}[\tilde{w}^{np}]\gamma_{lm}[\tilde{w}^{qr}]dx. \tag{2.8}$$

The tensor $\{A_{npqr}(\cdot,\lambda)\} \in L_\infty(\Omega)$ is positively defined as $\lambda > 0$ and symmetric $A_{iklm} = A_{kilm} = A_{lmik} = A_{ikml}$. It is analytical on λ in the half-plane $\mathrm{Re}\lambda > 0$ and $\overline{A_{npqr}(y,\lambda)} = A_{npqr}(y,\bar{\lambda})$. The following estimate holds [**7**]:

$$|A_{npqr}(y,\lambda)| \le C|\lambda|^{-1}. \tag{2.9}$$

Therefore, there exists the inverse Laplace transform:

$$A_{npqr}(y,t) = \frac{1}{2\pi i}\int_{\sigma-i\infty}^{\sigma+i\infty} A_{npqr}(y,\lambda)e^{\lambda t}d\lambda, \tag{2.10}$$

such that $A_{npqr}(y,t) = 0$ as $t < 0$, $A_{npqr}(y,t) \in L_2([0,T]; L_\infty(\Omega))$ ($\forall T < \infty$) and $\mathrm{Im}A_{npqr}(y,t) = 0$.

In the case of $0 < \alpha < 2$, the cell problem has the form:

$$\sum_{i,k,l,m=1}^{3} \frac{\partial}{\partial x_i}\left(a_{iklm}\gamma_{lm}[\tilde{w}^{np}]\right) e^k = 0, \quad x \in \Pi \setminus G^y, \tag{2.11}$$

$$\sum_{i,k,l,m=1}^{3} a_{iklm}\gamma_{lm}[\tilde{w}^{np}]\nu_i e^k = \tilde{p}\nu, \quad \tilde{p} = const, \quad x \in \partial G^y, \tag{2.12}$$

$$\int_{\partial G_\varepsilon} \tilde{w}^{np}\cdot\nu dS = 0, \tag{2.13}$$

$$\hat{w}^{np} - \phi^{np} \in H^1_{\mathrm{per}}(\Pi), \tag{2.14}$$

where $\hat{w}^{np}$ is an extension of $\tilde{w}^{np}$ on G^y such that $\hat{w}^{np} \in H^1(\Pi)$.

There exists a unique solution to the problem (2.11)–(2.14) (up to an additive constant vector). We define the tensor $\{A^0_{npqr}(y)\}$:

$$A^0_{npqr}(y) = \int\limits_{\Pi \setminus G^y} \sum_{i,k,l,m=1}^{3} a_{iklm} \gamma_{ik}[\tilde{w}^{np}] \gamma_{lm}[\tilde{w}^{qr}] dx. \tag{2.15}$$

The tensor $\{A^0_{npqr}(y)\} \in L_\infty(\Omega)$ is symmetric $A_{iklm} = A_{kilm} = A_{lmik} = A_{ikml}$ and positively defined.

It can be shown that for sufficiently smooth boundary ∂G_ε and functions $a(y)$, $d(y)$, $P(y)$ tensors $\{A_{npqr}(y, \lambda)\}$ and $A^0_{npqr}(y)$ are continuously differentiable in y.

To formulate the main result, we introduce the displacement vector-function of the medium

$$\hat{u}_\varepsilon(x,t) = u_\varepsilon(x,t)\chi_{\Omega_\varepsilon} + \left(\int\limits_0^t v_\varepsilon(x,\tau) d\tau + P_\varepsilon U^0_\varepsilon \right) \chi_{G_\varepsilon}, \tag{2.16}$$

where $\chi_{\Omega_\varepsilon}$, χ_{G_ε} are the characteristic functions of domains Ω_ε and G_ε respectively, and P_ε is an extension operator on G_ε defined in Lemma 3.1 in Section 3.

THEOREM 2.1. *Let $\mu_\varepsilon = \mu\varepsilon^\alpha$, $0 < \alpha < 2$ and the initial data of the problem (1.1)–(1.8) converge to the vector-functions U^0, U^1 and $V^1 \in W^1_2(\Omega)$ in the following sense*

(2.17)
$$\lim_{\varepsilon \to 0} \int\limits_{\Omega_\varepsilon} |U^0_\varepsilon - U^0|^2 dx = 0, \quad \lim_{\varepsilon \to 0} \int\limits_{\Omega_\varepsilon} |U^1_\varepsilon - U^1|^2 dx = 0, \quad \lim_{\varepsilon \to 0} \int\limits_{G_\varepsilon} |V^1_\varepsilon - V^1|^2 dx = 0.$$

Then the displacement vector function $\hat{u}_\varepsilon(x,t)$ (2.16) converges in $L_2(\Omega_T)$ ($\Omega_T = \Omega \times [0,T]$) to the vector-function $u(x,t)$ that is the solution of the following initial boundary value problem

$$\rho \frac{\partial^2 u}{\partial t^2} - \sum_{n,p,q,r=1}^{3} \frac{\partial}{\partial x_n} \left(A^0_{npqr}(x) \gamma_{qr}[u(x,t)] \right) e^p = 0, \tag{2.18}$$

$$u(x,t) = 0, \quad x \in \partial\Omega, \tag{2.19}$$

$$u(x,0) = U^0(x), \quad \frac{\partial u}{\partial t}(x,0) = V(x), \tag{2.20}$$

where

$$\rho = \rho(x) = \rho_f \frac{|G^x|}{|\Pi|} + \rho_s \frac{|\Pi \setminus G^x|}{|\Pi|}, \quad V = \frac{\rho_f}{\rho} \frac{|G^x|}{|\Pi|} V^1 + \frac{\rho_s}{\rho} \frac{|\Pi \setminus G^x|}{|\Pi|} U^1. \tag{2.21}$$

The coefficients $A^0_{npqr}(x)$ are defined by (2.15).

THEOREM 2.2. *Let $\mu_\varepsilon = \mu > 0$ and the initial displacement of the problem (1.1)–(1.8) satisfy the condition*

$$U^0_\varepsilon = a^j_\varepsilon + b^j_\varepsilon \times (x - x^j_\varepsilon), \quad x \in \partial G^j_\varepsilon, \tag{2.22}$$

where a^j_ε, b^j_ε are the constant vectors in $\mathbb{R}^3$. The initial data vector functions converge to the vector functions U^0, U^1, and $V^1 \in W^1_2(\Omega)$ as in (2.17) as $\varepsilon \to 0$.

Then the displacement vector function $\hat{u}_\varepsilon(x,t)$ (2.16) converges in $L_2(\Omega_T)$ to the vector function $u(x,t)$ that is the solution of the initial boundary value problem

$$\rho\frac{\partial^2 u}{\partial t^2} - \sum_{n,p,q,r=1}^{3} \frac{\partial^2}{\partial x_n \partial t}\left(\int_0^t A_{npqr}(x,t-\tau)\gamma_{qr}[u(x,\tau)]d\tau\right) e^p = 0, \tag{2.23}$$

$$u(x,t) = 0, \quad x \in \partial\Omega, \tag{2.24}$$

$$u(x,0) = U^0(x), \quad \frac{\partial u}{\partial t}(x,0) = V(x), \tag{2.25}$$

where coefficients $A_{npqr}(x,t)$, $\rho(x)$ and $V(x)$ are defined by (2.8) and (2.21).

REMARK 2.3. Theorem 2.1 is proved for $0 < \alpha < 2$. There is a question: What would happen if $\alpha \geq 2$? It can be expected that in the case $\alpha > 2$ the homogenized equation would have the same form (2.18) with $\rho = \rho_s \frac{|\Pi \setminus G^x|}{|\Pi|}$.

REMARK 2.4. Theorem 2.2 is proved under the assumption that the initial displacement $U^0_\varepsilon(x)$ satisfies the condition (2.22). There is a question: Is this condition essential, or the type of homogenized equation (2.23) will be the same under more general condition (1.9)?

3. Abstract analog of the problem (1.1)–(1.8). The existence of a weak solution

We introduce the displacement vector of fluid $w_\varepsilon = P_\varepsilon U^0_\varepsilon(x) + \int_0^t v_\varepsilon(x,\tau)d\tau$, where P_ε is the extension operator of the vector-functions $u \in W^1_2(\Omega_\varepsilon)$ to the set G_ε constructed according to the following lemma.

LEMMA 3.1. *Let $U_\varepsilon \in W^1_2(\Omega_\varepsilon)$ satisfy the following condition: $\int_{\partial G_\varepsilon} U_\varepsilon \nu = 0$.*

Then there exists an extension operator $P_\varepsilon : W^1_2(\Omega_\varepsilon) \to W^1_2(\Omega)$ such that $\hat{U}_\varepsilon = P_\varepsilon U_\varepsilon \in W^1_2(\Omega)$, $\mathrm{div}\hat{U}_\varepsilon(x) = 0$ for $x \in G_\varepsilon$, $\hat{U}_\varepsilon(x) = U_\varepsilon(x)$ for $x \in \Omega_\varepsilon$ and

$$\|\hat{U}_\varepsilon\|_{W^1_2(\Omega)} \leq C\|U_\varepsilon\|_{W^1_2(\Omega_\varepsilon)}.$$

The constant C does not depend on ε. Moreover, if $U_\varepsilon \in W^2_2(\Omega_\varepsilon)$, then $\hat{U}_\varepsilon \in W^2_2(\Omega)$.

This lemma can be proved by the method of construction of divergence-free fields (see, for example [**8**], Chapter 1, §2) and similarity considerations formulated in [**10**] (Chapter 3, §2).

Using displacement vectors u_ε and w_ε we rewrite the problem (1.1)–(1.8) as follows

$$\rho_s\frac{\partial^2 u_\varepsilon}{\partial t^2} - L[u_\varepsilon] = 0, \quad x \in \Omega_\varepsilon, t > 0, \tag{3.1}$$

$$\rho_f\frac{\partial^2 w_\varepsilon}{\partial t^2} - \mu_\varepsilon\Delta\frac{\partial w_\varepsilon}{\partial t} - \nabla p_\varepsilon = 0, \quad \mathrm{div} w_\varepsilon = 0, \quad x \in G_\varepsilon, t > 0, \tag{3.2}$$

$$w_\varepsilon = u_\varepsilon, \quad x \in \partial G_\varepsilon, t > 0, \tag{3.3}$$

$$T_s[u_\varepsilon] = T_f[\frac{\partial w_\varepsilon}{\partial t}] + p_\varepsilon \nu, \quad x \in \partial G_\varepsilon, t > 0, \tag{3.4}$$

$$u_\varepsilon = 0, \quad x \in \partial\Omega, t > 0, \tag{3.5}$$

$$u_\varepsilon(x,0) = U_\varepsilon^0(x), \quad \frac{\partial u_\varepsilon}{\partial t}(x,0) = U_\varepsilon^1(x), \quad x \in \Omega_\varepsilon, \tag{3.6}$$

$$w_\varepsilon(x,0) = \hat{U}_\varepsilon^0(x), \quad \frac{\partial w_\varepsilon}{\partial t} = V_\varepsilon^1(x), \quad x \in G_\varepsilon, \tag{3.7}$$

where $\hat{U}_\varepsilon^0 = P_\varepsilon U_\varepsilon^0$.

We will rewrite this problem in an operator form. To this end, we introduce the following spaces

$$\mathcal{H}_\varepsilon = L_2(\Omega_\varepsilon) \times L_2(G_\varepsilon) \times L_2(S_\varepsilon), \quad S_\varepsilon = \partial G_\varepsilon.$$

The elements u_ε of the space $\mathcal{H}_\varepsilon$ are the systems of three vector functions $u_\varepsilon = \{u_\varepsilon^\Omega, u_\varepsilon^G, u_\varepsilon^S\}^\top$, $u_\varepsilon^\Omega \in L_2(\Omega_\varepsilon)$, $u_\varepsilon^G \in L_2(G_\varepsilon)$, $u_\varepsilon^S \in L_2(S_\varepsilon)$. The scalar product in $\mathcal{H}_\varepsilon$ is introduced in the standard way:

$$(u_\varepsilon, w_\varepsilon)_{\mathcal{H}_\varepsilon} = (u_\varepsilon^\Omega, w_\varepsilon^\Omega)_{L_2(\Omega_\varepsilon)} + (u_\varepsilon^G, w_\varepsilon^G)_{L_2(G_\varepsilon)} + (u_\varepsilon^S, w_\varepsilon^S)_{L_2(S_\varepsilon)}. \tag{3.8}$$

Let us introduce two subspaces in $\mathcal{H}_\varepsilon$:

$$\mathcal{H}_\varepsilon^{sol} = clos\{u_\varepsilon \in \mathcal{H}_\varepsilon, u_\varepsilon^\Omega \in L_2(\Omega_\varepsilon), u_\varepsilon^G \in J(G_\varepsilon), u_\varepsilon^S = u_{\varepsilon,\nu}^G\},$$

where $J(G_\varepsilon)$ is a space of divergence free in G_ε vector functions $u_\varepsilon \in W_2^1(G_\varepsilon)$, $u_{\varepsilon,\nu}^G$ is a normal component of u_ε^G on S_ε,

$$\mathcal{H}_\varepsilon^{grad} = clos\{u_\varepsilon \in \mathcal{H}_\varepsilon, u_\varepsilon^\Omega = 0, u_\varepsilon^G = \nabla\varphi, u_\varepsilon^S = -\varphi\cdot\nu + \psi\cdot\tau\},$$

where $\varphi \in W^1(G_\varepsilon)$, $\psi \in L_2(S_\varepsilon)$, ν, τ are the normal and tangent vectors on S_ε respectively, and *clos* denotes the closure in $\mathcal{H}_\varepsilon$.

LEMMA 3.2. *The orthogonal decomposition is valid:*

$$\mathcal{H}_\varepsilon = \mathcal{H}_\varepsilon^{\text{sol}} \oplus \mathcal{H}_\varepsilon^{\text{grad}}.$$

PROOF. It is not difficult to verify that subspaces $\mathcal{H}_\varepsilon^{\text{sol}}$ and $\mathcal{H}_\varepsilon^{\text{grad}}$ are orthogonal, i.e. $(u,v)_{\mathcal{H}_\varepsilon} = 0$, $\forall u \in \mathcal{H}_\varepsilon^{\text{sol}}$ and $\forall v \in \mathcal{H}_\varepsilon^{\text{grad}}$. Let us show that $\mathcal{H}_\varepsilon^{\text{sol}} \oplus \mathcal{H}_\varepsilon^{\text{grad}} = \mathcal{H}_\varepsilon$.

Let $w = (w^\Omega, w^G, w^S) \neq 0$ and $w \in \mathcal{H}_\varepsilon \ominus (\mathcal{H}_\varepsilon^{\text{sol}} \oplus \mathcal{H}_\varepsilon^{\text{grad}})$. Thus
$(w,u)_{\mathcal{H}_\varepsilon} = 0$, $\forall u \in \mathcal{H}_\varepsilon^{\text{sol}}$ and $(w,v)_{\mathcal{H}_\varepsilon} = 0$, $\forall v \in \mathcal{H}_\varepsilon^{\text{grad}}$.

Let us choose $u \in \mathcal{H}_\varepsilon^{\text{sol}}$ of the form: $u = (u^\Omega, 0, 0)$. Then according to (3.8) we get $w^\Omega = 0$.

Now we choose $u = (u^\Omega, 0, u_\tau^S) \in \mathcal{H}_\varepsilon^{\text{sol}}$, $\forall u^\Omega \in L_2(\Omega_\varepsilon)$, $\forall u_\tau^S \in L_2(S_\varepsilon)$. Since $(w,u)_{\mathcal{H}_\varepsilon} = 0$, then, taking into account what we discussed above, we obtain $w_\tau^S = 0$.

At last, we choose $u = \{0, \nabla\varphi, -\varphi\cdot\nu + \psi\tau\} \in \mathcal{H}_\varepsilon^{\text{grad}}$. Then, since $w = (0, w^G, w_\nu^S)$ is orthogonal to $\mathcal{H}_\varepsilon^{\text{grad}}$, from (3.8) we have

$$0 = (w,v)_{\mathcal{H}_\varepsilon} = \int_{G_\varepsilon} w^G grad\varphi dx - \int_{S_\varepsilon} w_\nu^S \varphi\nu ds, \quad \forall\varphi \in W_2^1(G_\varepsilon).$$

It follows that $\text{div} w^G = 0$ and $w_\nu^G = w_\nu^S$; consequently, $w \in \mathcal{H}_{\text{sol}}$. Since $w \in \mathcal{H}_\varepsilon \ominus (\mathcal{H}_\varepsilon^{\text{sol}} \oplus \mathcal{H}_\varepsilon^{\text{grad}})$, we get $w = 0$. □

Let us introduce the following operators in the space $\mathcal{H}_\varepsilon^{\rm sol}$

$$\mathcal{A} = \mathcal{P}_\varepsilon^{\rm sol}\tilde{\mathcal{A}}, \quad \mathcal{B} = \mathcal{P}_\varepsilon^{\rm sol}\tilde{\mathcal{B}}, \quad \mathcal{C} = \mathcal{P}_\varepsilon^{\rm sol}\tilde{\mathcal{C}},$$

where $\mathcal{P}_\varepsilon^{\rm sol}$ is an orthogonal projector to subspace $\mathcal{H}_\varepsilon^{\rm sol}$, and $\tilde{\mathcal{A}}$, $\tilde{\mathcal{B}}$, $\tilde{\mathcal{C}}$ are the operators, which are defined as follows

$$\tilde{\mathcal{A}} = \begin{pmatrix} \rho_s & 0 & 0 \\ 0 & \rho_f & 0 \\ 0 & 0 & 0 \end{pmatrix}, \quad \tilde{\mathcal{B}} = \begin{pmatrix} 0 & 0 & 0 \\ 0 & -\mu_\varepsilon\Delta & 0 \\ 0 & T_f & 0 \end{pmatrix}, \quad \tilde{\mathcal{C}} = \begin{pmatrix} -L & 0 & 0 \\ 0 & 0 & 0 \\ -T_s & 0 & 0 \end{pmatrix}.$$

Operator $\tilde{\mathcal{A}}$ is bounded, and operators $\tilde{\mathcal{B}}$, $\tilde{\mathcal{C}}$ are defined on the dense set $D_\varepsilon \subset \mathcal{H}_\varepsilon^{\rm sol}$

$$D_\varepsilon = \mathcal{H}_\varepsilon^{\rm sol} \bigcap \{u_\varepsilon^\Omega \in W_2^2(\Omega_\varepsilon) : u_\varepsilon = 0, x \in \partial\Omega; u_\varepsilon^G \in W_2^2(G_\varepsilon), u_\varepsilon^S = u_\varepsilon^G|_S = u_\varepsilon^\Omega|_S\}$$

Then $\mathcal{A}$, $\mathcal{B}$, and $\mathcal{C}$ are symmetric and non-negative in the space $\mathcal{H}_\varepsilon^{\rm sol}$.

It is easy to see that in these notations, the original problem (1.1)–(1.8) can be written in the space $\mathcal{H}_\varepsilon^{\rm sol}$ in the form

$$\mathcal{A}\ddot{u}_\varepsilon + \mathcal{B}\dot{u}_\varepsilon + \mathcal{C}u_\varepsilon = 0, \quad u_\varepsilon \in D_\varepsilon, \tag{3.9}$$
$$u_\varepsilon(0) = u_\varepsilon^0, \quad \dot{u}_\varepsilon(0) = u_\varepsilon^1,$$

where

$$u_\varepsilon^0 = (U_\varepsilon^0, \hat{U}_\varepsilon^0, U_\varepsilon^0|_{S_\varepsilon})^\top, \quad \tilde{u}_\varepsilon^1 = (U_\varepsilon^1, V_\varepsilon^1, U_\varepsilon^1|_{S_\varepsilon})^\top.$$

Applying the Laplace transform

$$\tilde{u}_\varepsilon = \int_0^\infty u_\varepsilon(x,t)e^{\lambda t}dt,$$

we pass to the quadratic spectral bundle in space $\mathcal{H}_\varepsilon^{\rm sol}$:

$$\lambda^2\mathcal{A}\tilde{u}_\varepsilon + \lambda\mathcal{B}\tilde{u}_\varepsilon + \mathcal{C} = \tilde{f}_\varepsilon, \tag{3.10}$$

where $\tilde{f}_\varepsilon(x) = \mathcal{A}(u_\varepsilon^1 + \lambda u_\varepsilon^0) + \mathcal{B}u_\varepsilon^0$.

We will prove that there exists a unique solution to the problem (3.10) for $\mathrm{Re}\lambda > 0$. For this purpose, we consider the operator $\mathcal{E} = \mathcal{A} + \mathcal{B} + \mathcal{C}$, which is symmetric and positively defined on $D_\varepsilon \subset \mathcal{H}_\varepsilon^{\rm sol}$. Therefore, we can introduce the energy space $H_\mathcal{E}$ generated by this operator [**12**]. The scalar product $(u,v)_\mathcal{E}$ in $H_\mathcal{E}$ is generated by sesquilinear form $(\mathcal{E}u,v)_{\mathcal{H}_\varepsilon}$ as follows:

$$(u,v)_\mathcal{E} = \int\limits_{\Omega_\varepsilon} \rho_s(u^\Omega,\bar{v}^\Omega)dx + \int\limits_{G_\varepsilon} \rho_f(u^G,\bar{v}^G)dx + 2\mu_\varepsilon \int\limits_{G_\varepsilon} \sum_{i,k=1}^3 \gamma_{ik}[u^G]\gamma_{ik}[\bar{v}^G]dx$$
$$+ \int\limits_{\Omega_\varepsilon} \sum_{i,k,l,m=1}^3 a_{iklm}\gamma_{ik}[u^\Omega]\gamma_{lm}[\bar{v}^\Omega]dx \quad (u,v \in H_\mathcal{E}).$$

The energy space $H_\mathcal{E}$ is isomorphic to the subspace $\overset{\circ}{J}_\varepsilon(\Omega)$ of the vector functions $u(x) \in \overset{\circ}{W}{}_2^1(\Omega)$, which are divergence free on G_ε: $u^\Omega = u|_{\Omega_\varepsilon}$, $u^G = u|_{G_\varepsilon}$, $u^S = u|_{S_\varepsilon}$. The norm $\overset{\circ}{J}_\varepsilon(\Omega)$ is induced by the norm in $\overset{\circ}{W}{}_2^1(\Omega)$.

Let us define the sesquilinear form in $\overset{\circ}{J}_\varepsilon(\Omega)$

$$\Phi_{\varepsilon\lambda}[u_\varepsilon, v_\varepsilon] = \lambda \int\limits_\Omega \rho_\varepsilon(u_\varepsilon, \bar{v}_\varepsilon)dx + 2\mu_\varepsilon \int\limits_{G_\varepsilon} \sum_{i,k=1}^3 \gamma_{ik}[u_\varepsilon]\gamma_{ik}[\bar{v}_\varepsilon]dx \tag{3.11}$$

$$+\frac{1}{\lambda}\int\limits_{\Omega_\varepsilon}\sum_{i,k,l,m=1}^{3} a_{iklm}\gamma_{ik}[u_\varepsilon]\gamma_{lm}[\bar{v}_\varepsilon]dx$$

and the antilinear functional

$$F[v_\varepsilon]=\int\limits_{\Omega}F_\varepsilon\cdot\bar{v}_\varepsilon dx+\frac{2\mu_\varepsilon}{\lambda}\int\limits_{G_\varepsilon}\sum_{i,k=1}^{3}\gamma_{ik}[\hat{U}^0_\varepsilon]\gamma_{ik}[\bar{v}_\varepsilon]dx, \tag{3.12}$$

where $\mathrm{Re}\lambda>0$ and

$$\rho_\varepsilon=\rho_s\chi_{\Omega_\varepsilon}+\rho_f\chi_{G_\varepsilon},\quad F_\varepsilon(x)=\rho_s(\frac{1}{\lambda}U^1_\varepsilon+U^0_\varepsilon)\chi_{\Omega_\varepsilon}+\rho_f(\frac{1}{\lambda}V^1_\varepsilon+\hat{U}^0_\varepsilon)\chi_{G_\varepsilon}. \tag{3.13}$$

Then (3.10) can be formulated in a variational form as follows:

Find the vector function $u_\varepsilon\in\overset{\circ}{J}_\varepsilon(\Omega)$ *that satisfies the equality*

$$\Phi_{\varepsilon\lambda}[u_\varepsilon,v_\varepsilon]=F[v_\varepsilon],\quad\forall v_\varepsilon\in\overset{\circ}{J}_\varepsilon(\Omega). \tag{3.14}$$

To prove the existence and uniqueness of the solution (3.14), we will show that the form $\Phi_{\varepsilon\lambda}[u_\varepsilon,v_\varepsilon]$ is bounded and coercive in $\overset{\circ}{J}_\varepsilon(\Omega)$. Using the Cauchy-Schwartz inequality and taking into account the boundedness of the elasticity tensor $\{a_{iklm}\}$, we obtain from (3.11)

$$|\Phi_{\varepsilon\lambda}[u_\varepsilon,v_\varepsilon]|\le C\|u_\varepsilon\|_{W^1_2(\Omega)}\|v_\varepsilon\|_{W^1_2(\Omega)}.$$

So $\Phi_{\varepsilon\lambda}[u_\varepsilon,v_\varepsilon]$ is bounded in $\overset{\circ}{J}_\varepsilon(\Omega)$.

Let us prove that it is coercive. Separating in (3.11) the real part, we obtain

$$|\Phi_{\varepsilon\lambda}[u_\varepsilon,u_\varepsilon]|\ge\mathrm{Re}\lambda\int\limits_{\Omega}\rho|u_\varepsilon|^2dx+2\mu_\varepsilon\int\limits_{G_\varepsilon}\sum_{i,k=1}^{3}|\gamma_{ik}[u_\varepsilon]|^2dx \tag{3.15}$$

$$+\frac{\mathrm{Re}\lambda}{|\lambda|^2}\int\limits_{\Omega_\varepsilon}\sum_{i,k,l,m=1}^{3}a_{iklm}\gamma_{lm}[u_\varepsilon]\gamma_{ik}[\bar{u}_\varepsilon]dx.$$

Taking into account that the tensor $\{a_{iklm}\}$ is positively defined and using the Korn's inequality for $u_\varepsilon\in\overset{\circ}{W}{}^1_2(\Omega)$, we get

$$|\Phi_{\varepsilon\lambda}[u_\varepsilon,u_\varepsilon]|\ge C_1\mu_\varepsilon\|u_\varepsilon\|^2_{W^1_2(\Omega)},$$

where the constant $C_1>0$ does not depend on ε. Since $\mu_\varepsilon>0$, the form $\Phi_{\varepsilon\lambda}[u_\varepsilon,u_\varepsilon]$ is coercive.

The functional (3.12) is bounded in $\overset{\circ}{J}_\varepsilon(\Omega)$, according to the conditions of Theorem 2.1 on the initial data U^0_ε, U^1_ε, V^1_ε

Thus, according to the Lax-Milgram theorem [**5**], the problem (3.14) has a unique solution for any λ ($\mathrm{Re}\lambda>0$).

Taking into account the analytical dependence of the form (3.11) and functional (3.12) on λ ($\mathrm{Re}\lambda>0$), we conclude that the solution to the problem (3.14) is an analytical function with respect to λ ($\mathrm{Re}\lambda>0$) [**13**]. Moreover, from (3.11)–(3.15) it follows the estimates

$$\|u_\varepsilon\|_{L_2(\Omega)}\le\frac{C_1}{\mathrm{Re}\lambda},\quad\|u_\varepsilon\|_{W^1_2(\Omega)}\le\frac{C_2|\lambda|}{\mu_\varepsilon\mathrm{Re}\lambda}. \tag{3.16}$$

According to (3.11)–(3.14), $u_\varepsilon(x,\lambda)$ satisfies the equality

$$\lambda^2 \int_\Omega \rho_\varepsilon (u_\varepsilon, \bar{v}_\varepsilon) dx + 2\lambda\mu_\varepsilon \int_{G_\varepsilon} \sum_{i,k=1}^{3} \gamma_{ik}[u_\varepsilon]\gamma_{ik}[\bar{v}_\varepsilon] dx$$

$$+ \int_{\Omega_\varepsilon} \sum_{i,k,l,m=1}^{3} a_{iklm}\gamma_{ik}[u_\varepsilon]\gamma_{lm}[\bar{v}_\varepsilon] dx \tag{3.17}$$

$$= \int_\Omega \rho(U^1_\varepsilon + V^1_\varepsilon)\cdot \bar{v}_\varepsilon dx + \lambda \int_\Omega \rho \hat{U}^0_\varepsilon \cdot \bar{v}_\varepsilon dx + 2\mu_\varepsilon \int_{G_\varepsilon} \sum_{i,k=1}^{3} \gamma_{ik}[\hat{U}^0_\varepsilon]\gamma_{ik}[\bar{v}_\varepsilon] dx, \quad v_\varepsilon \in \overset{\circ}{J}_\varepsilon(\Omega).$$

We consider the function

$$\hat{u}_\varepsilon(x,t) = \frac{1}{2\pi i} \int_{\sigma - i\infty}^{\sigma + i\infty} \frac{u_\varepsilon(x,\lambda)}{\lambda^2} e^{\lambda t} d\lambda. \tag{3.18}$$

It is clear that $\hat{u}_\varepsilon(x,t) \in C(-T,T; L_2(\Omega)) \bigcap L_2(-T,T; \overset{\circ}{J}_\varepsilon(\Omega))$, and $\hat{u}_\varepsilon(x,t) = 0$ for $t < 0$.

Let us multiply (3.17) by $\frac{e^{\lambda t}}{2\pi i \lambda^4}\eta^{(4)}(t)$, where $\eta(t) \in C^4_0(-T,T)$, and integrate with respect to $\lambda \in (\sigma - i\infty, \sigma + i\infty)$ and to $t \in [-T,T]$. Then we obtain:

$$\int_0^T \int_\Omega \rho \hat{u}_\varepsilon(x,t) \bar{v}_\varepsilon(x) \eta^{(4)}(t) dx dt - 2\mu_\varepsilon \int_0^T \int_{G_\varepsilon} \sum_{n,p,q,r=1}^{3} \gamma_{np}[\hat{u}_\varepsilon]\gamma_{qr}[\bar{v}]\eta^{(3)} dx dt$$

$$+ \int_0^T \int_{\Omega_\varepsilon} \sum_{n,p,q,r=1}^{3} a_{npqr}\gamma_{np}[\hat{u}_\varepsilon]\gamma_{qr}[\bar{v}]\eta'' dx dt \tag{3.19}$$

$$= \int_\Omega \rho(U^1_\varepsilon + V^1_\varepsilon)\cdot \bar{v}_\varepsilon \eta(0) dx + \int_\Omega \rho \hat{U}^0_\varepsilon \cdot \bar{v}_\varepsilon \eta'(0) dx + 2\mu_\varepsilon \int_{G_\varepsilon} \sum_{i,k=1}^{3} \gamma_{ik}[\hat{U}^0_\varepsilon]\gamma_{ik}[\bar{v}_\varepsilon]\eta(0) dx.$$

This equality means that the generalized function

$$u_\varepsilon(x,t) = \frac{d^2}{dt^2}\hat{u}_\varepsilon(x,t) \tag{3.20}$$

is a solution to the Cauchy problem (3.9) in the sense of distribution.

On the other hand, it follows from (3.9) that the solution $u_\varepsilon(x,t)$ satisfy the equality

$$\int_\Omega \rho_\varepsilon \left(\frac{\partial u_\varepsilon}{\partial t}\right)^2 dx + \int_{\Omega_\varepsilon} \sum_{n,p,q,r=1}^{3} a_{npqr}\gamma_{np}[u_\varepsilon]\gamma_{qr}[u_\varepsilon] dx$$

$$+ 2\mu_\varepsilon \int_0^t \int_{G_\varepsilon} \sum_{n,p=1}^{3} \gamma^2_{np}[u_\varepsilon] dx d\tau$$

$$= \int_{\Omega_\varepsilon} \rho_s (U^1_\varepsilon)^2 dx + \int_{G_\varepsilon} \rho_f (V^1_\varepsilon)^2 dx + \int_{\Omega_\varepsilon} \sum_{n,p,q,r=1}^{3} a_{npqr}\gamma_{np}[U^0_\varepsilon]\gamma_{qr}[U^0_\varepsilon] dx. \tag{3.21}$$

Hence, $u_\varepsilon(x,t) \in W_2^1(\Omega_T) \bigcap L_2(0,T;\overset{\circ}{J}_\varepsilon(\Omega))$ and according to (3.19) $u_\varepsilon(x,t)$ satisfies the equality

$$\int_0^T\int_\Omega \rho u_\varepsilon(x,t)\bar{v}_\varepsilon(x)\eta''(t)dxdt - 2\mu_\varepsilon\int_0^T\int_{G_\varepsilon}\sum_{n,p,q,r=1}^{3}\gamma_{np}[u_\varepsilon]\gamma_{qr}[\bar{v}]\eta' dxdt$$

$$+\int_0^T\int_{\Omega_\varepsilon}\sum_{n,p,q,r=1}^{3} a_{npqr}\gamma_{np}[u_\varepsilon]\gamma_{qr}[\bar{v}]\eta dxdt \tag{3.22}$$

$$= \int_\Omega \rho(U_\varepsilon^1+V_\varepsilon^1)\cdot\bar{v}_\varepsilon\eta(0)dx + \int_\Omega \rho\hat{U}_\varepsilon^0\cdot\bar{v}_\varepsilon\eta'(0)dx + 2\mu_\varepsilon\int_{G_\varepsilon}\sum_{i,k=1}^{3}\gamma_{ik}[\hat{U}_\varepsilon^0]\gamma_{ik}[\bar{v}_\varepsilon]\eta(0)dx.$$

Since a linear span of functions $v_\varepsilon(x)\eta(t)$ is dense $W_2^1(\Omega_T)$, $u_\varepsilon(x,t)$ is a weak solution to the boundary value problem (3.1), and the pair of functions $(u_\varepsilon(x,t), v_\varepsilon(x,t) = \frac{\partial u_\varepsilon}{\partial t})$ is a weak solution to the original problem (1.1)–(1.8). It follows from (3.21) that this solution is unique.

4. Asymptotic behavior of the problem (3.14) as $\varepsilon \to 0$

The proof of the main result (Theorems 2.1 and 2.2) is based on a study of the asymptotic behavior of the solutions of the stationary problem (3.14) as $\varepsilon \to 0$. The proof consists of several stages.

First, we study the asymptotic behavior of the solutions with positive λ. In this case the problem (3.14) is equivalent to the variational problem

$$\Phi_{\varepsilon\lambda}[u_\varepsilon] \to \inf_{u_\varepsilon\in\overset{\circ}{J}_\varepsilon(\Omega)}, \tag{4.1}$$

where

$$\Phi_{\varepsilon\lambda}[u_\varepsilon] = \lambda\int_\Omega \rho_\varepsilon|u_\varepsilon|^2dx + 2\mu_\varepsilon\int_{G_\varepsilon}\sum_{i,k=1}^{3}\gamma_{ik}^2[u_\varepsilon]dx \tag{4.2}$$

$$+\frac{1}{\lambda}\int_{\Omega_\varepsilon}\sum_{i,k,l,m=1}^{3}a_{iklm}\gamma_{ik}[u_\varepsilon]\gamma_{lm}[u_\varepsilon]dx + 2\int_\Omega F_\varepsilon\cdot u_\varepsilon dx$$

$$-2\mu_\varepsilon\int_{G_\varepsilon}\sum_{i,k=1}^{3}\gamma_{ik}[\hat{U}_\varepsilon^0]\gamma_{ik}[u_\varepsilon]dx.$$

The functions ρ_ε and F_ε are defined by equalities (3.13).

We will prove that the solutions to the problem (4.1) converge in $L_2(\Omega)$ to the solution to the problem

$$\Phi_\lambda[u] \to \inf_{u\in\overset{\circ}{W}{}_2^1(\Omega)}, \tag{4.3}$$

where

$$\Phi_\lambda[u] = \lambda\int_\Omega \rho|u|^2dx + \int_\Omega\sum_{i,k,l,m=1}^{3}A_{iklm}(x,\lambda)\gamma_{ik}[u]\gamma_{lm}[u]dx + 2\int_\Omega F\cdot udx. \tag{4.4}$$

Here $F(x) = \rho U^0(x) + \frac{1}{\lambda}\rho V$, and functions $\rho(x)$ and $V(x)$ are defined by equalities (2.21).

This result we prove in subsection 4.1 is for the general location of the caverns, not necessarily locally periodic. We introduce special characteristics of the elastic medium with caverns (mesoscopic characteristics). For these characteristics we establish the sufficient conditions for the convergence of the problem (4.1) to the problem (4.3) and give the description of the tensor $\{A_{iklm}\}$.

In subsection 4.2, we verify that these conditions are fulfilled in the case of the locally periodic structures. We calculate the tensor $\{A_{iklm}\}$ by solving cell problems (2.3)–(2.7) and (2.11)–(2.14). Finally, in subsection 4.3, taking into account the analyticity of the solution of (4.1)with respect to λ, we prove the convergence of the solution of the problem (4.1) to the solution of the homogenized problem (4.3) for $\mathrm{Re}\lambda > 0$.

In Section 5, we complete the proof of Theorems 2.1 and 2.2, applying the inverse Laplace transform.

4.1. Asymptotic behavior of the variational problem (4.1) as $\varepsilon \to 0$. We will consider domains Ω_ε of a more general type. Specifically, we assume that

(1) caverns G_ε^j are homothetic contractions of a fixed domain $G \in \mathbb{R}^3$ (or a finite number of such domains), that is homeomorphic to a sphere $S \subset \mathbb{R}^3$ and has a smooth boundary;

(2) the distance between the minimal ball B_ε^j, containing the G_ε^j and neighboring balls or the boundary $\partial\Omega$ $r_\varepsilon^j = \mathrm{dist}(B_\varepsilon^j, \bigcup_{l,l\neq j} B_\varepsilon^l \bigcup \partial\Omega)$ and the diameters d_ε^j of the minimal balls B_ε^j, satisfy the following conditions

$$d_\varepsilon^j < C r_\varepsilon^j, \quad r_\varepsilon^j = O(\varepsilon), \tag{4.5}$$

where C is a positive constant does not depend on ε.

Let us introduce the special mesoscopic characteristics of the medium.

Let $K_h^j = K(h, x^j)$ be a cube with a side h ($\varepsilon \ll h \ll 1$) and a center in the point $x^j \in \Omega$; $R = \{R_{np}\}$ be an arbitrary tensor of the second rank in $\mathbb{R}^3$. Consider the minimizing problem

$$E^{\varepsilon,\gamma}_{x^j,h,\lambda}[v_{\varepsilon j}, R] \to \inf_{v_{\varepsilon j} \in J_\varepsilon(K_h^i)},$$

where

$$\begin{aligned} E^{\varepsilon,\gamma}_{x^j,h,\lambda}[v_{\varepsilon j}, R] = \Bigg\{ 2\mu_\varepsilon & \int_{K_h^j \cap G_\varepsilon} \sum_{i,k=1}^3 \gamma_{ik}^2[v_{\varepsilon\alpha}]dx \\ & + \frac{1}{\lambda} \int_{K_h^j \cap \Omega_\varepsilon} \sum_{i,k,l,m=1}^3 a_{iklm}\gamma_{ik}[v_{\varepsilon j}]\gamma_{lm}[v_{\varepsilon j}]dx \\ & + h^{-2-\gamma} \int_{K_h^j} \left| v_{\varepsilon j} - \sum_{k,l=1}^3 \phi^{kl}(x-z)R_{kl} \right|^2 \Bigg\}. \end{aligned} \tag{4.6}$$

Here the infimum is taken over all vector functions

$$v_{\varepsilon j} \in J_\varepsilon(K_h^j) = \{v_{\varepsilon j} \in W_2^1(K_h^j), \mathrm{div} v_{\varepsilon j} = 0, x \in G_\varepsilon\};$$

γ is an arbitrary number $0 < \gamma < 2$, and the vector function $\phi^{kl}(x)$ is defined by

$$\phi^{kl}(x) = \frac{1}{2}(x_k e^l + x_l e^k).$$

There exists a unique vector function $v_{\varepsilon j}(x, \lambda)$ that minimizes (4.6). Denote by $v_{\varepsilon j}^{np}$ the minimizer of (4.6) for $R = R^{np} \stackrel{\text{def}}{=} \frac{1}{2}(e^n \otimes e^p + e^p \otimes e^n)$. Then $v_{\varepsilon j}(x, \lambda) = \sum_{n,p=1}^{3} R_{np} v_{\varepsilon j}^{np}(x, \lambda)$ and

$$\tilde{E}_{x^j,h,\lambda}^{\varepsilon,\gamma}[R] = \sum_{n,p,q,r=1}^{3} \tilde{A}_{npqr}(x^j, \varepsilon, h, \lambda, \gamma) R_{np} R_{qr},$$

where

$$\tilde{A}_{npqr}(x^j, \varepsilon, h, \lambda, \gamma) = 2\mu_\varepsilon \int\limits_{K_h^j \cap G_\varepsilon} \sum_{i,k=1}^{3} \gamma_{ik}[v_{\varepsilon j}^{np}] \gamma_{ik}[v_{\varepsilon j}^{qr}] dx$$

$$+\frac{1}{\lambda} \int\limits_{K_h^j \cap \Omega_\varepsilon} \sum_{i,k,l,m=1}^{3} a_{iklm} \gamma_{ik}[v_{\varepsilon j}^{np}] \gamma_{lm}[v_{\varepsilon j}^{qr}] dx$$

$$+h^{-2-\gamma} \int\limits_{K_h^j} (v_{\varepsilon j}^{np} - \phi^{np}(x - x^j)) \cdot (v_{\varepsilon j}^{qr} - \phi^{qr}(x - x^j)) dx \tag{4.7}$$

are the components of symmetric and positively defined tensor of the 4th rank $\{\tilde{A}_{npqr}(x^j, \varepsilon, h, \lambda, \gamma)\}$ in $\mathbb{R}^3$.

The following theorem holds.

THEOREM 4.1. *Let $\forall x \in \Omega$ the following conditions are valid*

1) $\lim\limits_{h\to 0} \overline{\lim\limits_{\varepsilon\to 0}} \frac{\tilde{A}_{npqr}(x,\varepsilon,h,\lambda,\gamma)}{h^3} = \lim\limits_{h\to 0} \underline{\lim\limits_{\varepsilon\to 0}} \frac{\tilde{A}_{npqr}(x,\varepsilon,h,\lambda,\gamma)}{h^3} = \tilde{A}_{npqr}(x, \lambda)$, $\forall x \in \Omega$, $\exists \gamma$, *(*$0 < \gamma < 2$*).*

2) $\rho_\varepsilon(x) \to \rho(x)$, $F_\varepsilon(x) \to F(x)$ *weakly in* $L_2(\Omega)$ *as* $\varepsilon \to 0$,
where $\tilde{A}_{npqr}(x, \varepsilon, h, \lambda, \gamma)$, $\rho(x) \in L_\infty(\Omega)$, $f(x) \in L_2(\Omega)$,

3) the vector function $\hat{U}_\varepsilon^0$ *is uniformly bounded in* $W_2^1(\Omega)$ *with respect to* ε*; it is defined by equality (2.22) on* G_ε, *if* $\mu_\varepsilon \to \mu > 0$.

Then the sequence of solutions $\{u_\varepsilon(x, \lambda)\}$ *of the problem (4.1) converge in* $L_2(\Omega)$ *to the vector function* $u(x, \lambda)$ *that is the solution of the variational problem (4.3).*

PROOF. First, we consider the case $\mu_\varepsilon \to 0$, namely $\mu_\varepsilon = \mu\varepsilon^\alpha$ $(0 < \alpha < 2)$.

Since the solution $u_\varepsilon = u_\varepsilon(x, \lambda)$ to the problem (4.1) minimizes the functional $\Phi_{\varepsilon\lambda}[u_\varepsilon]$ in the class $u_\varepsilon \in \overset{\circ}{J}_\varepsilon(\Omega)$, then $\Phi_{\varepsilon\lambda}[u_\varepsilon] \le \Phi_{\varepsilon\lambda}[0] = 0$ and so the inequality holds

$$\lambda \int\limits_{\Omega} \rho_\varepsilon |u_\varepsilon|^2 dx + 2\mu_\varepsilon \int\limits_{G_\varepsilon} \sum_{i,k=1}^{3} \gamma_{ik}^2[u_\varepsilon] dx + \frac{1}{\lambda} \int\limits_{\Omega_\varepsilon} \sum_{n,p,q,r=1}^{3} a_{npqr} \gamma_{np}[u_\varepsilon] \gamma_{qr}[u_\varepsilon] dx$$

$$\le \frac{2\mu_\varepsilon}{\lambda} \int\limits_{G_\varepsilon} \sum_{i,j,k,l=1}^{3} \gamma_{ik}[\hat{U}_\varepsilon^0] \gamma_{lm}[u_\varepsilon] dx - 2 \int\limits_{\Omega} F_\varepsilon u_\varepsilon.$$

Hence, taking into account the boundedness and positive definiteness of tensor $\{a_{npqr}\}$, the boundedness of F_ε and the form of the extension of $\hat{U}^0_\varepsilon$ (Lemma 3.1), we get for $\lambda > 0$

$$\int_\Omega |u_\varepsilon|^2 dx \le C_1, \quad \int_{\Omega_\varepsilon} \sum_{n,p=1}^3 \gamma^2_{np}[u_\varepsilon] dx \le C_2, \tag{4.8}$$

$$\mu_\varepsilon \int_{G_\varepsilon} \sum_{n,p=1}^3 \gamma^2_{np}[u_\varepsilon] dx \le C_3, \tag{4.9}$$

where the constants C_1, C_2, C_3 do not depend on ε.

Now, we use the following lemma [**9**].

LEMMA 4.2. *For any vector function $u \in W^1_2(\Pi \setminus G)$ there exists its extension to the vector function $\tilde{u} = \tilde{P}_G u \in W^1_2(\Pi)$ and the inequality holds*

$$\int_\Pi \sum \gamma^2_{ik}[\tilde{u}] dx \le C \int_{\Pi\setminus G} \sum_{i,k=1}^3 \gamma^2_{ik}[u] dx,$$

where the constant C does not depend on u.

This lemma implies that the solution u_ε of the problem (4.1) can be extended to vector function $\tilde{u}_\varepsilon \in \overset{\circ}{W}{}^1_2(\Omega)$, satisfying the inequality

$$\int_\Omega \sum_{n,p=1}^3 \gamma^2_{np}[\tilde{u}_\varepsilon] dx \le C \int_{\Omega_\varepsilon} \sum_{n,p=1}^3 \gamma^2_{np}[u_\varepsilon] dx, \tag{4.10}$$

where C does not depend on ε.

Using Korn's inequality and (4.8), we obtain from (4.10)

$$\|\tilde{u}_\varepsilon\|_{W^1_2(\Omega)} \le C,$$

where C does not depend on ε.

It follows that the set of vector functions $\{\tilde{u}_\varepsilon(x,\lambda), \varepsilon > 0\}$ is weakly compact in $W^1_2(\Omega)$. Hence, we can extract a subsequence $\{\tilde{u}_\varepsilon, \varepsilon = \varepsilon_k \to 0\}$ weakly converging in $W^1_2(\Omega)$ and strongly in $L_2(\Omega)$ to a function $u(x) \in \overset{\circ}{W}{}^1_2(\Omega)$.

We will show that the solution u_ε of the problem (4.1) converges to the same function as $0 < \alpha < 2$.

Since $\tilde{u}_\varepsilon(x) = u_\varepsilon(x)$ as $x \in \Omega_\varepsilon$, then $\tilde{u}_\varepsilon - u_\varepsilon \in \overset{\circ}{W}{}^1_2(G^j_\varepsilon)$. Using Friedreich's inequality and taking into account that $\operatorname{diam} G_\varepsilon = O(\varepsilon)$, we get

$$\int_\Omega |\tilde{u}_\varepsilon - u_\varepsilon|^2 dx = \sum_j \int_{G^j_\varepsilon} |\tilde{u}_\varepsilon - u_\varepsilon|^2 dx \le C \sum_j \varepsilon^2 \int_{G^j_\varepsilon} |\nabla(\tilde{u}_\varepsilon - u_\varepsilon)|^2 dx.$$

Hence, due to Korn's inequality, (4.8) and (4.9) we obtain

$$\int_\Omega |\tilde{u}_\varepsilon - u_\varepsilon|^2 dx \le 2C_1 \sum_j \varepsilon^2 \int_{G^j_\varepsilon} \sum_{n,p=1}^3 |\gamma_{np}[\tilde{u}_\varepsilon - u_\varepsilon]|^2 dx \le C_2 \varepsilon^{2-\alpha}.$$

The constants in all inequalities are independent on ε.

Since $\tilde{u}_\varepsilon \to u$ in $L_2(\Omega)$ as $\varepsilon \to 0$, then this inequality implies that the solution u_ε to the problem (4.1) strongly converges in $L_2(\Omega)$ to the function $u(x,\lambda)$ as $\alpha < 2$.

The remaining part of the proof of Theorem 4.1 is analogous to the proof of the same theorem in [**7**]. The only difference is that the functional (4.2) $\Phi_{\varepsilon\lambda[u_\varepsilon]}$ contains a term with initial displacement $\hat{U}^0_\varepsilon$. However, taking into account that the vector functions $\hat{U}^0_\varepsilon$ and u_ε are bounded in $W^1_2(\Omega)$ and $\mu_\varepsilon \to 0$ as $\varepsilon \to 0$ this term tends to zero:

$$\frac{2\mu_\varepsilon}{\lambda}\int\limits_{G_\varepsilon}\sum_{i,j,k,l=1}^{3}\gamma_{ik}[\hat{U}^0_\varepsilon]\gamma_{lm}[u_\varepsilon]dx \to 0.$$

Thus, we can follow the scheme of [**7**] to prove Theorem 4.1 in the case $\mu_\varepsilon = \mu\varepsilon^\alpha$ $(0 < \alpha < 2)$.

In the case $\mu_\varepsilon = \mu > 0$ $(\alpha = 0)$ the proof of Theorem 4.1 follows from [**7**], because due to the condition (2.22) of Theorem 2.2, $\gamma_{ik}[\hat{U}^0_\varepsilon] = 0$ and, consequently, the term with initial displacement is absent in the functional $\Phi_{\varepsilon\lambda[u_\varepsilon]}$. □

4.2. Calculation of effective tensors in the case of locally periodic structures. Now we consider the locally periodic location of caverns, as described in §2. We will show that in this case condition 1) of Theorem 4.1 is valid and the limiting tensor is described by formula (2.8) as $\alpha = 0$ and formula (2.15) as $0 < \alpha < 2$.

In the case $\alpha = 0$ the proof of formula (2.8) is similar to [**7**]. Therefore we present the proof of (2.15) in the case of $0 < \alpha < 2$.

Let us construct the concentric to Π^ν_ε parallelepipeds

$$\Pi^{\pm\nu}_\varepsilon = \left\{x \in \mathbb{R}^3 : |x_i - x^\nu_i| \le (\theta_i\varepsilon)\frac{1 \pm h^\sigma}{2}, i = 1,2,3\right\},$$

where $0 < \sigma < \gamma/4$, $\varepsilon \ll h \ll 1$ and associate with them the partition of unity $\{\varphi^{\nu h}_\varepsilon(x)\}$ that satisfy the conditions: $\varphi^{\nu h}_\varepsilon(x) = 1$ as $x \in \Pi^{-\nu}_\varepsilon$, $\varphi^{\nu h}_\varepsilon(x) = 0$ as $x \notin \Pi^{+\nu}_\varepsilon$, $\sum_\nu \varphi^{\nu h}_\varepsilon = 1$, $|D^k\varphi^{\nu h}_\varepsilon(x)| \le \frac{C}{\varepsilon^{|k|}h^{\sigma|k|}}$ $(|k| = 0,1)$.

Without loss of generality, we can assume that G^ν_ε does not intersect with the cubes ∂K^j_h and are located within the K^j_h, together with parallelepipeds Π^ν_ε.

We seek the minimizer $v^{np}_{\varepsilon j}(x,\lambda)$ of the problem (4.6) as $R = R^{np}$ in the form:

$$v^{np}_{\varepsilon j}(x,\lambda) = V^{np}_\varepsilon(x,x^j) + \zeta_{\varepsilon j}(x),$$

where

$$V^{np}_\varepsilon(x,x^j) = \phi^{np}(x - x^j) + \varepsilon\sum_\nu\left(\hat{w}^{np}\left(\frac{x - x^\nu}{\varepsilon}\right) - \phi^{np}\left(\frac{x - x^\nu}{\varepsilon}\right)\right)\varphi^{\nu h}_\varepsilon(x).$$

Here $\hat{w}^{np}(x)$ is the extension of the solution of the cell problem (2.11)–(2.14) to G^j_ε in accordance with Lemma 3.1 and $\phi^{np} = \frac{1}{2}(x_n e^p + x_p e^n)$.

According to Lemma 3.1 $\operatorname{div}\hat{w}^{np}(x) = 0$. The definition of the function ϕ^{np} and the fact that $\varphi^{\nu h}_\varepsilon(x) = 1$, $x \in G_\varepsilon$, imply that $\operatorname{div}V^{np}_\varepsilon = 0$. Thus $V^{np}_\varepsilon \in J_\varepsilon(K^j_h)$. The minimizer $v^{np}_{\varepsilon j}(x,\lambda)$ of the problem (4.6) is of the same class. Therefore the vector function $\zeta_{\varepsilon j}(x)$ is of the same class and minimizes the following functional

$$\tilde{\Phi}_{\varepsilon\lambda}[\zeta_{\varepsilon j}] = 2\mu_\varepsilon\int\limits_{K^j_h\bigcap G_\varepsilon}\sum_{i,k=1}^{3}\gamma^2_{ik}[\zeta_{\varepsilon j}]dx$$

$$+\frac{1}{\lambda}\int\limits_{K_h^j\bigcap\Omega_\varepsilon}\sum_{i,k,l,m=1}^{3}a_{iklm}\gamma_{ik}[\zeta_{\varepsilon j}]\gamma_{lm}[\zeta_{\varepsilon j}]dx+h^{-2-\gamma}\int\limits_{K_h^j}|\zeta_{\varepsilon j}|^2dx$$

$$+\frac{2}{\lambda}\int\limits_{K_h^j\bigcap\Omega_\varepsilon}\sum_{i,k,l,m=1}^{3}a_{iklm}\gamma_{ik}[V_\varepsilon^{np}]\gamma_{lm}[\zeta_{\varepsilon j}]dx+\mu\int\limits_{K_h^j\bigcap G_\varepsilon}\sum_{i,k=1}^{3}\gamma_{ik}[\zeta_{\varepsilon j}]\gamma_{ik}[V_\varepsilon^{np}]dx$$

$$+h^{-2-\gamma}\int\limits_{K_h^j}(\zeta_{\varepsilon j},V_\varepsilon^{np}-\phi^{np})dx.$$

From here we obtain

$$\|\zeta_{\varepsilon\alpha}\|^2_{W_2^1(K_h^j)}\le Ch^{3+\gamma/2}.$$

It follows that function V_ε^{np} approximates the solution of the problem (4.6) for sufficiently small ε and h. Substituting $v_{\varepsilon j}^{np}(x)$ in (4.7) we get

$$\tilde{A}_{npqr}(y,\lambda)=\lim_{h\to 0}\underline{\lim}_{\varepsilon\to 0}\frac{\tilde{A}_{npqr}(y,\varepsilon,h,\lambda,\gamma)}{h^3}=\lim_{h\to 0}\overline{\lim_{\varepsilon\to 0}}\frac{\tilde{A}_{npqr}(y,\varepsilon,h,\lambda,\gamma)}{h^3}$$

$$=\frac{1}{\lambda}\int\limits_{\Pi\setminus G^y}\sum_{i,k,l,m=1}^{3}a_{iklm}\gamma_{ik}[w^{np}]\gamma_{lm}[w^{qr}]dx=\frac{1}{\lambda}A^0_{npqr}(y),$$

where tensor $A^0_{npqr}(y)$ is defined by (2.15).

Thus, the following theorem holds.

THEOREM 4.3. *In the case of locally periodic structures the tensor* $\{\tilde{A}_{npqr}(x,\lambda)\}$ *is defined by*

$$\tilde{A}_{npqr}(x,\lambda)=\begin{cases}A_{npqr}(x,\lambda), & \alpha=0,\\ \frac{1}{\lambda}A^0_{npqr}(x), & 0<\alpha<2,\end{cases}$$

where tensors $A_{npqr}(x,\lambda)$ *and* $A^0_{npqr}(x)$ *are given by (2.8) and (2.15), respectively.*

The tensor $\{\tilde{A}_{npqr}(x,\lambda)\}$ *is symmetric* $\tilde{A}_{npqr}=\tilde{A}_{pnqr}=\tilde{A}_{qrnp}=\tilde{A}_{nprq}$ *and positively defined as* $\lambda>0$*. It is analytical on* λ *in the half-plane* $\mathrm{Re}\lambda>0$ *and* $\overline{\tilde{A}_{npqr}(x,\lambda)}=A_{npqr}(x,\bar{\lambda})$*. Moreover, the following estimate is valid*

$$|\tilde{A}_{npqr}(x,\lambda)|\le C|\lambda|^{-1}.$$

Hence there exists a Laplace transform

$$\tilde{A}_{npqr}(x,t)=\frac{1}{2\pi i}\int\limits_{\sigma-i\infty}^{\sigma+i\infty}\tilde{A}_{npqr}(x,\lambda)e^{\lambda t}d\lambda,$$

and $\tilde{A}_{npqr}(x,t)=0$ *as* $t<0$, $\tilde{A}_{npqr}(x,t)\in L_2([0,T];L_\infty(\Omega))$, *and* $\mathrm{Im}\tilde{A}_{npqr}(x,t)=0$.

4.3. Passing to the complex λ.

THEOREM 4.4. *The solution to the problem (3.14) $u_\varepsilon(x,\lambda)$ converge in $L_2(\Omega)$ to the analytical vector function $w(x,\lambda)$ (for $\mathrm{Re}\lambda > 0$) which satisfies the identity*

$$\lambda^2 \int_\Omega \rho w(x,\lambda)\cdot\zeta(x)dx + \lambda \int_\Omega \sum_{n,p,q,r}^{3} (\tilde{A}_{npqr}(x,\lambda)\gamma_{np}[w(x,\lambda)]\gamma_{qr}[\zeta])dx = \tag{4.11}$$

$$= \int_\Omega \rho(\lambda U^0 + V(x))\cdot\zeta(x)dx,$$

and the following estimates hold:

$$\|w\|_{L_2(\Omega)} \le \frac{C}{\mathrm{Re}\lambda}, \quad \|w\|_{W_2^1(\Omega)} \le \frac{C|\lambda|}{\mathrm{Re}\lambda}. \tag{4.12}$$

PROOF. It follows from subsections 4.1–4.2 that for $\lambda > 0$ the solution to problem (4.1) converges to the solution of problem (4.3) which satisfies the identity (4.11). According to §3, the solution of the problem (4.1) is analytical with respect to λ ($\mathrm{Re}\lambda > 0$) and the estimates (3.16) are valid. Therefore, Vitally's theorem implies that for every compact set on the half-plane $\mathrm{Re}\lambda > 0$, the solution $u_\varepsilon(x,\lambda)$ of (3.14) converges to analytical with respect to λ ($\mathrm{Re}\lambda > 0$) vector function $w(x,\lambda)$, and $w(x,\lambda)$ satisfies identity (4.11) and estimates (4.12) hold. According to Theorem 4.3, tensor $\{\tilde{A}_{npqr}(x,\lambda)\}$ is analytical with respect to λ ($\mathrm{Re}\lambda > 0$), and we conclude that the vector function $w(x,\lambda)$ satisfies the identity (4.11) for complex λ. □

5. End of the proof of the main theorem

According to the equality (3.21), the solution $u_\varepsilon(x,t)$ of the problem (3.1) is uniformly bounded in $W_2^1(\Omega_T)\bigcap L_2(0,T;\overset{\circ}{J}_\varepsilon(\Omega)$ with respect to ε. Hence, we can extract a subsequence $\{u_\varepsilon, \varepsilon=\varepsilon_k\to 0\}$ which converges to some function $u(x,t)$ weakly in $W_2^1(\Omega)$ and strongly in $L_2(\Omega)$. Let us show that $u(x,t)$ is a weak solution of (2.23)–(2.25). Since the solution of this problem is unique (Lemma 5.1), it follows that the whole sequence $u_\varepsilon(x,t)$ converges to $u(x,t)$ in $L_2(\Omega)$ as $\varepsilon\to 0$.

LEMMA 5.1. *The problem (2.18)–(2.20) has a unique solution $w(x,t)$ satisfying the inequality $\|w\|_{W_2^1(\Omega_T)} \le CT$.*

PROOF. If we have two solutions $w^1(x,t)$ and $w^2(x,t)$, then their difference $\delta(x,t) = w^1(x,t)-w^2(x,t)$ is a solution of the problem (2.18)–(2.20)with initial data equal to zero and $\|\delta\|_{W_2^1(\Omega_T)} < CT$, $\forall T$. It follows that the Laplace transformation

$$\tilde{\delta}(x,\lambda) = \int_0^T \delta(x,t)e^{-\lambda t}dt \in \overset{\circ}{W}{}_2^1(\Omega)$$

is an analytical vector function with respect to λ in half-plane $\mathrm{Re}\lambda > 0$ and satisfies the identity

$$\lambda\int_\Omega(\tilde{\delta},v)dx + \int_\Omega \sum_{n,p,q,r=1}^{3} A_{npqr}(x,\lambda)\gamma_{qr}[\tilde{\delta}]\gamma_{np}[v]dx = 0, \quad \forall v\in W_2^1(\Omega). \tag{5.1}$$

According to Theorem 4.3, the tensor $A_{npqr}(x,\lambda)$ is analytical with respect to λ in half-plane Re$\lambda > 0$ and positively defined for $\lambda > 0$. It follows from (5.1) and analyticity of $\tilde{\delta}(x,\lambda)$ that $\tilde{\delta}(x,\lambda) \equiv 0$ and, consequently, $\delta(x,\lambda) \equiv 0$. □

We consider the vector function

$$\hat{w}(x,t) = \frac{1}{2\pi i} \int_{\sigma-i\infty}^{\sigma+i\infty} \frac{w(x,\lambda)}{\lambda^2} e^{\lambda t} d\lambda. \tag{5.2}$$

Accorrding to the conditions of $w(x,\lambda)$ (Theorem 4.4) $\hat{w}(x,t) \in C[-T,T; W_2^1(\Omega)]$, and $\hat{w}(x,t) = 0$ as $t < 0$.

Let us multiply the identity (4.11) by $\frac{e^{\lambda t}}{2\pi i \lambda^4}\eta^{(4)}(t)$, where $\eta(t) \in C_0^4(-T,T)$, and integrate with respect to $\lambda \in (\sigma - i\infty, \sigma + i\infty)$ and to $t \in [-T,T]$. So we get

$$\begin{aligned} &\int_{\Omega_T} \rho(x)\hat{w}(x,t)\cdot\zeta(x)\eta^{(4)}(t)dxdt \\ &+ \int_{\Omega_T} \sum_{n,p,q,r=1}^{3} \int_0^t (A_{npqr}(x,\tau)\gamma_{np}[\hat{w}(x,\tau)]\gamma_{qr}[\zeta(x)])d\tau\eta^{(3)}(t)dxdt \\ &= \int_{\Omega} \rho(U^0\eta'(0) + V(x)\eta(0))\cdot\zeta(x)dx. \end{aligned} \tag{5.3}$$

It follows from here that $w(x,t) = \frac{d^2\hat{w}(x,t)}{dt^2}$ is a solution to (2.23)–(2.25) in the sense of distributions. Let us show that $w(x,t)$ is equal to $u(x,t)$. Since the set of functions $u_\varepsilon(x,t)$ is compact in $L_2(\Omega_T)$, it is sufficient to show that $u_\varepsilon(x,t)$ converges weakly to $w(x,t)$.

According to definitions (3.20) and (3.18) we have

$$\int_0^t \int_0^s u_\varepsilon(x,\tau)d\tau ds = \frac{1}{2\pi i}\int_{\sigma-i\infty}^{\sigma+i\infty} \frac{u_\varepsilon(x,\lambda)}{\lambda^2}e^{\lambda t}d\lambda.$$

The following equalities are valid for arbitrary $\zeta(x) \in L_2(\Omega)$ and $\eta(t) \in C_0^2[0,T]$

$$\begin{aligned} \int_0^T \int_\Omega u_\varepsilon \cdot \zeta\eta dxdt &= \int_\Omega \int_0^T \int_0^t \int_0^s u_\varepsilon(x,\tau)\cdot\zeta(x)d\tau ds\eta'' dxdt \\ &= \frac{1}{2\pi i}\int_\Omega \int_0^T \int_{\sigma-i\infty}^{\sigma+i\infty} \frac{u_\varepsilon}{\lambda^2}\cdot\zeta e^{\lambda t}d\lambda\eta'' dtdx \\ &= \frac{1}{2\pi i}\int_{\sigma-i\infty}^{\sigma+i\infty} \frac{1}{\lambda^2}\int_\Omega \zeta\cdot u_\varepsilon(x,\lambda)dx \int_0^T e^{\lambda t}\eta'' dtd\lambda. \end{aligned} \tag{5.4}$$

Here the replacement of integrals is correct because the last integral is convergent absolutely.

According to Theorem 4.4, $u_\varepsilon(x,\lambda)$ converges to the vector function $w(x,\lambda)$. Passing to the limit in (5.4) as $\varepsilon \to 0$ and taking into account (5.2), we obtain

$$\lim_{\varepsilon\to 0}\int_0^T\int_\Omega u_\varepsilon\cdot\zeta\eta dxdt = \int_0^T\int_\Omega \hat{w}(x,t)\zeta\eta'' dxdt.$$

Since the linear span of functions $\zeta(x)\eta(t)$ is dense in $L_2(\Omega_T)$, it follows that the vector function $u_\varepsilon(x,t)$ converges weakly in $L_2(\Omega_T)$ to the vector function $\frac{d^2\hat{w}(x,t)}{dt^2} = w(x,t)$, and hence $w(x,t) = u(x,t)$.

Theorem 2.2 is proved.

References

[1] O. A. Oleinik, A. S.Shamaev and G. A. Yosifian, *Mathematical problems in elasticity and homogenization*, Studies in Mathematics and its Applications, 26, North-Holland Publishing Co., Amsterdam, 1992. MR1195131 (93k:35025)

[2] J. Sanchez-Hubert, *Asymptotic study of the macroscopic behaviour of a solid-fluid mixture*, Math. Meth. Appl. Sci. 2 (1980), 1–11. MR561374 (81a:76046)

[3] R. P. Gilbert and A. Mikelić, *Homogenizing the acoustic properties of the seabed*, J. Nonlinear Anal. **40** (2000), 185–212 . MR1768409 (2002k:74017)

[4] A. M. Meirmanov, *Nguetseng's two-scale convergence method for filtration and seismic acoustic problems in elastic porous media*, Sibirsk. Mat. Zh. **48** (2007), 645–667. MR2347913 (2008k:74017)

[5] E. Sanchez-Palencia, *Non homogeneous media and vibration theory*, Springer, Berlin, 1980.

[6] M. V. Goncharenko and N. K. Radyakin, *An averaged model of oscillations of the elastic medium with a lot of small caverns filled with a low-viscosity incompressible fluid*, Dopovidi NAN Ukraine (2011), no. 5, 7–11.

[7] M. V. Goncharenko and E. Ya. Khruslov, *Averaged model of vibration of a damped elastic medium*, Ukr. Mat. Zh. **62** (2010), no. 10, 1309–1329. MR2924566

[8] O. A. Ladizhenskaya, *The Mathematical Theory of Viscous Incompressible Flow*, New York-London-Paris: Gordon and Breach, Science Publishers, 1969. MR0254401 (40:7610)

[9] M. V. Goncharenko and E. Ya. Khruslov, *On the exponential decay of vibrations of damped elastic media*, Ukr. Math. Zh. **63** (2011), 1641–1659. MR3109679

[10] V. A. Marchenko and E. Ya. Khruslov, *Homogenization of Partial Differential Equations*, Springer, 2006. MR2182441 (2007b:35028)

[11] O. A. Ladizhenskaya, *The Boundary Value Problems of Mathematical Physics*, Applied Mathematical Sciences, 49, Berlin-Heidelberg-New York: Springer Verlag, 1985.

[12] S. G. Mikhlin, *The Problem of the Minimum of a Quadratic Functional*, Holden-Day Incorporated, 1965. MR0171196 (30:1427)

[13] T. Kato, *Perturbation theory for linear operators*, Springer, 1995. MR1335452 (96a:47025)

B. Verkin Institute for Low Temperature Physics and Engineering, 47 Lenin Avenue, Kharkiv 61103, Ukraine

E-mail address: khruslov@ilt.kharkov.ua

B. Verkin Institute for Low Temperature Physics and Engineering, 47 Lenin Avenue, Kharkiv 61103, Ukraine

E-mail address: goncharenko@ilt.kharkov.ua

B. Verkin Institute for Low Temperature Physics and Engineering, 47 Lenin Avenue, Kharkiv 61103, Ukraine

E-mail address: radyakin@ilt.kharkov.ua

Amer. Math. Soc. Transl.
Volume **233**, 2014

Double Scaling Limit in the Painlevé IV Equation and Asymptotics of the Okamoto Polynomials

V. Yu. Novokshenov and A. A. Shchelkonogov

Dedicated to the 90th birthday of Vladimir Aleksandrovich Marchenko.

Abstract. A set of Okamoto polynomials $Q_n(z)$ provides one-parameter rational solutions of the Painlevé IV equation (PIV). They satisfy nonlinear recurrence ODEs in z involving three neighboring polynomials Q_{n-1}, Q_n and Q_{n+1}. Similarly to semi-classical orthogonal polynomials arising in various aspects of mathematics and physics, we consider the scaling limit as $n \to \infty$ and $\sqrt{n}/|z| = \text{const}$ and establish the asymptotic distribution of zeros at infinity. In this limit, the zeroes of $Q_n(z)$ form a lattice with one-to-one correspondence to poles of $u(z) = -2z/3 + \partial_z \log Q_{n+1}/Q_n$ which is a rational solution to PIV. Each pole of $u(z)$ has residue ± 1 so that they are interpreted as positive or negative electric charges interacting with quadratic field. It turns out that in the limit above the lattice become regular at infinity; thus the equation of state with one-to-one Coulomb interaction is generalized to an infinite number of particles.

0. Introduction

The forth Painlevé equation equation (PIV)

$$u'' = \frac{(u')^2}{2u} + \frac{3}{2}u^3 + 4zu^2 + 2(z^2 - \alpha)u + \frac{\beta}{2u} \tag{0.1}$$

has a number of applications in modern mathematical physics. Besides being integrable in the sense of soliton theory, it has hierarchies of solutions coming from supersymmetric quantum mechanics [**1**], random matrix models [**11**], and nonlinear Schrödinger equations [**12**]. For example, in the Gaussian unitary matrix ensembles with prescribed spectral degeneracy, the averages of characteristic polynomials give a τ-function of the rational solutions of the PIV equation [**2**]. In the theory of orthogonal polynomials the PIV equation often arises for determination of recurrence coefficients of the polynomials from the weight function. In its turn, the recurrence coefficients are expressed in terms of special polynomials that form rational solutions of the PIV equation.

2010 *Mathematics Subject Classification.* Primary 30D35, 30E10, 33C75, 34M35, 34M55, 34M60.

Key words and phrases. Painlevé equations, meromorphic solutions, distribution of poles, double scaling limit, Okamoto polynomials, Riemann-Hilbert problem, Stiltjes relations, Coulomb gas.

The study of polynomials associated with the PIV is also motivated by equilibrium distributions of Coulomb charged particles. Indeed, any pole of a rational solution to the PIV has a residue equal to $+1$ or -1. The poles can be interpreted as positive and negative charges interacting by logarithmic potential (see §1) and external quadratic potential. Each pole coincides with zero of the related polynomial. What is the distribution of poles for large orders of polynomials? This question seems to be interesting for the dynamics of Coulomb log-gases [**5**] and approximations in logarithmic potential theory [**16**].

To answer this question, we use the integrability properties of the PIV equation. Namely, in §2 we establish invariants of the given set of rational solutions in terms of monodromy data. Then the same invariants are found for approximate solutions arising in double scaling limit $z \to \infty$ and $\sqrt{n}/|z| = \text{const}$ (§4). This gives the parameters of approximate solution which is an elliptic function modulated with respect to the "slow variable" $\varphi = \arg z$ (§3). Thus a distribution of poles (Coulomb particles) forms a regular lattice at infinity. This enables us to generalize the equation of equilibrium state (Stieltjes relations) for an infinite number of particles. Finally, we prove that Stieltjes relations in this limit coincide with modulation (Whitham) equations for the elliptic function.

1. Stieltjes relations and the PIV equation

All solutions of the PIV equation (0.1) are meromorphic functions [**6**]. For some choice of parameters α and β there are families of rational solutions which are generated recursively by a set of *Backlúnd transformations* [**3**] (see (3.6) below) in §2). All of them have an "electrostatic" interpretation as an equilibrium set of Coulomb charged particles.

THEOREM 1.1 ([**12**]). *Any rational solution to equation* (0.1) *has the form*

$$u(z) = \epsilon z + \sum_j \frac{c_j}{z - z_j}, \quad \epsilon = 0, -\frac{2}{3}, -2, \quad c_j = \pm 1, \quad j = 1, 2, \ldots, \tag{1.1}$$

and the Stieltjes relation *is true*

$$\sum_{j \neq k} \frac{c_j}{z_k - z_j} + (\epsilon + 1) z_k = 0, \quad k = 1, 2, \ldots. \tag{1.2}$$

PROOF. Take a Laurent series near a pole $z = z_j$

$$u(z) = \sum_{k=-l}^{\infty} c_k (z - z_j)^k, \quad z \to z_j.$$

Scaling the leading terms of the series in equation (0.1) yields $l = 1$, $c_{-1}^2 = 1$. The similar comparison at infinity proves that the rational solution $u(z)$ has at most linear growth $u(z) = \epsilon z + O(1)$ as $z \to \infty$ with $\epsilon = 0$, $-2/3$, -2. Expand a rational soliton $u(z)$ into simple fractions (1.1) and put it into equation (0.1), looking for terms of order $O(z - z_k)^{-2}$ as $z \to z_k$. Scaling these terms gives relations (1.2) for any pole z_k. □

The problem of the equilibrium state of Coulomb particles dates back to the end of the 19th century. It arose from the attempts to establish the inner atomic structure in the framework of G. G.Thompson's model. Strictly speaking, the log-potential (1.3) does not relate to particles, but to infinite uniformly charged rods in $\mathbb{R}^3$ space. However, the particles interacting by log-potential in the plane still have applications in physics [**5**].

Actually, in 1885 T. Stieltjes considered [**17**] n particles on the line interacting pairwise with repulsive logarithmic potential in the harmonic field. He proved that the equilibrium of this system is exactly the set of zeroes of Hermite polynomials $H_n(z) = (-1)^n e^{z^2} \frac{d^n}{dz^n} e^{-z^2}$. More precisely, the extremum condition for the potential function (1.3) with $\epsilon = -2$ which is the system of the *Stieltjes relations* determines exactly the roots of the equation $H_n(z) = 0$.

Following [**17**] it is natural to identify poles with residues $c_j = +1$ as positive charges and those with $c_j = -1$ as negative charges. Moreover, the Stieltjes relation (1.2) defines local extremum of potential function [**7**]

$$U(z_1, z_2, \ldots, z_n) = (\epsilon + 1) \sum_{j=1}^{n} c_j z_j^2 + \sum_{j \neq k}^{n} c_j c_k \log(z_j - z_k)^2. \tag{1.3}$$

Note that Stieltjes' result describes a particular distribution of poles to the PIV equation (0.1). Indeed, the Hermite polynomials form the family of rational solution

$$u(z) = -2z + \frac{H_n'(z)}{H_n(z)},$$

which has poles on the real axis with all positive charges $c_j = +1$.

In this paper, we address the questions [**18**]:

$Q1$: | *How can we generalize Stieltjes relations for an infinite number of Coulomb particles?*

or, equivalently

$Q2$: | *Does an equilibrium distribution of infinite number of Coulomb particles (Coulomb gas) exist?*

Note that an infinite set of charged particles in equilibrium has been known for a long time—those are poles of elliptic functions. For example, the series representation of Weierstrass function

$$\wp(z|\omega, \omega') = \frac{1}{z^2} + \sum_{(m,n)\neq(0,0)} \left\{ \frac{1}{(z - 2m\omega - 2n\omega')^2} - \frac{1}{(2m\omega - 2n\omega')^2} \right\},$$

where ω and ω' are basic periods and $2m\omega + 2n\omega'$, $m, n \in \mathbb{Z}$ is the lattice of poles.

This representation is very similar to equation (1.1). Putting $z = z_0$, $\wp(z_0|\omega, \omega') = 0$ here one gets an infinite analog of the Stiltjes relation (1.2). This gives a direction of generalization (1.2) on the infinite set of poles.

As a model we will consider *Okamoto polynomials* Q_n, $n = 0, 1, \ldots$, as generators of rational solutions of the PIV equation [**15**]. Indeed,

$$u_n(z) = -\frac{2}{3} z + \frac{d}{dz} \log \frac{Q_{n+1}(z)}{Q_n(z)}, \tag{1.4}$$

satisfies the PIV equation (0.1) with $\alpha = 2n$, $\beta = -\frac{2}{9}$ and Q_n defined by the recurrence [**3**]

(1.5)

$$Q_{n+1} Q_{n-1} = \frac{9}{2} \left[Q_n \frac{d^2 Q_n}{dz^2} - \left(\frac{dQ_n}{dz} \right)^2 \right] + \left[2z^2 + 3(2n - 1) \right] Q_n^2, \quad n = 1, 2, \ldots,$$

where

$$Q_0(z) = Q_1(z) = 1, \quad Q_2(z) = 2z^2 + 3.$$

All zeroes of Q_n are simple and they coincide with the poles of rational solution (1.4) with residues $c_j = +1$ as n is odd and $c_j = -1$ as n is even. Thus the distribution of zeroes of Q_n as $n \to \infty$ coincides with singularities of a PIV solution for $\alpha \to \infty$. The latter limit (see §2) yields an elliptic function equation, and the distribution of poles of an elliptic function should mimic the distribution of zeros of Q_n. Figure 1 illustrates this similarity for relatively small n. In fact, the correct approximation is achieved in the limit $n \to \infty$, $\sqrt{n}/|z| = O(1)$, while the module of the elliptic function depends on "slow variable" $\arg t$ (see §4).

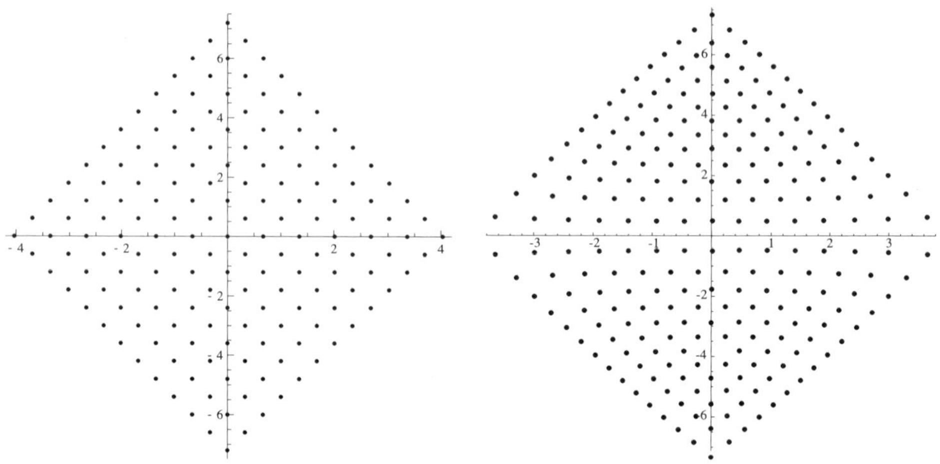

FIGURE 1. The poles of Weierstrass elliptic function $\wp(e^{\pi/4}z|\omega,\omega')$ (left) and zeroes of the Okamoto polynomials $Q_n(z)$ for $n = 15$ (right).

2. Elliptic asymptotics of the PIV equation

Following [9] let us first make the scaling in the PIV equation (0.1)

$$u(z) = \sqrt{2t}\, v(t), \quad t = \frac{z^2}{2}, \quad |t| \to \infty. \tag{2.1}$$

Then equation (0.1) takes the form ($' = d/dt$)

$$v'' = \frac{(v')^2}{2v} + \frac{3}{2}v^3 + 4v^2 + 2\left(1 - \frac{\alpha}{2t}\right)v - \frac{v'}{t} + \frac{v}{8t^2} + \frac{\beta}{8t^2 v}. \tag{2.2}$$

Since $\alpha = 2n \to \infty$, $\beta = -2/9$, assume the scaling limit

$$t = |t|e^{i\varphi}, \quad \frac{\alpha}{2t} = -A + 1 = O(1), \quad |t| \to \infty. \tag{2.3}$$

Thus equation (2.2) in leading order $v = V(t) + o(1)$ yields

$$V'' = \frac{(V')^2}{2V} + \frac{3}{2}V^3 + 4V^2 + 2AV.$$

Actually, this equation is for an elliptic function. Indeed, multiplying both parts by $2V'/V$ and integrating in t yields

$$(V')^2 = V^4 + 4V^3 + 4AV^2 + 4BV, \tag{2.4}$$

where B is an integration constant.

The solution $V = V(t\,|\,A, B)$ of equation (2.4) is defined by elliptic integral

$$\int_0^V \frac{d\xi}{\mu(\xi)} = t - t_0, \tag{2.5}$$

over the Riemann surface

$$\mu^2 = \xi^4 + 4\xi^3 + 4A\xi^2 + 4B\xi. \tag{2.6}$$

We will estimate the remainder terms of equation (2.2)

$$(v')^2 - v^4 - 4v^3 - 4v^2 = -4v \int_{t_0}^{t} \left((1-A)v' + \frac{(v')^2}{2v} \right) \frac{dt}{t}$$

$$+\frac{v}{4} \int_{t_0}^{t} \left(v' + \frac{\beta v'}{v^2} \right) \frac{dt}{t^2}. \tag{2.7}$$

We are going to apply approximation (2.5) at a large domain

$$\mathcal{L} = \{ t \,|\, -\pi/2 < \arg t < 0, \quad |t| \gg 1 \}.$$

Suppose that $v = V(t) + o(1)$ on $\mathcal{L}$ and integrate the right-hand side of (2.7) over $\mathcal{L}$ while $t = |t|e^{i\varphi}$

$$\int_{t_0}^{t} \frac{V'dt}{t} = V(t) - V(t_0) + \frac{i}{|t|} \int_{\varphi_0}^{\varphi} e^{-i\varphi} V(|t|e^{i\varphi}) d\varphi,$$

$$\int_{t_0}^{t} \left(V' + \frac{\beta V'}{V^2} \right) \frac{dt}{t^2} = \frac{i}{|t|} \int_{\varphi_0}^{\varphi} e^{-i\varphi} \left(V' + \frac{\beta V'}{V^2} \right) d\varphi + o(|t|^{-1}),$$

$$\int_{t_0}^{t} \frac{(V')^2 dt}{Vt} = i \int_{\varphi_0}^{\varphi} \left\langle \frac{(V')^2}{V} \right\rangle d\varphi + o(1),$$

where $\langle \cdot \rangle$ is an average of two-period elliptic function

$$\left\langle \frac{(V')^2}{V} \right\rangle = \int_\ell \frac{\mu(\xi)}{\xi} d\xi \Big/ \int_\ell \frac{d\xi}{\mu(\xi)},$$

and ℓ is some cycle passing over the Riemann surface (2.6) while t goes along $\mathcal{L}$.

The definition (2.3) and leading orders in $|t|$ of the above integrals yield equations for A and B

$$A(\varphi) = 1 - \frac{n}{|t|} e^{-i\varphi}, \tag{2.8}$$

$$B(\varphi) = -\frac{i}{2} \int_{\varphi_0}^{\varphi} \left\langle \frac{(V')^2}{V} \right\rangle d\varphi + B(\varphi_0). \tag{2.9}$$

Equation (2.9) can be transformed into more explicit form. Note that

$$\frac{\partial}{\partial B}\int \frac{\mu(\xi)}{\xi}d\xi = 2\int \frac{d\xi}{\mu(\xi)},$$

and get the solution of (2.9)

$$\int\limits_{\ell} \frac{\mu(\xi)}{\xi}d\xi = Ce^{-i\varphi}.$$

Since integration along $\mathcal{L}$ maps into large generic cycle $\ell = N_1 l_1 + N_2 l_1$ with N_1, $N_2 \gg 1$ and l_1, l_2 - basic cycles of (2.6), this equation takes form [**14**]

$$\arg\int\limits_{\ell_1} \frac{\mu(\xi)}{\xi}d\xi = \arg\int\limits_{\ell_2} \frac{\mu(\xi)}{\xi}d\xi = -\varphi + \arg C = -\varphi + \frac{\pi}{2}. \tag{2.10}$$

Here we put $\arg C = \pi/2$ because at the ends of the arc $\mathcal{L}$ the elliptic function (2.5) degenerates; for example, as $\varphi = -\pi/2$ we have $\ell_2 = 0$ and $\ell_1 = (0, -2\sqrt{i})$, so that $C = 8i/3$.

THEOREM 2.1. *Equation (2.10) has the unique solution $B = B(\varphi)$ in the domain $-\pi/2 < \arg\varphi < 0$.*

PROOF. Following [**9**], introduce a slightly modified elliptic Riemann surface (2.6)

$$\nu^2 = \xi(\xi-\xi_1)(\xi-\xi_3)(\xi-\xi_5) \equiv \xi^4 + 4e^{i\phi}\xi^3 + 4A(\phi)e^{2i\phi}\xi^2 + 4B(\phi)e^{3i\phi}.$$

Here the cuts are made over intervals (ξ_5, ξ_3) and $(0, \xi_1)$; on the upper sheet $\nu(\xi) \to \xi^2 + 2e^{i\phi}\xi$ as $\xi \to \infty^+$. The basic cycles a and b on the upper sheet are chosen as

$$\frac{1}{2}a = (\xi_5, \xi_3), \quad \frac{1}{2}b = (\xi_3, \xi_1).$$

Consider the following *Boutroux problem*:

> *Find a smooth function $B(\phi)$ together with branch points ξ_1, ξ_3 and ξ_5 at the intervals $\phi \in (-\pi/4, 0) \cup (0, \pi/4)$ such that*
>
> $$\xi_1 + \xi_3 + \xi_5 = -4e^{i\phi}, \qquad \mathrm{Re}\int\limits_{a,b} \frac{\nu(\xi)}{\xi}d\xi = 0, \tag{2.11}$$
>
> *with the boundary conditions*
>
> $$\xi_{3,1} \to -\frac{2}{3}, \quad \text{as} \quad \phi \to 0, \qquad \xi_1 \to 0, \quad \xi_{3,5} \to -2e^{\pm i\pi/4}, \quad \text{as} \quad \phi \to \pm\frac{\pi}{4}. \tag{2.12}$$

The existence of the unique solution to the Boutroux problem follows from Theorem 1 in [**9**]. Equation (2.11) is equivalent to (2.10) for

$$\phi = \frac{\varphi}{2} + \frac{\pi}{2}, \quad \xi \mapsto \xi e^{i\varphi/2}.$$

□

REMARK 2.2. Equations (2.8) and (2.10) describe modulation of elliptic function $V = V(t|A, B)$ with respect to the "slow variable" $\varphi = \arg t$. They are similar to Whitham equations for modulation of multi-periodic wave packets in nonlinear media. These equations arise in asymptotic solutions of Painlevé equations in the complex plane near essential singularities (e.g. infinity for PI, PII and PIV equations). Indeed, if $\alpha = O(1)$ in (2.3), then elliptic function (2.1) modulated by equations (2.10) ($A \equiv 1$) provides the asymptotics at infinity $z \to \infty$, $-\pi/2 < \arg z < 0$. (See [**9**] and [**4**], Chapter 9.)

The poles of the function (2.5) form a rectangular lattice in $t \in \mathbb{C}$. We choose parameters A, B and t_0 in a way to match the poles of

$$u(z) = zV\left(\frac{z^2}{2} - t_0(\varphi)\,\middle|\, A(\varphi), B(\varphi)\right), \tag{2.13}$$

with the poles of the rational solution (1.4)

$$u(z) = -\frac{2}{3}z + \frac{d}{dz}\log\frac{Q_{n+1}(z)}{Q_n(z)},$$

i.e. with the zeroes of Okamoto polynomial $Q_n(z)$ as $z \to \infty$, $\sqrt{n}/|z| = O(1)$.

3. Isomonodromic deformation method for the PIV equation

In order to match the above parameters we use the invariants of the PIV equation. Those are the monodromy data of the λ-equation of the Lax pair (see [**4**], Chapter 5)

$$\Psi_\lambda = A(\lambda, z)\Psi, \tag{3.1}$$

where $\Psi = \Psi(\lambda, z)$ is 2×2 matrix function rational in auxiliary parameter λ

$$A = \lambda\begin{pmatrix}1 & 0\\ 0 & -1\end{pmatrix} + \begin{pmatrix} z & y\\ 2(uw + \theta_0 - \theta_\infty)y^{-1} & -z\end{pmatrix}$$
$$+\frac{1}{\lambda}\begin{pmatrix} -uw + \theta_0 & -yw/2\\ 2(uw - 2\theta_0)y^{-1} & uw - \theta_0\end{pmatrix},$$

where u, w and y are functions of z satisfying nonlinear equations

$$\frac{d}{dz}\log y = -u - 2z,$$

$$\begin{cases} \dfrac{du}{dz} = -4uw + u^2 + 2zu + 4\theta_0,\\[2mm] \dfrac{dw}{dz} = 2w^2 - 2uw - 2zw + \theta_0 + \theta_\infty. \end{cases} \tag{3.2}$$

Equation (3.2) is equivalent to the PIV equation (0.1) on function $u = u(z)$ with parameters

$$\alpha = 2\theta_\infty - 1, \quad \beta = -8\theta_0^2.$$

The monodromy conditions for functions Ψ of equation (3.1) are given by the jump conditions

$$\Psi_{j+1}(\lambda, z) = \Psi_j(\lambda, z)S_j, \quad j = 1, 2, 3, 4. \tag{3.3}$$

Here four basic solutions Ψ_j of (3.1) are normalized at infinity in four sectors $\Omega_j = \{\lambda \mid -\pi/4 + \pi j/2 < \arg\lambda < \pi/4 + \pi j/2, \quad j = 1, 2, 3, 4\}$

$$\Psi_j(\lambda, z) = \left(I + O(\lambda^{-1})\right) \exp\left\{\left(\frac{\lambda^2}{2} + z\lambda\right)\sigma_3\right\} \lambda^{-\theta_\infty \sigma_3}, \tag{3.4}$$

$$\lambda \to \infty, \quad \lambda \in \Omega_j, \quad \sigma_3 = \begin{pmatrix} 1 & 0 \\ 0 & -1 \end{pmatrix}.$$

It is known [4], [9] that jump matrices S_j (3.3) are independent of z iff u, w and y satisfy equation (3.2), or, equivalently, $u = u(z)$ satisfies the PIV equation (0.1). Thus the scalar entries of these matrices s_1, s_2, s_3 and s_4

$$S_{2k} = \begin{pmatrix} 1 & s_{2k} \\ 0 & 1 \end{pmatrix}, \quad S_{2k-1} = \begin{pmatrix} 1 & 0 \\ s_{2k-1} & 1 \end{pmatrix}, \quad k = 1, 2,$$

are the integrals of motion (first integrals) of the PIV equation (0.1). Of course, they are not independent; they satisfy, for example, the algebraic relation [4]

$$(1 + s_2 s_3) e^{2\pi i \theta_\infty} + (s_1 s_4 + (1 + s_3 s_4)(1 + s_1 s_2))\, e^{2\pi i \theta_\infty} = 2\cos(2\pi\theta_0).$$

We calculate the invariants s_j both for the rational solution (1.4) and approximate solution (2.13).

We begin with the rational solutions (1.4) which are generated by recurrence equations (1.5). For $n = 0$ we have $Q_0 = Q_1 = 1$ and

$$u_0(z) = -\frac{2}{3}z, \quad \theta_\infty = \frac{n+1}{2}, \quad \theta_0 = -\frac{1}{6}. \tag{3.5}$$

In other terms, the set of solutions (1.4) is constructed by the Backlünd transformations [3]

$$\tilde{u} = \frac{(u' - u^2 - 2zu)^2 + 2\beta}{2u(u' - u^2 - 2zu + 2(\alpha + 1))},$$

where $u = u(z;\, \alpha, \beta)$ and $\tilde{u} = u(z;\, \tilde{\alpha}, \tilde{\beta})$ are solutions of the PIV equation (0.1) for coefficients $\tilde{\alpha} = \alpha + 2$ and $\tilde{\beta} = \beta$. For our case we have recurrence relation

$$u_{n+1} = \frac{9(u_n' - u_n^2 - 2zu_n)^2 - 4}{18u_n(u_n' - u_n^2 - 2zu_n + 4n + 2)}, \tag{3.6}$$

which is equivalent to equations (1.4) and (1.5).

It is known ([4], Chapter 6) that the Backlünd transformations form the left-hand side rational "dressing" of Ψ-functions of equation (3.1)

$$\tilde{\Psi} = R(\lambda)\Psi,$$

so they do not change the corresponding s_j invariants of $u_n(z)$.

Thus it is enough to calculate s_j for the case (3.5). From equations (3.2) and (3.5) we find that $w(z) = \frac{z}{3}$ and

$$u(0) = w(0) = 0, \quad y(0) = 1. \tag{3.7}$$

The values (3.7) together with $z = 0$ reduce the equation to a exactly solvable system. Indeed, putting

$$\xi = \lambda^2, \quad \begin{pmatrix} \Psi_{11} \\ \Psi_{21} \end{pmatrix} = \xi^{-1/4} \begin{pmatrix} \phi_1 \\ \phi_2 \end{pmatrix},$$

we get the scalar equations [8]

$$\frac{d^2\phi_1}{d\xi^2} = \left(-\frac{1}{4} + \frac{k}{\xi} + \frac{\frac{1}{4} - m^2}{\xi^2}\right)\phi_1,$$
$$\frac{d^2\phi_2}{d\xi^2} = \left(-\frac{1}{4} + \frac{k+\frac{1}{2}}{\xi} + \frac{\frac{1}{4} - \left(m+\frac{1}{2}\right)^2}{\xi^2}\right)\phi_2,$$

where $k = \frac{2\theta_\infty - 1}{4}$ and $m = \frac{2\theta_0 - 1}{4}$. This yields the exact solution of equation (3.1) with respect to (3.7) and normalization (3.4) (see [8], Theorem 4)

$$\Psi(\lambda, 0) = \begin{pmatrix} L_{k,m}(\lambda) & L_{k,-m}(\lambda) \\ -(2k+2m+1)L_{k+\frac{1}{2},m+\frac{1}{2}}(\lambda) & -4mL_{k+\frac{1}{2},-m-\frac{1}{2}}(\lambda) \end{pmatrix},$$

where $L_{k,m}$ are Whittaker functions

$$L_{k,m}(\lambda) = \lambda^{2m+\frac{1}{2}} e^{-\lambda^2/2} {}_1F_1\left(m - k + \frac{1}{2}, 2m+1;\, \lambda^2\right).$$

For the case (3.5) we have $k = 0$, $m = -\frac{1}{3}$. The jump conditions (3.3) now yield from [8], Theorem 5

$$s_j = -i, \quad j = 1, 2, 3, 4. \tag{3.8}$$

4. Distribution of poles in double scaling limit

In elliptic function ansatz (2.5) we need to find the phase shift t_0. This is achieved by comparing the invariants s_j found for (2.5) from the monodromy equation (3.1). First we formulate the result by A. Kapaev on the asymptotics of the PIV transcendents at infinity in the case of $\alpha, \beta = O(1)$.

THEOREM 4.1 ([9]). *If the invariants s_j satisfy inequalities*

$$(1 - s_1 s_2)(1 + s_2 s_3) \neq 1, \quad (1 - s_1 s_2)(1 + s_1 s_4) \neq 1, \quad 1 - s_1 s_2 \neq 0, \tag{4.1}$$

then the asymptotic solution of the PIV equation (0.1)

$$u(z) = \sqrt{2t}V(t\,|\,A, B) + o(t^{1/2}), \quad t = \frac{z^2}{2}, \quad t \to \infty, \quad -\pi/4 < \arg t < 0.$$

Here function V is given by inversion of the elliptic integral (2.5)

$$\int_0^V \frac{d\xi}{\mu(\xi)} = t - t_0,$$

where $\mu^2 = \xi^4 + 4\xi^3 + 4A\xi^2 + 4B\xi$, $A \equiv 1$ and B satisfy equations (2.10). The phase shift t_0 satisfies relations

$$\begin{aligned} t_0 &= p\Omega_a + q\Omega_b + O(t^{-1}), \\ p &= -\frac{1}{2\pi i}\log\left((1 - s_1 s_2)(1 + s_2 s_3) - 1\right), \\ q &= \frac{1}{2\pi i}\log(1 - s_1 s_2) + \frac{2}{3}\alpha - \frac{5}{2}i\sqrt{\beta}, \end{aligned} \tag{4.2}$$

where Ω_a, Ω_b are basic periods of (2.5)

$$\Omega_{a,b} = \int_{a,b} \frac{d\xi}{\mu(\xi)}. \tag{4.3}$$

REMARK 4.2. Note that **[9]** uses another monodromy equation by A. Kitaev instead of (3.1) by M. Jimbo and T. Miwa. Thus the function $\tilde{\Psi}$ and invariants $\tilde{s}_j$ in **[9]** differ from those used above:

$$\Psi(\lambda, z) = e^{\frac{z^2\sigma_3}{2}} \begin{pmatrix} 1 & iu \\ 0 & 1 \end{pmatrix} (2\lambda)^{\frac{\sigma_3}{8}} \tilde{\Psi}(\sqrt{2\lambda}, z),$$
$$s_1 = -\tilde{s}_1, \quad s_4 = -\tilde{s}_4.$$

We return to the double scaling limit case (2.3)

$$\alpha = 2n, \quad \beta = -2/9, \quad \frac{n}{|t|} = \text{const}, \quad t \to \infty.$$

Note that large parameter $\alpha = 2n$ enters only the phase shift of the periodic function V. Thus the asymptotics in Theorem 4.1 do not depend on it, i.e. t_0 can be made finite by subtracting a proper number of periods Ω_b.

The set of invariants s_j (3.8) satisfy conditions (4.1); thus the phase shift (4.2) has coefficients

$$p = \frac{i}{2}, \quad q = \frac{1}{2\pi i}\log 2 + \left(\frac{4n}{3} - \frac{5\sqrt{2}}{3}\right) \operatorname{mod} n. \tag{4.4}$$

THEOREM 4.3. *The zeroes z_k, $-\pi/2 < \arg z_k < 0$ of the Okamoto polynomials $Q_n(z)$ (1.5) in the double scaling limit $n \to \infty$, $\sqrt{n}/|z| = O(1)$ have the form*

$$z_k = \frac{1}{2}(pN_1\Omega_a + qN_2\Omega_b) + o(1), \quad N_1, N_2 \in \mathbb{Z}, \quad N_1, N_2 \to \infty,$$

where p and q are given by (4.4) and Ω_a, Ω_b are basic periods (4.3) on the Riemann surface (2.6) with parameters (see (2.8) and (2.9)).

The zeroes z_k of polynomial Q_n coincide with the poles of PIV transcendent (1.4) when residues (1.1) $c_j = +1$ and n is odd, or when residues $c_j = -1$ and n is even.

We return now to the Stieltjes relations (1.2) discussed above in §1. Let $z = z_k$ be a pole of solution $u(z)$. Consider the Cauchy integral over the large contour $\mathcal{C}$ which contains all poles of $u(z)$

$$I = \frac{1}{2\pi i}\int_{\mathcal{C}} \frac{u(\zeta)}{\zeta^2 - z_k^2} d\zeta, \tag{4.5}$$

where $\mathcal{C} = \{z \,||z^2 - z_k^2| = 2Cn\}$ and $\mathcal{C}$ is passed counter-clockwise. Because of (1.4) we have $u(z) = -\frac{2}{3}z + O(1)$ as $z \to \infty$; thus integral (4.5) has the pole at infinity and $I = -2/3$.

According to Theorem 4.1 we apply asymptotics $u(z) = \sqrt{2t}V(t) + o(t^{1/2})$, $t = z^2/2$

$$I = \frac{1}{4\pi i}\int_{|t-t_k|=Cn} \frac{V(t)}{t - t_k} dt = -\frac{1}{4\pi}\int_0^{2\pi} \langle V(Cne^{i\varphi} + t_k) \rangle d\varphi + o(1), \tag{4.6}$$

where $\langle\cdot\rangle$ is the average of the elliptic function (2.5) over the period. On the other hand, since the integrand has simple poles in the circle and double pole at $t = t_k$

$$V(t) = \frac{c_j}{t - t_j} - 1 + O(t - t_j), \quad t \to t_j,$$

we have the expansion

$$I = \frac{1}{2} \sum_{\substack{j \neq k \\ |t_j - t_k| < Cn}} \frac{c_j}{t_j - t_k} - \frac{1}{2} = \frac{1}{4z_k} \cdot \sum_{\substack{j \neq k \\ |z_j^2 - z_k^2| < 2Cn}} \left(\frac{c_j}{z_j - z_k} - \frac{c_{-j}}{z_{-j} - z_k} \right) - \frac{1}{2}.$$

Here we used the obvious symmetry $z_j = -z_{-j}$ and $c_j = -c_{-j}$.

Combining the above calculations we come to the Stieltjes relation

$$\sum_{j \neq k} \frac{c_j}{z_k - z_j} + \frac{1}{3} z_k = 0, \quad k = 1, 2, \dots .$$

We transform the integral (4.6) as we did in §2

$$\langle V \rangle = \int_{\ell} \frac{\xi}{\mu(\xi)} d\xi \Big/ \int_{\ell} \frac{d\xi}{\mu(\xi)},$$

where $\ell = N_1 a + N_2 b$, $N_1, N_2 \gg 1$ is a large contour on the Riemann surface (2.6). It arises while t goes along $\mathcal{C}$. Since $I = -2/3$ and due to asymtotics (4.6) the average $\langle V \rangle$ yields

$$\arg \int_{a,b} \frac{\xi}{\mu(\xi)} d\xi = \arg \int_{a,b} \frac{d\xi}{\mu(\xi)}. \tag{4.7}$$

It can be shown that equation (4.7) is equivalent to the Whitham equations (2.10). This will be done in our next publication.

References

[1] D. Bermudez and D. J. Fernandez, *Non-hermitian Hamiltonians and the Painlevé IV equation with real parameters*, Phys. Lett. A **375** (2011), 2974–2978. MR2818319 (2012k:81096)

[2] Y. Chen and M. V. Feigin, *Painlevé IV and degenerate Gaussian unitary ensembles*, J. Phys. A: Math. Gen. **39** (2006), 12381–12393. MR2265830 (2007m:82026)

[3] P. A. Clarkson, *Special Polynomials Associated with Rational Solutions of the Painlevé Equations and Applications to Soliton Equations*, Comp. Methods and Function Theory **6** (2) (2006), 329–401. MR2291140 (2008b:33046)

[4] A. S. Fokas, A. R. Its, A. A. Kapaev and V. Yu. Novokshenov, *Painlevé Transcendents. The Riemann-Hilbert Approach*, Mathematical Surveys and Monographs, **128**. American Mathematical Society, Providence, RI, 2006. MR2264522 (2010e:33030)

[5] P. J. Forrester, *Log-gases and random matrices*, London Math. Soc. Monographs, **34**. Princeton University Press, 2010. MR2641363 (2011d:82001)

[6] V. I. Gromak, I. Laine and S. Shimomura, *Painlevé equations in the complex plane*, de Gruytor Studies in Mathematics, vol. 28, Walter de Gruyter, Berlin, 2002. MR1960811 (2003m:34210)

[7] D. P. Hardin and E. B. Saff, *Discretizing manifolds via minimum energy points*, Notices of the American Mathematics Society **51**(10) (2004), 1186–1194. MR2104914 (2006a:41049)

[8] K. Kaneko, *A new solution of the fourth Painlevé equation with a solvable monodromy*, Proc. Japan Acad. Ser. A Math. Sci. **81**(5) (2005), 75–79. MR2143546 (2006a:34244)

[9] A. A. Kapaev, *Global asymptotics of the fourth Painlevé transcendent*, POMI Preprint **6** (1996), `ftp://ftp.pdmi.ras.ru/pub/publicat/preprint/1996/06-96.ps.gz`

[10] A. A. Kapaev and E. Hubert, *A note on the Lax pairs for Painlevé equations*, J. Phys, A: Math. Gen. **32**(46) (1999), 8145–8156. MR1731965 (2000i:34172)

[11] A. P. Magnus, *Painlevé-type differential equations for the recurrence coefficients of semi-classical orthogonal polynomials*, J. Comput. Appl. Math. **57** (1995), 215–237. MR1340938 (96f:42027)

[12] V. G. Marikhin, A. B. Shabat, M. Boiti and F. Pempinelli, *Self-similar solutions of NLS-type dynamical systems*, J. Exp. Theor. Phys. **117** (3) (2000), 634–643. MR1777679 (2001e:37099)

[13] V. Yu. Novokshenov, *Padé approximations of Painlevé I and II transcendents*, Theor. Math. Phys. **159** (3) (2009), 852–861. MR2568568

[14] V. Yu. Novokshenov, *The Boutroux ansatz for the second Painlevé equation in the complex domain*, Izv. Akad. Nauk SSSR, ser. matem. **54** (1990), 1229–1251. MR1098625 (92c:34006)

[15] K. Okamoto, *Studies on Painlevé equations. III, Second and fourth Painlevé equations P_{II} and P_{IV}*, Math. Ann. **275** (1986), 221–255. MR854008 (88m:58064)

[16] E. B. Saff, *Logarithmic Potential Theory with Applications to Approximation Theory*, Surveys in Approximation Theory **5** (2010), 165–200. MR2734198 (2012k:31001)

[17] T. J. Stieltjes, *Sur certains polynomes qui vérifient une équation différentielle linéaire du second ordre et sur la theorie des fonctions de Lamé*, Acta Mathematica **6** (1) (1885), 321–326. MR1554669

[18] A. P. Veselov, *On Stieltjes relations, Painlevé-IV hierarchy and complex monodromy*, J. Phys. A: Math. Gen. **34** (2001), 3511–3519. MR1838065 (2002e:34153)

Institute of Mathematics RAS, Chernyshevskii street 112, 450008, Ufa, Russia

Chelyabinsk State University, Brat'ev Kashirinykh street 129, 454000, Chelyabinsk, Russia

Amer. Math. Soc. Transl.
Volume **233**, 2014

Recovering a Potential of the Sturm–Liouville Problem from Finite Sets of Spectral Data

A. M. Savchuk and A. A. Shkalikov

To dear Vladimir Alexandrovich Marchenko with sincere respect and gratitude

ABSTRACT. It is well known that a potential q of the Sturm-Liouville operator $Ly = -y'' + q(x)y$ on the finite interval $[0,\pi]$ can be recovered by the two spectra $\{\mu_k\}_1^\infty$ and $\{\lambda_k\}_1^\infty$ of this operator generated by Dirichlet-Neumann and Dirichlet-Dirichlet boundary conditions, respectively. Given potential q belonging to a Sobolev space $W_2^\theta[0,\pi]$ with $\theta > -1$ we associate its $2N$-approximation q_N constructed by the final sets $\{\mu_k\}_1^N$ and $\{\lambda_k\}_1^N$. The main result claims that for $-1 \leqslant \tau < \theta$ the estimate $\|q - q_N\|_\tau \leqslant CN^{\tau-\theta}$ holds, where $\|\cdot\|_\tau$ is the norm in W_2^τ and the constant C depends on R but does not depend on q if $\|q\|_\theta \leqslant R$.

1. Introduction and results for a classical L_2-potential

In the present paper we deal with the classical inverse problem for the Sturm-Liouville operator originated by Borg [**2**]. The problem is to recover a potential given by two spectra of this operator with Dirichet-Neumann and Dirichlet-Dirichlet boundary conditions. It is well known that this problem has a unique solution. Given only finitely many, say $2N$, eigenvalues one can not recover a potential explicitly. However, one can do this with an accuracy depending on N. The accuracy will be understood in the norms of Sobolev spaces with different smoothness indices.

The approach of this paper can be applied to get similar results for other inverse problems on the finite interval (in particular, to the problem of recovering a potential q by the spectral function). However, in the present paper we concentrate our attention only on the Borg problem. Speaking on the historical background for inverse problems on the finite interval we mention only the most important papers of Borg [**2**], Levinson [**19**], Marchenko and Ostrovskii [**23**], [**24**], [**26**], Gelfand, Levitan and Gasymov [**6**], [**7**], [**20**], [**21**], Krein [**17**], [**18**]. Detailed information can be found in the books of Marchenko [**24**], Levitan [**21**], Pöshel and Trubowitz [**28**], and Freiling and Yurko [**5**]. Inverse problems for periodic potentials were considered in the papers of Korotyaev [**16**], Djakov and Mityagin [**3**], [**4**], Kappeler, Schaad and Topalov [**15**] and many others. Direct and inverse problems for the Sturm-Liouville operators with distribution potentials have been studied in our papers [**31**]–[**35**] and those of Hryniv and Mykytyuk [**10**]–[**14**].

2010 *Mathematics Subject Classification.* Primary 35A55; Secondary 34B24.

The stability for inverse problems was studied in a local form in the papers of Marchenko and Maslov [**25**], Yurko [**38**], Ryabushko [**29**], [**30**], and McLaughlin [**22**]. The problem on global or uniform stability was raised in our paper [**35**]. The results of that paper (see Theorem 2.4 below) are used as the main tool in this paper. The investigation of stability for the inverse problem with finite spectral data was initiated in a recent paper by Marletta and Weikard [**27**] (see Remark 1 below). The main novelty of the present paper is that we evaluate precisely the speed of convergence in different norms, for potentials q belonging to different Sobolev spaces (not only for $q \in L_2$) and with uniform constants.

To make the understanding of this paper easier we decided in the first section to formulate the results for the classical case, when a potential q belongs to the space $L_2[0,1]$, and no additional concepts or notations are required. In the second section we will prove the results in a general form.

Consider the Sturm-Liouville operator

$$Ly = -y'' + q(x)y$$

on the finite interval $[0,\pi]$. At the beginning we assume that the potential q is real and $q \in L_2[0,1]$. Denote by L_D and $L_{DN}(h)$ the restrictions of the operator L on the domains

$$\mathcal{D}(L_D) = \{y \in W_2^2[0,1] |\ y(0) = y(\pi) = 0\}$$

and

$$\mathcal{D}(L_{DN}(h)) = \{y \in W_2^2[0,1] |\ y(0) = y'(\pi) - hy(\pi) = 0\}.$$

Here h is a real number and $W_2^\theta[0,1]$, $\theta \in \mathbb{R}$, are Sobolev spaces. (Below we use these spaces with $\theta \geqslant -1$.) L_D and $L_{DN}(h)$ are named Dirichlet and Dirichlet-Neumann operators. (Formally in the second case in addition we have to assume $h = 0$; however, later we consider a problem in a more general setup and the given name will be approved.) It is known that they are self-adjoint and have simple eigenvalues $\{\lambda_k\}_1^\infty$ and $\{\mu_k\}_1^\infty$. Moreover, (see [**20**] or the more general Proposition 2 in [**33**]); these sequences interlace

$$\mu_1 < \lambda_1 < \mu_2 < \lambda_2 \cdots \mu_n < \lambda_n \dots . \tag{1.1}$$

It is also known that that the eigenvalues $\{\lambda_k\}_1^\infty$ and $\{\mu_k\}_1^\infty$ admit the following representations

$$\rho_{2k-1} = \sqrt{\mu_k} = k - \frac{1}{2} + \frac{g_0}{2k-1} + \frac{\alpha_{2k-1}}{k} \tag{1.2}$$

$$\rho_{2k} = \sqrt{\lambda_k} = k + \frac{q_0}{2k} + \frac{\alpha_{2k}}{k}, \tag{1.3}$$

where $\{\alpha_k\}_1^\infty \in \ell_2$ and

$$q_0 = \frac{1}{\pi}\left(\sigma(\pi) - \sigma(0)\right), \quad g_0 = q_0 - \frac{2}{\pi}h, \quad \sigma(x) = \int q(t)\,dt. \tag{1.4}$$

Here σ is an arbitrary primitive of q (below we will consider particular primitives as well).

Notice that asymptotic formulae (1.2) and (1.3) are equivalent to

$$\mu_k = (k-1/2)^2 + g_0 + \beta_{2k-1},$$
$$\lambda_k = k^2 + q_0 + \beta_{2k},$$

with $\sum_1^\infty |\beta_n|^2 < \infty$. Moreover, if $q'(x) \in L_2$, then $\beta_n = O(n^{-2})$. In particular, in the case $q = q_0 = \text{const}$ we have $\lambda_k^0 = k^2 + q_0, \quad \mu_k^0 = (k - 1/2)^2 + g_0 + O\left(k^{-2}\right)$. (We do not need precise information on the remainder $O\left(k^{-2}\right)$ in this asymptotic.) Now, given a potential $q \in L_2$ let us take a finite set of its eigenvalues $\{\mu_k\}_1^N$ and $\{\lambda_k\}_1^N$, assuming that N is sufficiently large, namely, such that $\lambda_N < \mu_{N+1}^0$. Then

$$\mu_1 < \lambda_1 < \mu_2 < \lambda_2 \cdots \mu_N < \lambda_N < \mu_{N+1}^0 < \lambda_{N+1}^0 \cdots, \tag{1.5}$$

and it follows from the results of [**26**] or [**23**] that there is a unique potential q_N such that the first N eigenvalues of the operators $L_{DN}(h)$ and L_D coincide with $\{\mu_k\}_1^N$ and $\{\lambda_k\}_1^N$, while the other eigenvalues are the same as in the case of constant potential $q = q_0$, i.e. coincide with the sequences $\{\mu_k^0\}_{N+1}^\infty$ and $\{k^2 + q_0\}_{N+1}^\infty$. Moreover, using the methods of [**28**], starting from $q(x) = q_0$, one can effectively construct the potential q_N solving sequentially $2N$ differential equations (see details in [**33**] and [**35**]). We say that q_N *is $2N$-approximation of the potential q by its finite set $\{\mu_k\}_1^N$ and $\{\lambda_k\}_1^N$* of the spectral data. We would like to emphasize that the construction og q_n involves not only finite sets of the eigenvalues but also the numbers q_0 and h which are assumed to be known from asymptotic formulae.

One may expect that the functions q_N approximate the potential q in L_2 norm as $N \to \infty$. This is true; namely, the following statement holds.

THEOREM 1.1. *Assume that $q \in L_2(0, 1)$, the numbers q_0 and h are known as well as finite sets of spectral data $\{\mu_k\}_1^N$ and $\{\lambda_k\}_1^N$ of q. Let q_N be the $2N$-approximations of the potential q by these finite sets of the spectral data. Then the asymptotic formulae* (1.2) *and* (1.3) *hold, where $\{\alpha_k\}_1^\infty \in \ell_2$ and*

$$\|q_N - q\|_{L_2} \leqslant C(q) \sum_{k=N+1}^{\infty} |\alpha_k|^2 \to 0, \quad \text{as } N \to \infty,$$

with a constant $C = C(q)$ depending on q. Moreover, a more precise estimate holds; namely, for all potentials q such that $\|q\|_{L_2} \leqslant R$,

$$\|q_N - q\|_{L_2} \leqslant C(R) \sum_{k=N+1}^{\infty} |\alpha_k|^2, \tag{1.6}$$

where C depends only on R.

We can not give a reference where such results were formulated. However, the first assertion of this theorem in the case $q_0 = 0, \quad h = 0$ one can obtain from the results of Marchenko and Ostrovskii [**26**] and Rjabushko [**29**], [**30**]. Apparently the same methods can be applied to get this assertion for $h \neq 0$. The second statement is new.

The limitation of Theorem 1.1 is that it gives no information on how fast the functions q_N approximates the potential q. Actually, we can not get such information in L_2 norm. However, the speed of convergence can be approximated in a weaker norm. Let us consider the space $W_2^{-1}[0, 1]$ which consists of the functions q such that their primitives $u(x) = \int q(t)\, dt$ (here the primitives are understood in the sense of distributions) belong to $L_2[0, 1]$. The norm in $W_2^{-1}[0, 1]$ is defined by $\|q\|_{-1} = \inf \|u(x) - c\|_{L_2}$, where u is a primitive of q and infimum is taken over all constants c. The spaces $W_2^\theta[0, 1]$ for $\theta \in (-1, 0)$ we define by the method of real or complex interpolation (both methods for Hilbert spaces coincide [**37**]). Often we will use abbreviation W_2^θ instead of $W_2^\theta[0, 1]$ for all $\theta \geqslant -1$. It can be easily obtained from the definitions that for all $\theta \geqslant -1$ the estimate $\|q\|_\theta \leqslant \|\int q\|_{\theta+1}$ holds for all primitives $\int q$ of the function q. Certainly, for all $-1 \leqslant \tau < \theta$ the space W_2^θ is continuously and compactly embedded in the space W_2^τ with the norm estimate $\|q\|_\tau \leqslant \|q\|_\theta$.

THEOREM 1.2. *Assume that $q \in L_2[0,\pi]$ and $\|q\|_{L_2} \leqslant R$. Let the numbers h and q_0 be known as well as the first eigenvalues $\{\mu_k\}_1^N$ and $\{\lambda_k\}_1^N$ of the operators $L_{DN}(h)$ and L_D. Let q_N be the $2N$-approximations of the potential q by these finite sets of the spectral data. Denote*

$$(1.7)\quad \sigma(x) = h - (\pi - x)q_0 + \int_0^x q(t)\,dt, \quad \sigma_N(x) = h - (\pi - x)q_0 + \int_0^x q_N(t)\,dt.$$

Then for all $\theta \in [0,1)$ the estimate

$$(1.8)\qquad \|q_N - q\|_{\theta-1} \leqslant \|\sigma_N(x) - \sigma(x)\|_\theta \leqslant \frac{C(R)}{N^{1-\theta}}$$

holds, where the constant $C(R)$ depends only on R. In particular, for all $\theta \in (1/2,1)$ the estimate

$$(1.9)\qquad \left|\int_0^x (q_N(t) - q(t))\,dt\right| = |\sigma_N(x) - \sigma(x)| \leqslant \frac{C(R,\theta)}{N^{1-\theta}}$$

holds with the constant $C(R,\theta)$ depending only on R and θ.

Here we remark only that the estimate (1.9) follows directly from (1.8), taking into account the embedding $W_2^\theta \hookrightarrow C[0,1]$ with the norm estimate $\|q\|_C \leqslant C(\theta)\|q\|_\theta$, provided that $\theta > 1/2$ (see [**1**, Chapter 5]), for example).

Now, assume that we have no information on a potential q, but we possess the information on the asymptotic behavior (see (1.2) and (1.3)) for the eigenvalues μ_k and λ_k and from this information we have

$$(1.10)\qquad \sum_{n=1}^N |\alpha_n|^2 \leqslant r \quad \text{for all } N < \infty.$$

We also calculate the number

$$(1.11)\quad d = \min_{1\leqslant k<\infty} d_k, \qquad \text{where } d_{2k-1} = \lambda_k - \mu_k,\ d_{2k} = \mu_{k+1} - \lambda_k,\ k = 1,2,\dots.$$

Since $d_n \to \infty$ as $n \to \infty$, one has to find only finitely many number d_n to evaluate d. Now we can formulate another version of Theorem 1.2.

THEOREM 1.3. *Let asymptotic formulae* (1.2) *and* (1.3) *hold, and the estimate* (1.10) *be valid. Then the estimate* (1.8) *holds, provided that the constant $C(R)$ on the right-hand side is replaced by a constant $C = C(r,d)$, depending only on the numbers r in* (1.10) *and d in* (1.11).

The next theorem can be of considerable interest in numerical analysis. In practical realization of the recovery procedure one cannot calculate the eigenvalues $\{\mu_k\}_1^N$ and $\{\lambda_k\}_1^N$ explicitly. It is natural to assume that we have these spectral data up to a given accuracy ε, i.e.

$$(1.12)\qquad |\mu_k - \tilde{\mu_k}| \leqslant \varepsilon, \quad |\lambda_k - \tilde{\lambda_k}| \leqslant \varepsilon, \quad k = 1,2.\dots,N,$$

where $\{\widetilde{\mu}_k\}_1^N$ and $\{\widetilde{\lambda}_k\}_1^N$ are the data obtained by measurements. Then, we cannot guarantee that the approximation of the potential q is improving as $N \to \infty$. The left-hand side of (1.8) may diverge if we replace each q_N by $\widetilde{q}_N$ which is constructed in the same way as q_N, but with the help of approximate finite sets $\{\widetilde{\mu}_k\}_1^N$ and $\{\widetilde{\lambda}_k\}_1^N$. Naturally we have problems: What accuracy (depending on ε) can we achieve for recovering the potential q by approximate finite spectral data and how can we choose the number N in an optimal way? The answers are given in the next theorems.

THEOREM 1.4. *Assume that $q \in L_2[0,\pi]$ and $\|q\|_{L_2} \leqslant R$. Let $\{\mu_k\}_1^N$ and $\{\lambda_k\}_1^N$ be the first N eigenvalues of the operators $L_{DN}(h)$ and L_D corresponding to the potential q and let approximate spectral data $\{\widetilde{\mu}_k\}_1^N$ and $\{\widetilde{\lambda}_k\}_1^N$ be subject to (1.12). Denote by $\widetilde{q}_N$ the $2N$-approximation of q constructed by the approximate spectral data and by $\widetilde{\sigma}_N$ its primitive, such that $\widetilde{\sigma}_N(0) = \sigma(0) = h - \pi q_0$. Then the following estimates hold.*

- *For $\theta \in [0, 1/2)$:*

$$\|q_0 + q_N - q\|_{\theta-1} \leqslant \|\sigma_N(x) - \sigma(x)\|_\theta \leqslant C(R)\left(\frac{\varepsilon}{1-2\theta} + \frac{1}{N^{1-\theta}}\right);$$

- *For $\theta = 1/2$:*

$$\|q_0 + q_N - q\|_{-1/2} \leqslant \|\sigma_N(x) - \sigma(x)\|_{1/2} \leqslant C(R)\left(\varepsilon(\ln N)^{1/2} + \frac{1}{N^{1/2}}\right); \tag{1.13}$$

- *For $\theta \in (1/2, 1)$:*

$$\|q_0 + q_N - q\|_{\theta-1} \leqslant \|\sigma_N(x) - \sigma(x)\|_\theta \leqslant C(R)\left(\varepsilon N^{2\theta-1} + \frac{1}{N^{1-\theta}}\right),$$

where $C(R)$ depends only on R.

As a corollary we get

THEOREM 1.5. *Let the conditions and notations of Theorem 1.4 hold and $0 < \varepsilon < e^{-1}$. Then the following estimates hold.*

- *For $\theta \in [0, 1/2)$:*

$$\|\widetilde{\sigma}_N(x) - \sigma(x)\|_\theta \leqslant C(R)\frac{\varepsilon}{1-2\theta}, \quad \textit{provided that} \quad N \geqslant \left(\frac{1-2\theta}{\varepsilon}\right)^\gamma, \ \gamma = \frac{1}{1-\theta}.$$

- *For $\theta = 1/2$:*

$$\|\sigma_N(x) - \sigma(x)\|_{1/2} \leqslant \left(1+\sqrt{2}\right) C(R)\varepsilon|\ln\varepsilon|^{1/2}, \quad \textit{provided that} \quad N = \varepsilon^{-2};$$

- *For $\theta \in (1/2, 1)$:*

$$\|\sigma_N(x) - \sigma(x)\|_\theta \leqslant 2C(R)\left(\frac{\varepsilon}{2\theta-1}\right)^{-1+1/\theta}, \quad \textit{provided that} \quad N = \left(\frac{2\theta-1}{\varepsilon}\right)^{1/\theta}.$$

REMARK 1.6. For the case $q_0 = h = 0$, Marletta and Weikard [**27**] obtained the following estimate

$$|\widetilde{\sigma}_N(x) - \sigma(x)| \leqslant C(q)\left(\varepsilon\ln N + N^{-1/2}\right),$$

with a constant $C(q)$ depending on q. This estimate is "close" to our estimate (1.13) because the space $W_2^{1/2}$ is "close" to be embedded in the space of continuous functions $C[0,\pi]$ (actually, $W_2^{1/2+\delta} \hookrightarrow C[0,\pi]$ with the norm estimates for all $\delta > 0$).

Since neither $W_2^{1/2} \hookrightarrow C[0,\pi]$ nor $W_2^{1/2} \hookleftarrow C[0,\pi]$ holds, the last estimate and (1.13) are independent.

2. Main results and their proofs

Now let us recall the definition of the Sturm-Liouville operator with a distribution potential. Suppose that $q(x) \in W_2^{-1}[0,1]$. Then there exists (in sense of distributions) a primitive $\sigma(x) = \int q(t)\,dt$ which belongs to $L_2[0,1]$. It is known [**8**, Chapter 1] that a distribution primitive is defined up to a constant. Denote

$$y^{[1]}(x) = y'(x) - \sigma(x)y(x).$$

Following our paper [**31**] the operator L_D we define by the differential expression

$$Ly = -\left(y^{[1]}\right)' - \sigma(x)y^{[1]} - \sigma^2(x)y$$

on the domain

$$\mathcal{D}(L_D) = \{y, y^{[1]} \in W_1^1[0,1] \mid Ly \in L_2[0,1],\ \ y(0) = y(\pi) = 0\}.$$

Similarly, the operator L_{DN} is defined by the same differential expression on the domain

$$\mathcal{D}(L_{DN}) = \{y, y^{[1]} \in W_1^1[0,1] \mid Ly \in L_2[0,1],\ \ y(0) = y^{[1]}(\pi) = 0\}.$$

We remark that the choice of a constant when taking a primitive σ plays no role in the definition of the operator L_D. However, it does play a role in the definition of L_{DN}. It is easily seen for the functions $q \in W_2^\theta[0,1]$ with $\theta > -1/2$. In this case $\sigma \in W_2^{\theta+1}[0,1] \hookrightarrow C[0,1]$ and the value $\sigma(\pi)$ is well defined. The boundary condition in the definition of L_{DN} takes the form $y'(\pi) - \sigma(\pi)y(\pi) = 0$. Therefore, the addition of a constant to σ does not change the differential expression but does change the boundary condition. In the case $q \in W_2^\theta[0,1]$ with $\theta < -1/2$ a primitive $\sigma = \int q(t)\,dt$ is not point-wise defined and the mean value of q does not exist. However, both operators L_D and L_{DN} are self-adjoint (provided that q is a real distribution), possess discrete spectrum, and the interlacing property (1.1) remains valid (see [**33**]). Certainly, the eigenvalues of these operators do not preserve the asymptotic behavior (see (1.2) and (1.3)), since the numbers q_0 and h are not defined if the singularity of a potential $> 1/2$, i.e. $\theta < -1/2$. Two spectra $\{\mu_k\}_1^\infty$ and $\{\lambda_k\}_1^\infty$ of the operators L_{DN} and L_D determine a function σ uniquely (see [**11**], [**33**]). This means that in the case of continuous σ we define simultaneously not only the potential $q = \sigma'$ but also the constant $h = \sigma(\pi)$ which determines the boundary condition.

Given two spectra $\{\mu_k\}_1^\infty$ and $\{\lambda_k\}_1^\infty$ of the operators L_{DN} and L_D we associate the numbers

$$s_{2k-1} = \sqrt{\mu_k} - (k - 1/2), \quad s_{2k} = \sqrt{\lambda_k} - k, \quad k = 1, 2, \dots,. \tag{2.1}$$

Certainly, specifying two spectra is equivalent to specifying the sequence $\{s_k\}_1^\infty$. Further it will be more convenient to deal with the last sequence. We say that such a sequence defines *the regularized spectral data* of the Borg problem.

To use the language of the theory of mappings in what follows, we should understand in what spaces the regularized spectral data can be placed in (depending on θ) as the primitive σ of the defined operators ranges over $W_2^\theta, \theta \geqslant 0$. One may guess that the spaces for the sequences $\{s_k\}_1^\infty$ of the regularized spectral data have to be associated with ℓ_2 weighted spaces, i.e. with the spaces ℓ_2^θ consisting of the sequences $\mathbf{x} = \{x_1, x_2, \dots,\}$ of complex numbers such that

$$\|\mathbf{x}\|_\theta^2 := \sum_1^\infty |x_k|^2 k^{2\theta} < \infty.$$

However, the spaces we are looking for have a more complicated structure. Later we will formally define these spaces and denote by ℓ_B^θ (the index B means that we associate this space with the Borg problem), but first we would like to present some explanations about their structure. To do this we have to recall the results on the eigenvalue asymptotic formulae for the operators L_D and L_{DN}.

THEOREM 2.1. *Let m be any integer $\geqslant 1$. Suppose that $\sigma(x) \in W_2^m$ and $\|\sigma\|_m \leqslant R$. Then the following formulae for the eigenvalues μ_k and λ_k of the operators L_{DN} and L_D hold:*

- *For odd $m = 2s + 1$,*

$$\begin{aligned}\sqrt{\mu_k} =& k - \frac{1}{2} + \frac{g_0}{(2k-1)} + \frac{g_1}{(2k-1)^3} + \dots + \frac{g_s}{(2k-1)^{2s+1}} - (-1)^s \frac{a_{2k-1}}{(2k-1)^{2s+1}} \\ &+ \alpha_{2k},\\ \sqrt{\lambda_k} =& k + \frac{q_0}{(2k)} + \frac{q_1}{(2k)^3} + \dots + \frac{q_s}{(2k)^{2s+1}} - (-1)^s \frac{a_{2k}}{(2k)^{2s+1}} + \alpha_{2k-1}.\end{aligned}$$

- *For even $m = 2s$,*

$$\begin{aligned}\sqrt{\mu_k} =& k - \frac{1}{2} + \frac{g_0}{(2k-1)} + \frac{g_1}{(2k-1)^3} + \dots + \frac{g_s}{(2k-1)^{2s+1}} - (-1)^s \frac{b_{2k-1}}{(2k-1)^{2s}} \\ &+ \alpha_{2k-1},\\ \sqrt{\lambda_k} =& k + \frac{q_0}{(2k)} + \frac{q_1}{(2k)^3} + \dots + \frac{q_s}{(2k)^{2s+1}} - (-1)^s \frac{b_{2k}}{(2k)^{2s}} + \alpha_{2k-1}.\end{aligned}$$

In these formulae the remainders form a sequence $\{\alpha_k\}_1^\infty$ belonging to the space ℓ_2^{m+1}, and its norm in ℓ_2^{m+1} is estimated by a constant C depending only on R. The numbers $a_n = a_n(\sigma)$ and $b_n = b_n(\sigma)$ are defined by

$$a_n = \frac{1}{\pi}\int_0^\pi \sigma^{(m)}(t)\cos nt\,dt, \quad b_n = \frac{1}{\pi}\int_0^\pi \sigma^{(m)} \sin nt\,dt, \qquad n = 1, 2, \dots.$$

The numbers $q_j, g_j, 0 \leqslant j \leqslant s$, are continuous (nonlinear for $j \geqslant 1$) functionals from $\sigma \in W_2^m$, and their linear parts h_j^0, g_j^0 are expressed by the formulae

$$h_j^0 = (-1)^j \pi^{-1}\left[\sigma^{(2j)}(\pi) - \sigma^{(2j)}(0)\right], \quad g_j^0 = (-1)^{j+1}\pi^{-1}\left[\sigma^{(2j)}(\pi) + \sigma^{(2j)}(0)\right].$$

PROOF. The proof of this theorem in the case $h = \sigma(\pi) = 0$ with the remainder estimate $\sum_{n=1}^\infty |\alpha|^2 k^{2(m+1)} < C(\sigma)$ belongs to V. A. Marchenko [**23**]. A shorter proof for arbitrary h with the uniform remainder estimate was found by the authors [**34**]. □

THEOREM 2.2. *Suppose that $\sigma(x) \in W_2^\theta$, $0 \leqslant \theta \leqslant 1$, and $\|\sigma\|_\theta \leqslant R$. Then*

$$\sqrt{\lambda_k} = k - b_{2k} + \alpha_{2k}, \tag{2.2}$$

$$\sqrt{\mu_k} = k - \frac{1}{2} - b_{2k-1} + \alpha_{2k-1}, \tag{2.3}$$

where

$$b_n = \frac{1}{\pi}\int_0^\pi \sigma(t)\sin nt\,dt, \quad n = 1, 2, \dots$$

and

$$\sum_{n=1}^{\infty} |\alpha_n|^2 \left(n^{2\theta}\right)^2 < C, \quad i.e.\ \|\{\alpha_k\}_1^\infty\|_{2\theta} < C,$$

where the constant C depends only on R. The numbers $s_k = -b_k + \alpha_k$ form the sequence $\{s_k\}_1^\infty \in \ell_2^\theta$, provided that $\theta \in [0, 1/2)$. In the case $\theta \in [1/2,\, 1]$

$$s_{2k-1} = \frac{g_0}{2k-1} + \gamma_{2k-1}$$

and

$$s_{2k} = \frac{q_0}{2k} + \gamma_{2k},$$

where $\{\gamma_k\}_1^\infty \in \ell_2^\theta$.

PROOF. The first statement of this theorem is proved by the authors [**32**], [**34**]. To prove the second statement we consider the sequences

$$\mathbf{e}_1 = \left\{\frac{1}{k}\right\}_1^\infty \quad \text{and} \quad \mathbf{e}_2 = \left\{\frac{(-1)^k}{k}\right\}_1^\infty.$$

Construct the space $\ell_B^1 := \ell_2^1 \oplus \operatorname{span}\{\mathbf{e}_k\}_{k=1}^2$, which is a 2-dimensional extension of the space ℓ_2^1. Integrating by parts we remark that

$$b_n = \frac{1}{\pi}\int_0^\pi \sigma(t)\,\sin nt\,dt = \frac{1}{\pi n}\sigma(0) - \frac{(-1)^n}{\pi n}\sigma(\pi) + \frac{1}{\pi n}\int_0^\pi \sigma'(t)\cos nt\,dt.$$

Therefore,

(2.4) $\qquad T\sigma = \{b_n\}_{n=1}^\infty = \sigma(0)\mathbf{e}_1 - \sigma(\pi)\mathbf{e}_2 + \{\alpha_n\}_{n=1}^\infty, \quad \text{where } \{\alpha_n\}_{n=1}^\infty \in \ell_2^1$

is a bounded operator from W_2^1 into ℓ_B^1; moreover, T is an isomorphism. Obviously, $T : L_2 \to \ell_2$ is an isomorphism, too. By virtue of the interpolation theorem (see [**37**, Chapter 1],for example) $T : W_2^\theta \to [\ell_B^1, \ell_2]_\theta$ is an isomorphism for all $\theta \in (0, 1)$. Now, according to the claim proved in Lemma 1 of [**33**] we have

$$[\ell_B^1, \ell_2]_\theta = \begin{cases} \ell_2^\theta, & \text{if } 0 \leqslant \theta < 1/2, \\ \ell_2^\theta \oplus \operatorname{span}\{\mathbf{e}_k\}_{k=1}^2, & \text{if } 1/2 \leqslant \theta \leqslant 1. \end{cases}$$

Comparing the coefficients at even and odd coordinates in (2.4), (2.2), and (2.3), we find $q_0 = \pi^{-1}(\sigma(\pi) - \sigma(0))$, $g_0 = -\pi^{-1}(\sigma(\pi) + \sigma(0)) = q_0 - (2/\pi)h$. This proves Theorem 2.2. □

Now let us consider the nonlinear map

$$F(\sigma) = \{s_k\}_{k=1}^\infty.$$

Theorem 2.2 shows that F is well defined as a map from the space W_2^θ into $\ell_B^\theta = [\ell_B^1, \ell_2]_\theta$, provided that $\theta \in [0, 1]$. Analogously, Theorem 2.1 suggests the construction of the spaces ℓ_B^θ for $\theta > 1$ such that the map $F : W_2^\theta \to \ell_B^\theta$ will be well defined.

Consider the sequences

$$\mathbf{e}_{2k-1} = \{k^{-(2s-1)}\}_{k=1}^\infty, \qquad \mathbf{e}_{2k-1} = \{(-1)^k k^{-(2s-1)}\}_{k=1}^\infty, \quad s = 1, 2. \dots.$$

Let $m = [\theta/2 + 3/4]$, where $[a]$ is the integer part of the number a. We set

$$\ell_B^\theta = \ell_2^\theta \oplus \operatorname{span}\{\mathbf{e}_k\}_{k=1}^{2m}.$$

Here we have taken into account that the sequence $\mathbf{e}_k$ does not belong to the space ℓ_2^θ for $k \leqslant 2m$ and belongs to it for $k > 2m$. Thus, ℓ_B^θ consists of the elements $\mathbf{x}+\sum_{k=1}^m c_k\mathbf{e}_k$, where $\mathbf{x} \in \ell_2^\theta$ and $\{c_k\}_1^m$ are arbitrary complex numbers. The inner product of elements of ℓ_B^θ is determined by the formula

$$(\mathbf{x}+\sum_{k=1}^m c_k\mathbf{e}_k,\ \mathbf{y}+\sum_{k=1}^m d_k\mathbf{e}_k) = (\mathbf{x},\mathbf{y})_\theta + \sum_{k=1}^m c_k\overline{d_k}.$$

Although the space ℓ_B^θ is defined as a finite-dimensional extension of the weighed space ℓ_2^θ, it is convenient to write its elements in the form of usual sequences. For example, if $1/2 \leqslant \theta < 5/2$, then ℓ_B^θ consists of sequences $\mathbf{x} = \{x_k\}_{k=1}^\infty$ with the coordinates

$$x_k = y_k + c_1k^{-1} + c_2(-1)^k k^{-1}, \quad \text{where } \{y_k\}_1^\infty \in l_2^\theta, \quad c_1, c_2 \in \mathbb{C}.$$

It readily follows from this representation that ℓ_B^η is compactly embedded into ℓ_B^θ for $\eta > \theta$. (Here we take into account the compactness of the embedding $\ell_2^\eta \hookrightarrow \ell_2^\theta, \eta > \theta$.) Now we are ready to reformulate Theorems 2.1 and 2.2 in the abstract form, taking from this a significant advantage.

THEOREM 2.3. *For all $\theta \geqslant 0$ the operator $F(\sigma) = \{s_k\}_{k=1}^\infty$ is well defined as an operator from W_2^θ into ℓ_B^θ and admits a representation of the form*

$$F(\sigma) = T\ \sigma + \Phi(\sigma),$$

where $T : \sigma \to \{b_n\}_1^\infty$ is a linear isomorphism from W_2^θ into ℓ_B^θ and Φ maps W_2^θ onto ℓ_B^τ with

$$\tau = \begin{cases} 2\theta, & 0 \leqslant \theta \leqslant 1, \\ \theta+1, & 1 \leqslant \theta < \infty. \end{cases}$$

Moreover, the map $\Phi : W_2^\theta \to l_B^\tau$ is bounded in any ball of W_2^θ, i.e.

$$\|\Phi(\sigma)\|_\tau \leqslant C(R)\|\sigma\|_\theta,$$

where the constant $C(R)$ depends only on the radius R of the ball $\|\sigma\|_\theta \leqslant R$.

PROOF. The proof of this theorem can be found in the paper by the authors [**34**]. First, the proof was given for integer values $\theta = m$ and for any fixed $\theta_0 \in (0, 1/4)$. Then for intermediate values $\theta \in [\theta_0, 1]$ and $\theta \in [m,\ m+1]$ the claim was obtained by applying the interpolation theorem for the nonlinear map Φ. Here we have to make a stipulation. Our reference in [**34**] to Tartar's interpolation theorem for nonlinear maps was incorrect. The essence of the matter is explained in our recent paper [**36**]. Actually, one has to apply our interpolation theorem proved in [**36**] for analytic maps instead of Tartar's theorem. In this case one has to keep in mind that for $\theta > 0$ the map $\Phi : W_2^\theta \to \ell_B^\theta$ is analytic in any complex ball $\|\sigma\|_\theta < R$ after finite-dimensional correction of Φ (see Theorem 5.6 in [**36**]). Here it is important to prove independently this theorem not only for $\theta = 0$ but also for some $\theta_0 > 0$. This was done in [**36**]. □

Theorem 2.3 plays a key role in the proof of the uniform stability theorem which is formulated below. It is not convenient in the sequel to deal with negative eigenvalues μ_k or λ_k, because the corresponding numbers s_k representing regularized spectral data may take non-real values. By shifting the spectral parameter we can avoid this situation. Let us denote by $W^\theta_{2,\mathbb{R}}$ the set of all real functions in W^θ_2 and by $W^\theta_{2,+}$ the set of all functions σ from $W^\theta_{2,\mathbb{R}}$ for which $\mu_1(\sigma) > 1/4$. By $\mathcal{B}^\theta_+(R)$ we denote the intersection of $W^\theta_{2,+}$ with the ball of radius R in W^θ_2. For the functions $\sigma \in W^\theta_{2,+}$ all the eigenvalues μ_k and λ_k are positive and $s_1 > 0$. In this case the interlacing condition is equivalent to the inequalities $s_1 > 0, s_k - s_{k+1} < 1/2, k = 1, 2 \dots.$ Denote by $\Omega^\theta(r, d)$ the set of all real sequences $\{s_k\}_1^\infty$ belonging to the ball of radius r in ℓ^θ_B and satisfying the conditions

$$s_1 > 0, \quad s_k - s_{k+1} \leqslant 1/2 - d, \quad k = 1, 2, \dots, \quad d \in (0, 1/2).$$

Obviously, the last condition is equivalent to (1.11) (with a different number d), provided that $\mu_1 > 1/4$.

THEOREM 2.4. *For any $\theta \geqslant 0$ the map $F : W^\theta_{2,+} \to \Omega(\infty, 0)$ is a bijection. Given any $\mathbf{s}$ and $\mathbf{s}_1$ belonging to $\Omega^\theta(r, d)$ the preimages $\sigma = F^{-1}\mathbf{s}$ and $\sigma_1 = F^{-1}\mathbf{s}_1$ lie in $\mathcal{B}^\theta_+(R)$ and one has the estimates*

$$C_1\|\mathbf{s} - \mathbf{s}_1\|_\theta \leqslant \|\sigma - \sigma_1\|_\theta \leqslant C_2\|\mathbf{s} - \mathbf{s}_1\|_\theta. \tag{2.5}$$

Here the number R and the constants C_1, C_2 depend only on r and h. The numbers R and the constants C_2 and C_1^{-1} increase as $r \to \infty$ or $d \to 0$. Conversely, if σ and σ_1 lie in the set $\mathcal{B}^\theta_+(R)$, then the sequences $\mathbf{s}$ and $\mathbf{s}_1$ of regularized spectral data of these functions lie in $\Omega^\theta(r, d)$, and one has the estimates

$$C_1\|\mathbf{s} - \mathbf{s}_1\|_\theta \leqslant \|\sigma - \sigma_1\|_\theta \leqslant C_2\|\mathbf{s} - \mathbf{s}_1\|_\theta. \tag{2.6}$$

Here the numbers $r > 0, d \in (0,\ 1/2)$ and the constants C_1, C_2 depend only on R. The numbers r, d^{-1}, C_2 and C^{-1} may increase as $R \to \infty$.

PROOF. In the case $\theta > 0$ the proof is given in [**35**] and [**36**]. □

We will exploit this important theorem to estimate the speed of the recovery procedure by a finite set of spectral data. Further we restrict ourselves to the consideration of the case $\theta < 9/2$. The changes which one has to make in the case of arbitrary θ are obvious, but the formulations become much more cumbersome. We start with the following lemma.

LEMMA 2.5. *The functionals q_1 and g_1 in asymptotic formulae (1.2) have the form*

$$q_1 = \frac{q'(0) - q'(\pi)}{\pi} + \frac{1}{\pi}\int_0^\pi q^2(t)dt - 2h_0^2 - \frac{1}{\pi}\int_0^\pi\int_0^t\int_0^s q(v)q(s)q(t)dv\,ds\,dt$$
$$+ h_0\int_0^\pi\int_0^t q(s)q(t)ds\,dt - \frac{\pi^2}{3}h_0^3;$$

$$g_1 = \frac{q'(0) + q'(\pi)}{\pi} + \frac{1}{\pi}\int_0^\pi q^2(t)dt - \frac{4}{\pi}\sigma(\pi)q(\pi) - 2g_0^2$$
$$+ \frac{1}{\pi}\int_0^\pi\int_0^t\int_0^s q(v)q(s)q(t)dv\,ds\,dt$$
$$+ \frac{1}{\pi}\int_0^\pi q(t)dt\int_0^\pi\int_0^t q(s)q(t)ds\,dt - \frac{\pi^2}{3}g_0^3 + \frac{2}{\pi}\sigma(\pi)(\sigma^2(\pi) - \sigma^2(0)).$$

Given $\sigma \in W_2^\theta$ in the case $5/2 \leqslant \theta < 9/2$, there is an infinitely differentiable function $\sigma_(x)$ such that

$$q_0(\sigma_0) = q_0(\sigma), \quad g_0(\sigma_0) = g_0(\sigma), \quad q_1(\sigma_0) = q_1(\sigma), \quad g_1(\sigma_0) = g_1(\sigma). \tag{2.7}$$

PROOF. The representation of the functionals q_1 and g_1 can be obtained by straightforward calculations using the formulae for the solutions of the equation $Ly = \rho^2 y$ obtained in [**34**]. However, the calculations are cumbersome, and we do not present them (the concrete form of these functionals is not used in what follows). The second assertion of Lemma 2.5 follows from the fact that $F : W_{2,+}^\theta \to \Omega(\infty, 0)$ is a bijection. □

Now, assume that $\sigma \in W_{2,\mathbb{R}}^\theta, 0 \leqslant \theta < 9/2$ and let $\{s_k\}_1^{2N}$ be a finite set of the regularized spectral data corresponding to σ. Assume that asymptotic formulae for the eigenvalues μ_k and λ_k are known; namely: the numbers q_0, g_0 in the case $1/2 \leqslant \theta < 5/2$ and the numbers q_0, g_0, q_1, g_1 in the case $5/2 \leqslant \theta < 9/2$ are known. Take $\sigma_0(x) = 0$ if $0 < \theta < 1/2$ and $\sigma_0 = h + q_0(x - \pi)$ if $1/2 \leqslant \theta < 5/2$. In the case $5/2 \leqslant \theta < 9/2$ we assume that σ_0 is any smooth function satisfying conditions (2.7). Denote by σ_N the function such that $s_k(\sigma_N) = s_k(\sigma)$ for $k = 1, 2, \ldots, 2N$ and $s_k(\sigma_N) = s_k(\sigma_0)$ for $k > 2N$, i.e. the eigenvalues μ_k and λ_k of the operators L_{DN} and L_D with primitives σ_{2N} and σ_0 coincide for all $k > N$. We say that the functions σ_N are the $2N$-approximations of σ by finite sets of spectral data. The $2N$-approximations in the case $\theta \in [5/2, 9/2)$ do depend on the choice of a function σ_0, but the estimates we obtain below do not. In the case $\theta = 1$ the $2N$-approximations for the primitives coincide with those given in section 1 (cf. (1.7)).

Recall that a sequence $\mathbf{s} = \{s_k\}_1^\infty \in \ell_B^\theta$ of regularized spectral data can be written in the form (see the proof of Theorem 2.3)

$$\mathbf{s} = \begin{cases} c_1\mathbf{e}_1 + c_2\mathbf{e}_2 + \mathbf{a} & \text{if } 1/2 \leqslant \theta < 5/2, \\ c_1\mathbf{e}_1 + c_2\mathbf{e}_2 + c_3\mathbf{e}_3 + c_4\mathbf{e}_4 + \mathbf{a} & \text{if } 5/2 \leqslant \theta < 9/2, \end{cases} \tag{2.8}$$

where

$$c_1 = \frac{\pi}{2}(q_0 + g_0), \; c_2 = \frac{\pi}{2}(q_0 - g_0), \; c_3 = \frac{\pi}{2}(q_1 + g_1), \; c_4 = \frac{\pi}{2}(q_1 - g_0),$$
$$\text{and } \mathbf{a} = \{a_k\}_1^\infty, \|\mathbf{a}\|_\theta \leqslant \|\mathbf{s}\|_\theta < \infty.$$

.

THEOREM 2.6. *Let $\sigma \in W_{2,\mathbb{R}}^\theta$, $0 < \theta < 9/2$ and $\|\sigma\|_\theta \leqslant R$. Let σ_{2N} be $2N$-approximations of the functions σ by finite set of spectral data. Then*

$$\|\sigma(x) - \sigma_N(x)\|_\theta \to 0, \quad \text{as } N \to \infty.$$

Given any $\tau \in [0, \theta)$ the following estimate holds

$$\|\sigma(x) - \sigma_{2N}(x)\|_\tau \leqslant C(R)\frac{1}{N^{\theta-\tau}},$$

where $C(R)$ depends only on R.

PROOF. Let $\mathbf{s} = \{s_k(\sigma)\}_1^\infty$, $\mathbf{s}_N = \{s_k(\sigma_N)\}_1^\infty$. It follows from the construction of σ_N and representation (2.8) that

$$s_k(\sigma) - s_k(\sigma_N) = a_k(\sigma) - a_k(\sigma_N) = a_k(\sigma) - a_k(\sigma_0) \quad \text{for } k \geqslant N + 1.$$

Then by virtue of Theorem 2.4 we have

$$\|\sigma(x)-\sigma_N(x)\|_\tau^2 \leqslant C(R)\|\mathbf{s}-\mathbf{s}_N\|_\tau^2 = C(R)\sum_{2N+1}^{\infty}\frac{|\alpha_k(\sigma)-\alpha_k(\sigma_0)|^2k^{2\theta}}{k^{2(\theta-\tau)}}$$
$$\leqslant C(R)\frac{1}{N^{2(\theta-\tau)}}\sum_{k=2N+1}^{\infty}(|\alpha_k^2(\sigma)+|\alpha_k^2(\sigma_0)|)k^{2\theta}.$$

We see that the right-hand side $= o(1)N^{2(\tau-\theta)}$ as $N\to\infty$. We also see that it can be estimated by $C(R)N^{2(\tau-\theta)}(R+o(1))$ uniformly with respect to σ in the ball $\|\sigma\|_\theta \leqslant R$. □

REMARK 2.7. It follows from Theorem 2.4 that we can replace the constant $C(R)$ in the last theorem by a constant $C(r,d)$ if we do not possess a priori information on the function σ and have only the information on the spectral data.

THEOREM 2.8. *Let $\sigma \in W^\theta_{2,\mathbb{R}}, 0<\theta<9/2$ and $\|\sigma\|_\theta \leqslant R$. Let $\{\mu_k\}_1^N$ and $\{\lambda_k\}_1^N$ be the first N eigenvalues of the operators L_{DN} and L_D corresponding to $\sigma=\int q$. Fix a small number $\varepsilon>0$. Assume that numbers $\{\widetilde{\mu}_k\}_1^N$ and $\{\widetilde{\lambda}_k\}_1^N$ are subject to the estimates*

$$|\mu_k-\widetilde{\mu}_k|\leqslant\varepsilon,\quad |\lambda_k-\widetilde{\lambda}_k|\leqslant\varepsilon,\quad k=1,2,\dots,N.$$

Let σ_N be $2N$-approximation of the function σ, and $0\leqslant\tau<\theta$. Then

$$\|\sigma(x)-\sigma_N(x)\|_\tau \leqslant C(R)\frac{1}{N^{\theta-\tau}}+\varepsilon C(R)\begin{cases}N^{\tau-1/2} & \text{if } \tau>1/2,\\ (\ln N)^{1/2} & \text{if } \tau=1/2,\\ 1, & \text{if } \tau<1/2,\end{cases}\tag{2.9}$$

where $C(R)$ depends only on R. In particular,

(2.10)

$$\|\sigma(x)-\sigma_N(x)\|_\tau \leqslant 2C(R)\begin{cases}\varepsilon^\gamma,\ \gamma=\frac{2(\theta-\tau)}{2\theta-1}, & \text{if } \tau>1/2 \text{ and } N=\varepsilon^{-2/(2\theta-1)},\\ \varepsilon|\ln\varepsilon| & \text{if } \tau=1/2 \text{ and } N=\varepsilon^{-1/(\theta-\tau)},\\ \varepsilon & \text{if } \tau<1/2 \text{ and } N\geqslant\varepsilon^{-1/(\theta-\tau)}.\end{cases}$$

PROOF. We have $\lambda_k=(k+s_{2k})^2$. Since $\|\sigma\|_0\leqslant\|\sigma\|_\theta\leqslant R$ and $\sum|s_{2k}|^2 < C^2(R)$ we have $s_{2k}>-k/2$ for all indices k with the possible exception of some first, say p, indices. The number p depends only on R and does not depend on σ. Representing $\widetilde{\lambda}_k$ in the form $\widetilde{\lambda}_k=(k+\widetilde{s}_{2k})^2$ we also get $\widetilde{s}_{2k}>-k/2$ with the possible exception of first $\widetilde{p}$ indices. If we need to shift the spectral parameter, we may assume that both inequalities hold for all indices. Then for all $k\geqslant1$ we have

$$|\lambda_k-\widetilde{\lambda}_k| = |(k+s_{2k})^2-(k+\widetilde{s}_{2k})^2| = |2k+s_{2k}+\widetilde{s}_{2k}||s_{2k}-\widetilde{s}_{2k}|\geqslant k|s_{2k}-\widetilde{s}_{2k}|.$$

Similarly we get

$$|\mu_k-\widetilde{\mu}_k|\geqslant k|s_{2k-1}-\widetilde{s}_{2k-1}|.$$

Therefore

$$|s_k-\widetilde{s}_k|\leqslant\varepsilon k^{-1}.$$

Denote by $\widetilde{\sigma}_{2N}$ the $2N$-approximation constructed by the finite set $\{\widetilde{\mu}_k\}_1^N$ and $\{\widetilde{\lambda}_k\}_1^N$. Using the same arguments as in the proof of Theorem 2.6, by virtue of

Theorem 2.4 we obtain

$$\|\sigma(x)-\widetilde{\sigma}_{2N}(x)\|_\tau^2 \leqslant C(R)\|\mathbf{s}-\mathbf{s}_N\|_\tau^2 = C(R)\sum_1^{2N}|s_k-\tilde{s}_k|^2k^{2\tau}$$

$$+C(R)\sum_{2N+1}^{\infty}\frac{|\alpha_k(\sigma)-\alpha_k(\sigma_0)|^2k^{2\theta}}{k^{2(\theta-\tau)}} \leqslant C(R)\sum_{k=1}^{2N}\varepsilon^2k^{2\tau-2}+C(R)\frac{1}{N^{2(\theta-\tau)}}.$$

Obviously, this estimate implies (2.9). The estimate (2.10) follows readily from (2.9). □

References

[1] O. V. Besov, V. P. Il'in, and S. M. Nikol'skii, *Integral Expressions for Functions and Embedding Theorems* (in Russian), Nauka, Moscow, 1975. MR0430771 (55:3776)

[2] G. Borg, *Eine Umkehrung der Sturm-Liouvilleschen Eigenwertaufgabe. Bestimmung der Differentialgleichung durch die Eigenwerte*, Acta Math. **78** (1946), 1–96. MR0015185 (7:382d)

[3] P. B. Djakov and B. S. Mityagin, *Instability zones of one-dimensional periodic Schrödinger and Dirac operators*, Uspekhi Mat. Nauk, *61*:4 (2006), 77–182; English transl.: Russian Math. Surveys, **61**:4 (2006), 663–766. MR2279044 (2007i:47054)

[4] P. Djakov and B. Mityagin, *Fourier Method for One-Dimensional Schrödinger Operators with Singular Periodic Potentials*, http://arxiv.org/abs/0710.0237v1. MR2683242 (2011k:47083)

[5] G. Freiling and V. Yurko, *Inverse Sturm-Liouville Problems and Their Applications*, Nova Sci. Publ. Corporation, 2001. MR2094651 (2005f:34001)

[6] M. G. Gasymov and B. M. Levitan, *Determination of a differential equation by two of its spectra*, Uspekhi Mat. Nauk **19**:2 (1964), 3–63; English transl.: Russian Math. Surveys, **19**:2 (1964), 1-63. MR0162996 (29:299)

[7] I. M. Gelfand and B. M. Levitan, *On the determination of a differential equation from its spectral function*, Izv. Akad. Nauk SSSR Ser. Mat. **15**:4 (1951), 309–360. MR0045281 (13:558f)

[8] I. M. Gel'land and G. E. Shilov, *The Generalized Functions and Operations with Them* (in Russian), Fizmatgiz, Moscow, 1959.

[9] R. O. Hryniv, *Analyticity and uniform stability in the inverse singular Sturm-Liouville spectral problem*, Inverse Problems **27** (2011), 1–25. MR2805768 (2012k:34035)

[10] R. O. Hryniv and Ya. V. Mykytyuk, *Inverse spectral problems for Sturm-Liouville operators with singular potentials*, Inverse Problems **19**:3 (2003), 665–684. MR1984883 (2004e:34016)

[11] R. O. Hryniv and Ya. V. Mykytyuk, *Inverse spectral problems for Sturm-Liouville operators with singular potentials, II. Reconstruction by two spectra*, in: Functional Analysis and Its Applications, North-Holland Mathematical Studies, vol. 197 (V. Kadets and W. Zelazko, eds.), North-Holland Publishing Co., Amsterdam, 2004, pp. 97–114. MR2098874 (2005h:34020)

[12] R. O. Hryniv and Ya. V. Mykytyuk, *Transformation operators for Sturm-Liouville operators with singular potentials*, Math. Phys. Anal. Geom. **7**:2 (2004), 119-149. MR2057122 (2005i:34048)

[13] R. O. Hryniv and Ya. V. Mykytyuk, *Eigenvalue asymptotics for Sturm-Liouville operators with singular potentials*, J. Funct. Anal. **238**:1 (2006), 27–57. MR2234122 (2007h:34155)

[14] R. O. Hryniv and Ya. V. Mykytyuk, *Inverse spectral problems for Sturm-Liouville operators with singular potentials. IV. Potentials in the Sobolev space scale*, Proc. Edinb. Math. Soc. (2) **49**:2 (2006), 309–329. MR2243789 (2007c:34013)

[15] T. Kappeler, B. Schaad, and P. Topalov, *Asymptotics of spectral quantities of Schrödinger operators*, Proc. of Symposia in Pure Math., vol. 84, AMS, 2012, pp. 243–284. MR2985321

[16] E. Korotyaev, *Characterization of the spectrum of Schrödinger operators with periodic distributions*, Int. Math. Res. Not. (2003) no. 37, 2019–2031. MR1995145 (2004e:34134)

[17] M. G. Krein, *Solution of the inverse Sturm-Liouville problem*, Dokl. Akad. Nauk SSSR **76** (1951), 21–24. MR0039895 (12:613c)

[18] M. G. Krein, *On a method of effective solution of an inverse boundary problem*, Dokl. Akad. Nauk SSSR **94** (1954), 987–990. MR0062904 (16:38h)

[19] N. Levinson, *The inverse Sturm-Liouville problem*, Mat. Tidsskr. B., 1949, pp. 25–30. MR0032067 (11:248e)
[20] B. M. Levitan, *Determination of a Sturm-Liouville differential equation in terms of two spectra*, Izv. Akad. Nauk SSSR Ser. Mat. **28**:1 (1964), 63–78. MR0159980 (28:3194)
[21] B. M. Levitan, *Inverse Sturm-Liouville Problems*, VNU Science Press, Utrecht, 1987. MR933088 (89b:34001)
[22] J. R. Mclaughlin, *Stability theorems for two inverse spectral problems*, Inverse Problems **4** (1988), 529–540. MR954908 (89f:34034)
[23] V. A. Marchenko, *Some questions of the theory of one-dimensional linear differential operators of the second order*, Trudy Moskov. Mat. Obshch. **1** (1951), 328–420. MR0058064 (15:315b)
[24] V. A. Marchenko, *Sturm-Liouville Operators and Their Applications* (in Russian), Naukova Dumka, Kiev, 1977. MR0481179 (58:1317)
[25] V. A. Marchenko and K. V. Maslov, *Stability of the problem of reconstructing the Sturm–Liouville operator from the spectral function*, Mat. Sb. **81**:4 (1970), 525–551; English transl.: Russian Acad. Sci. Sb. Math. **10**:4 (1970), 475–502. MR0264154 (41:8750)
[26] V. A. Marchenko and I. V. Ostrovskii, *A characterization of the spectrum of Hill's operator*, Mat. Sb. **97**:4 (1975), 540–606; English transl.: Russian Acad. Sci. Sb. Math. **26**:4 (1975), 493–554. MR0409965 (53:13717)
[27] M. Marletta and R. Weikard, *Weak stability for an inverse Sturm-Liouville problem with finite spectral data and complex potential*, Inverse Problems **21**:4 (2005), 1275–1290. MR2158108 (2006b:34023)
[28] J. Pöschel and E. Trubowitz, *Inverse Spectral Theory*, Academic Press, Boston, 1987. MR894477 (89b:34061)
[29] T. I. Ryabushko, *Stability of the reconstruction of a Sturm-Liouville operator from two spectra*, Teor. Funkts., Funkts. Anal. Prilozh. **18** (1973), 176–185. MR0335934 (49:712)
[30] T. I. Ryabushko, *Estimates of the norm of the difference of two potentials in a Sturm-Liouville boundary value problem*, Teor. Funkts., Funkts. Anal. Prilozh. **39** (1983), 114–117. MR734693 (85k:34054)
[31] A. M. Savchuk and A. A. Shkalikov, *Sturm–Liouville operators with singular potentials*, Mat. Zametki **66**:6 (1999), 897–912; English transl.: Math. Notes **66**:6 (1999), 741–753. MR1756602 (2000m:34061)
[32] A. M. Savchuk and A. A. Shkalikov, *Sturm-Liouville operators with distribution potentials*, Trudy Moskov. Mat. Obshch. **64** (2003), 159–219; English transl.: Trans. Moscow Math. Soc., 2003, pp. 143–192. MR2030189 (2004j:34198)
[33] A. M. Savchuk and A. A. Shkalikov, *Inverse problem for Sturm-Liouville operators with distribution potentials: Reconstruction from two spectra*, Russian J. Math. Phys. **12**:4 (2005), 507–514. MR2201315 (2006i:34024)
[34] A. M. Savchuk and A. A. Shkalikov, *On the eigenvalues of the Sturm-Liouville operator with potentials from Sobolev spaces*, Mat. Zametki **80**:6 (2006), 864–884; English transl.: Math. Notes **80**:6 (2006), 814–832. MR2311614 (2008d:34160)
[35] A. M. Savchuk and A. A. Shkalikov, *Inverse Problems for Sturm-Liouville Operators with Potentials in Sobolev Spaces: Uniform Stability*, Funktsionalnyi Analiz i Ego Prilozheniya, Vol. 44, No. 4, 2010, pp. 34–53; Engl. Transl.: Functional Analysis and Its Applications, Vol. 44, No. 4, 2010, pp. 270–285. MR2768563 (2012b:34033)
[36] A. M. Savchuk and A. A. Shkalikov, *On the interpolation of analytic maps*, Mat. Zametki, **94**:4 (2013), 678–281. English transl.: Math. Notes **94**:4 (2013).
[37] H. Triebel, *Interpolation Theory, Function Spaces, Differential Operators*, North-Holland, Amsterdam, 1978. MR503903 (80i:46032b)
[38] V. A. Yurko, *On the stability of the reconstruction of the Sturm-Liouville operator*, Differential Equations and Function Theory (Saratov University), vol. 3, 1980, 113–124.

Moscow State University
E-mail address: artem_savchuk@mail.ru

Moscow State University
E-mail address: ashkalikov@yahoo.com

Amer. Math. Soc. Transl.
Volume **233**, 2014

High-Velocity Estimates, Inverse Scattering and Topological Effects

Ricardo Weder

To Vladimir A. Marchenko on the occasion of his 90th birthday

Abstract. We review the results in a time-dependent method to study the high-velocity behaviour of scattering solutions to Schrödinger equations. This approach gives precise high-velocity estimates with error bounds. To introduce the method we first consider the simple case of scattering by a potential. Then we give a detailed discussion of the magnetic Aharonov-Bohm effect.

1. Introduction

In this paper we consider the problem of obtaining high-velocity estimates for scattering solutions to the Schrödinger equation. To state our problem in a precise way we first consider potential scattering for the Schrödinger equation in n dimensions. The Hilbert space of states is $\mathcal{H} = L^2(\mathbf{R}^n)$.

The free Schrödinger equation without potential is given by

$$i\hbar\frac{\partial}{\partial t}\psi = \frac{\mathbf{P}^2}{2M}\psi, \tag{1.1}$$

where $\mathbf{P} := -i\hbar\nabla$ is the momentum operator and M is the mass of the particle.

The interacting Schrödinger equation is

$$i\hbar\frac{\partial}{\partial t}\psi = \left(\frac{\mathbf{P}^2}{2M} + q\mathbf{V}\right)\psi, \tag{1.2}$$

where q is the charge of the particle and $\mathbf{V}$ is the potential, a real-valued function.

For the simplicity of the presentation we divide by $\hbar$ and we write (1.1) and (1.2) as

$$i\hbar\frac{\partial}{\partial t}\psi = H_0\psi \tag{1.3}$$

where H_0 is the free Hamiltonian,

$$H_0 := \frac{\mathbf{p}^2}{2m}, \qquad m := \frac{M}{\hbar}, \ \mathbf{p} := -i\nabla, \tag{1.4}$$

2010 *Mathematics Subject Classification.* Primary 81U40, 35P25, 35Q40, 35R30.

Key words and phrases. High velocity, inverse scattering, topological effects.

The author's research was partially supported by CONACYT under Project CB-2008-01-99100.

and

$$i\frac{\partial}{\partial t}\psi = H\psi, \tag{1.5}$$

where H is the interacting Hamiltonian

$$H := H_0 + V, \qquad V := q\mathbf{V}/\hbar. \tag{1.6}$$

We state some standard definitions. We say that an operator is Kato–small in $L^2(\mathbf{R}^n)$ if it is bounded with respect to the Laplacian with relative bound zero [**44**]. For any set $O \subset \mathbf{R}^n$ we denote by $F(\mathbf{x} \in O)$ the operator of multiplication by the characteristic function of O.

To avoid technical complications we suppose that V is of short range. Our short-range class is given by

(1.7)

$$\mathcal{V}_{SR} = \left\{ V \middle| V \text{ is Kato–small in } L^2(\mathbf{R}^n), \int_0^\infty dR \left\| V(\mathbf{y})\,(-\Delta + I)^{-1} F(|\mathbf{y}| \geq R) \right\| < \infty \right\}.$$

For operators of multiplication by a function the decay property in (1.7) is equivalent [**45**] to the following condition that is more intuitive

$$\int_0^\infty dR \left\| F(|\mathbf{y}| \geq R)\; V(\mathbf{y})\; (-\Delta + I)^{-1} \right\| < \infty.$$

Actually, we prefer to use (1.7) because it is more convenient technically.

When the time is large and negative the particle is far away in space, where the potential is negligible. For this reason we can approximate its evolution by an incoming asymptotic solution to the free Schrödinger equation

$$e^{-itH_0}\,\phi_- \tag{1.8}$$

where $\phi_- \in \mathcal{H}$ is the incoming asymptotic state. As time increases the particle is close to the origin, where the potential is strong. In this situation we have to take into consideration the potential and the particle dynamics that is given by a scattering solution to the Schrödinger equation

$$e^{-itH}\,\phi, \tag{1.9}$$

where $\phi \in \mathcal{H}$ is the scattering state. For large positive times the particle is again far away from the origin, where the potential is small, and its evolution is once more well approximated by an outgoing asymptotic solution to the free Schrödinger equation

$$e^{-itH_0}\,\phi_+, \tag{1.10}$$

where $\phi_+ \in \mathcal{H}$ is the outgoing asymptotic state. The states ϕ, and $\phi_\pm$ have to be chosen in such a way that

$$\lim_{t \to \pm\infty} \left\| e^{-itH_0}\,\phi_\pm - e^{-itH}\,\phi \right\| = 0. \tag{1.11}$$

In the direct scattering problem the potential and the incoming asymptotic state, ϕ_- are known, and one has to calculate the scattering state, ϕ, and the outgoing asymptotic state, ϕ_+. For this purpose it is useful to introduce the wave operators

$$W_\pm = \mathrm{s} - \lim_{t \to \pm\infty} e^{itH}\,e^{-itH_0}. \tag{1.12}$$

It is a standard result [**45**] that, under assumption (1.7) the wave operators exist and are asymptotically complete, i.e. they are partially isometric with initial subspace $\mathcal{H}$ and final subspace the subspace of absolute continuity of H.

Then, if

$$\phi = W_- \phi_-, \tag{1.13}$$

and

$$\phi_+ = W_+^* \phi, \tag{1.14}$$

the conditions (1.11) are satisfied because e^{-itH} are unitary . It follows that the scattering solution is given by

$$\psi(x,t) := e^{-itH} W_- \phi_-, \tag{1.15}$$

and that

$$\lim_{t \to \pm\infty} \left\| \psi(x,t) - e^{-itH_0} \phi_\pm \right\| = 0. \tag{1.16}$$

The scattering operator, S, is defined as follows:

$$S = W_+^* W_-. \tag{1.17}$$

Clearly, it sends the incoming asymptotic state to the outgoing asymptotic state

$$S\phi_- := \phi_+. \tag{1.18}$$

To make our notion of high-velocity estimates precise, we first introduce our high-velocity asymptotic states.

Let us denote by $B_{m\eta}(m\mathbf{v})$ the open ball in $\mathbf{R}^n$, of center $m\mathbf{v}$ and radius $m\eta$. In the case where $\mathbf{v} = 0$ we just use the notation $B_{m\eta}$. By $\mathcal{F}$ we designate the Fourier transform

$$\mathcal{F}\varphi(p) := \frac{1}{(2\pi)^{n/2}} \int e^{-ip\cdot x} \varphi(x)\, dx.$$

For simplicity we also denote by $\hat{\varphi}$ the Fourier transform of φ.

Let $\Phi_0 \in \mathcal{H}$ be an asymptotic state with momentum space wave function $\hat{\Phi}_0 \in C_0^\infty(B_{m\eta})$.

The boosted configuration

$$\Phi_{\mathbf{v}} = e^{im\mathbf{v}\cdot\mathbf{x}}\, \Phi_0 \Leftrightarrow \hat{\phi}_{\mathbf{v}}(\mathbf{p}) = \hat{\phi}_0(\mathbf{p} - m\mathbf{v}) \tag{1.19}$$

has compact velocity support contained in the open ball around $\mathbf{v}$ with radius η.

By high-velocity estimates of scattering solutions we mean precise estimates, uniform in time, and with error bound that decays as a power of the velocity, of the following scattering solutions

$$\psi_{\mathbf{v}}(x,t) := e^{-itH} W_- \Phi_{\mathbf{v}},$$

when the velocity $\mathbf{v}$ is large.

Also by high-velocity estimates of the scattering operator we mean precise estimates with error bound that decays as a power of the velocity of the following outgoing asymptotic states,

$$S\, \Phi_{\mathbf{v}},$$

for large velocity $\mathbf{v}$.

These high-velocity estimates are important in many problems, for example, in inverse scattering problems and in topological effects in quantum mechanics. For a detailed discussion of the interest of our high-velocity estimates and the reasons to consider time-dependent methods instead of stationary ones, see [**82**].

In [**23, 24, 25**] we introduced a time-dependent method to estimate the high-velocity behavior of the scattering solutions and of the scattering operator. In this method all the physical properties of the scattering process, such as localization and propagation, are used in the proofs. This gives our method a lot of flexibility, thus allowing us to solve many problems, with proofs that closely follow the physical intuition. We briefly review some of the results

In [**23, 24, 25**] the non-relativistic Schrödinger equation for N-body systems with long-range potentials was studied. The case of the Stark effect (scattering by a decaying electric potential and an electric field that is constant in space) for N-body systems was considered in [**61**], [**40**], [**1**], [**59**]. Moreover, [**41**] studied the Stark effect with a time-periodic electric field, and [**2**] the case of electric fields that are asymptotically zero in time. In [**26**] the case of two-cluster scattering was studied. The papers [**6, 7**] and [**37**] considered two-body systems with decaying electromagnetic potentials in all space. Moreover, [**8**] considers the case of decaying electromagnetic potentials and a constant magnetic field. The article [**62**] considers N-body systems in all space with time-dependent potentials. The relativistic case of the Dirac equation was studied by [**35, 36**] for time independent potentials and by [**33, 34**] for time-dependent potentials. The contributions [**64, 65, 66, 67, 70, 71, 72**] consider the non-linear Schrödinger equation and [**68, 74**] the non-linear Klein-Gordon equation. In [**79, 80**] the forced nonlinear Schrödinger equation in the half-line is studied. The paper [**46**] considers a non-linear Schrödinger equation with a potential and a non-local nonlinearity. In [**42**] the problem of the Schrödinger equation with a strong repulsive potential is considered, and in [**43**] the Schrödinger equation in a semiclassical process is considered. The papers [**16, 17**] study scattering in a black-hole metric. The articles [**39, 75, 9, 10, 11**] consider the magnetic Aharonov-Bohm effect, and [**81**] studies the electric Aharonov-Bohm effect [**5**]. For review papers see [**63, 69, 73, 76, 78**].

This paper is organized as follows. In Section 2 we consider the case of potential scattering. We obtain high-velocity estimates for the solution to the Schrödinger equation and for the scattering operator. We also give a method for the unique reconstruction of the potential from the high-velocity limit of the scattering operator. In Section 3 we consider high-velocity estimates for the solution to the Schrödinger equation in the exterior of a handle-body that contains a magnetic flux inside, and we discuss the Aharonov-Bohm effect in a general situation. Furthermore, we discuss precise quantitative high-velocity estimates for the solution to the Schrödinger equation under the conditions of the fundamental experiments of Tomonura et al. [**48, 54, 55**], which are considered the only convincing experimental verification of the magnetic Aharonov-Bohm effect.

2. Potential Scattering

In this section we review the results in the case of scattering by a short-range potential obtained in [**25**] . We give the proofs from [**25**] to be able to explain the main ideas of our method in a simple case.

For any Borel function f we define the operator $f(\mathbf{p})$ by functional calculus, or equivalently, as $f(\mathbf{p}) := \mathcal{F}^{-1} f(\cdot)\mathcal{F}$. It is an elementary result that under translation in configuration or momentum space generated, respectively, by $\mathbf{p}$ and x we have that

$$e^{i\mathbf{p}\cdot\mathbf{v}t}\, f(x)\, e^{-i\mathbf{p}\cdot\mathbf{v}t} = f(x+\mathbf{v}t), \tag{2.1}$$

$$e^{-im\mathbf{v}\cdot x}\, f(\mathbf{p})\, e^{im\mathbf{v}\cdot x} = f(\mathbf{p}+m\mathbf{v}), \tag{2.2}$$

and in particular,

$$e^{-im\mathbf{v}\cdot x}\, e^{-itH_0}\, e^{im\mathbf{v}\cdot x} = e^{-imv^2t/2} e^{-i\mathbf{p}\cdot\mathbf{v}t} e^{-itH_0}, \tag{2.3}$$

for any $\mathbf{v}\in\mathbf{R}^n, n=1,2,\cdots$, where $v := |\mathbf{v}|$.

A first basic point in our method is that physical intuition tells us that for high velocity the scattering solution is close to the incoming asymptotic free solution, eventually multiplied by a phase. A second basic point is that for high velocity the free quantum evolution follows the classical trajectory given by Newton's equation, with tails that decay rapidly. In other words, the incoming asymptotic free solution is concentrated along the classical trajectory and it decays very fast in the classically forbidden region, i.e. in the region where, according to Newton's equation, the solution is zero. The following propagation estimate expresses this second point in the operator norm sense, in a way that is convenient for our purposes.

We denote by $F(x\in\mathcal{M})$ the operator of multiplication by the characteristic function of $\mathcal{M}$.

LEMMA 2.1. *For any $f\in C_0^\infty(B_{m\eta})$, for some $\eta>0$, and any $l=1,2,\cdots$, there is a constant C_l such that the following estimate holds:*

$$\left\| F(x\in\tilde{\mathcal{M}})\, e^{-itH_0}\, f\left(\frac{\mathbf{p}-m\mathbf{v}}{v^\mu}\right) F(x\in\mathcal{M})\right\| \le C_l(1+rv^\mu+\eta\, v^{2\mu}|t|)^{-l}, \tag{2.4}$$

for every $\mathbf{v}\in\mathbf{R}^n$, $t\in\mathbf{R}$, $\mu\in\mathbf{R}$ and any measurable sets $\mathcal{M}$ and $\tilde{\mathcal{M}}$ in $\mathbf{R}^n$ such that $r := \operatorname{dist}\left(\tilde{\mathcal{M}},\mathcal{M}+\mathbf{v}t\right)-\eta v^\mu|t|\ge 0$.

PROOF. We follow the proof of [**75**]. By (2.1)–(2.3) it is enough to establish the estimate

$$\left\| F(x\in\tilde{\mathcal{M}})\, e^{-itH_0}\, f\left(\frac{\mathbf{p}}{v^\mu}\right) F(x\in\mathcal{M})\right\| \le C_l\,(1+rv^\mu+\eta\, v^{2\mu}|t|)^{-l}, \tag{2.5}$$

provided that $r := \operatorname{dist}\left(\tilde{\mathcal{M}},\mathcal{M}\right)-\eta v^\mu|t|\ge 0$. It follows from a simple calculation that

$$\begin{aligned}\left\| F(x\in\tilde{\mathcal{M}})\, e^{-itH_0}\, f\left(\frac{\mathbf{p}}{v^\mu}\right) F(x\in\mathcal{M})\phi\right\|^2 \\ \le \tfrac{1}{(2\pi)^n}\textstyle\int_{\tilde{\mathcal{M}}}dx\int_{\mathcal{M}}dy\int_{\mathcal{M}}dz\, \tilde{f}_t(x-y)\,\phi(y)\,\overline{\tilde{f}_t(x-z)\,\phi(z)},\end{aligned} \tag{2.6}$$

where

$$\tilde{f}_t(x) := \frac{1}{(2\pi)^{n/2}}\int e^{ip\cdot x}e^{-ip^2t/2m}\, f(p/v^\mu)\, dp. \tag{2.7}$$

Since, $2|\phi(y)|\,|\phi(z)|\le|\phi(y)|^2+|\phi(z)|^2$, it follows that

$$\left\| F(x\in\tilde{\mathcal{M}})\, e^{-itH_0}\, f\left(\frac{\mathbf{p}}{v^\mu}\right) F(x\in\mathcal{M})\phi\right\|^2 \le \frac{1}{(2\pi)^n}\left[\int_{|x|\ge r+\eta v^\mu|t|}|\tilde{f}_t(x)|\,dx\right]^2\|\phi\|^2. \tag{2.8}$$

By the corollary to Theorem XI.14 in [**45**], for any $N = 1, 2, \ldots$, there is a constant C_N such that

$$|\tilde{f}_t(x)| \leq C_N\, v^{n\mu}\, (1 + |x|v^{\mu} + v^{2\mu}|t|)^{-N}, \tag{2.9}$$

for $x \geq \eta v^{\mu}|t|$. We complete the proof of the lemma, inserting (2.9) in (2.8). □

By Corollary 2.4 in [**20**], condition (1.7) implies the integrable decay of

$$\|V^S(\mathbf{x})\, g(\mathbf{p})\, F(|\mathbf{x}| \geq R)\|$$

for all $g \in C_0^{\infty}(\mathbf{R}^n)$. It follows that any potential in $\mathcal{V}_{SR}$ satisfies the assumption (2.10) in Lemma 2.2 with $\rho = 0$. Clearly, larger ρ implies faster decay.

The following lemma is our basic dynamical estimate. It allows us to prove our high-velocity estimate of the scattering solution in Theorem 2.4.

LEMMA 2.2. *Suppose that $V^S \in \mathcal{V}_{SR}$ and that for some $0 \leq \rho \leq 1$ and all $g \in C_0^{\infty}(\mathbf{R}^n)$*

$$(1 + R)^{\rho} \left\| V^S(\mathbf{x})\, g(\mathbf{p})\, F(|\mathbf{x}| \geq R) \right\| \in L^1((0, \infty),\, dR). \tag{2.10}$$

Then for any high-velocity state $\Phi_{\mathbf{v}}$, as in (1.19), there is a function h with $(1 + \tau)^{\rho} h(\tau) \in L^1((0, \infty))$ such that for every $\mathbf{v} \in \mathbf{R}^n$ satisfying $v \geq 4\eta$ we have

$$\left\| V^S(\mathbf{x})\, e^{-itH_0} \Phi_{\mathbf{v}} \right\| \leq h(|vt|). \tag{2.11}$$

PROOF. Since Φ_0 belongs to the Schwartz class

$$\left\| (1 + \mathbf{x}^2)^{3/2}\, \Phi_{\mathbf{v}} \right\| \leq C,$$

with the constant C uniform in $\mathbf{v}$. That is to say, the asymptotic state $\Phi_{\mathbf{v}}$ is well localized in space uniformly with respect to $\mathbf{v}$. Note that

$$\left\| V^S(\mathbf{x})\, g(\mathbf{p} - m\mathbf{v}) \right\| = \left\| V^S(\mathbf{x})\, g(\mathbf{p}) \right\|$$

is independent of $\mathbf{v}$. Let us take $g \in C_0^{\infty}(B_{m\eta})$ with $g \equiv 1$ on the support of $\hat{\Phi}_0$. Then

$$\begin{aligned} \left\| V^S(\mathbf{x})\, e^{-itH_0} \Phi_{\mathbf{v}} \right\| &= \left\| V^S(\mathbf{x})\, e^{-itH_0} (g(\mathbf{p} - m\mathbf{v}))^2 \Phi_{\mathbf{v}} \right\| \\ &\leq C \left\| V^S(\mathbf{x})\, e^{-itH_0} (g(\mathbf{p} - m\mathbf{v}))^2\, (1 + \mathbf{x}^2)^{-3/2} \right\|, \end{aligned} \tag{2.12}$$

where we used (2.2). It follows that

$$\left\| V^S(\mathbf{x})\, e^{-itH_0} \Phi_{\mathbf{v}} \right\| \leq T_1 + T_2 + T_3, \tag{2.13}$$

where

(2.14)

$$\begin{aligned} T_1 := &\Big\| V^S\, g(\mathbf{p} - m\mathbf{v})\, F(|\mathbf{x} - \mathbf{v}t| > |vt|\, 5/8)\, e^{-itH_0}\, g(\mathbf{p} - m\mathbf{v})\, F(|\mathbf{x}| < |vt|/8) \\ &\times (1 + \mathbf{x}^2)^{-3/2} \Big\|, \end{aligned}$$

(2.15)

$$\begin{aligned} T_2 := &\Big\| V^S\, g(\mathbf{p} - m\mathbf{v})\, F\,(|\mathbf{x} - \mathbf{v}t| > |vt|\, 5/8)\, e^{-itH_0}\, f(\mathbf{p} - m\mathbf{v})\, F(|\mathbf{x}| \geq |vt|/8) \\ &\times (1 + \mathbf{x}^2)^{-3/2} \Big\| \\ &\leq C \left\| F(|\mathbf{x}| \geq |vt|/8)\, (1 + \mathbf{x}^2)^{-3/2} \right\| \leq C\, (1 + |vt|)^{-3}, \end{aligned}$$

(2.16) $T_3 := \left\| V^s\, g(\mathbf{p} - m\mathbf{v})\, F(|\mathbf{x} - \mathbf{v}t| \le |vt|\, 5/8)\, e^{-itH_0}\, f(\mathbf{p} - m\mathbf{v})\, (1 + \mathbf{x}^2)^{-3/2} \right\|.$

The term T_1 decays when $|vt|$ is large because it is localized in the classical forbidden region: taking in Lemma 2.1

$$\mathcal{M} = \{|\mathbf{x}| \le |vt|/8\}, \quad \mathcal{M}' = \{|\mathbf{x} - \mathbf{v}t| > |vt|\, 5/8\},$$

$\mu = 0$, and $r \ge |vt|/4$, we obtain that

$$T_1 \le C\, (1 + |vt|/4)^{-3}. \tag{2.17}$$

Note that the decay of T_2 is a consequence of the term $(1 + \mathbf{x}^2)^{-3/2}$ that is there because of the localization in space of the asymptotic state $\Phi_{\mathbf{v}}$, which is uniform in v. Moreover, the term T_3 decays as $|vt|$ is large because it is localized where the particle is allowed to be according to the classical evolution, namely, far away in space where the potential V is small

$$T_3 \le C \left\| V^s\, g(\mathbf{p})\, F(|\mathbf{x}| \ge |vt|\, 3/8) \right\| =: h_1(|vt|), \tag{2.18}$$

where $(1 + \tau)^\rho\, h_1(\tau) \in L^1((0, \infty))$ by (2.10). The proof of the lemma follows from (2.13), (2.15), (2.17), and (2.18). □

COROLLARY 2.3. *Suppose that $V^s \in \mathcal{V}_{SR}$. Then,*

$$\left\| (W_\pm - I)\, e^{-itH_0}\, \Phi_{\mathbf{v}} \right\| = O(v^{-1}), \tag{2.19}$$

uniformly for $t \in \mathbf{R}$.

PROOF. The corollary is a simple consequence of Duhamel's formula and Lemma 2.2. By Duhamel's formula,

$$(W_+ - I)\, e^{-itH_0}\, \Phi_{\mathbf{v}} = i \int_0^\infty dt'\, e^{it'H}\, V^s\, e^{-it'H_0}\, e^{-itH_0}\, \Phi_{\mathbf{v}}.$$

Then, by Lemma 2.2,

$$\begin{aligned} \left\| (W_+ - I)\, e^{-itH_0}\, \Phi_{\mathbf{v}} \right\| &\le C \int_{-\infty}^{\infty} dt'' \left\| V^s\, e^{-it''H_0}\, \Phi_{\mathbf{v}} \right\| \\ &\le \frac{C}{v} \int_{-\infty}^{\infty} d(vt'')\, h(|vt''|). \end{aligned} \tag{2.20}$$

This completes the proof for W_+. The case of W_- follows in a similar way. □

The theorem below gives our high-velocity estimate.

THEOREM 2.4. *Assume that $V^s \in \mathcal{V}_{SR}$. Then, uniformly for $t \in \mathbf{R}$,*

$$\left\| \psi_{\mathbf{v}}(x, t) - e^{-itH_0}\, \Phi_{\mathbf{v}} \right\| = O(v^{-1}). \tag{2.21}$$

PROOF. By the intertwining relations $e^{-itH}\, W_\pm = W_\pm\, e^{-itH_0}$. Then, the scattering solution $\psi_{\mathbf{v}}(x, t) = e^{-itH}\, W_- \Phi_{\mathbf{v}}$ satisfies, $\psi_{\mathbf{v}}(x, t) = W_-\, e^{-itH_0}\, \Phi_{\mathbf{v}}$. Hence, equation (2.21) follows from Corollary 2.3. □

Theorem 2.4 shows that our first basic point above is actually a mathematical fact. Namely, that in the limit $v \to \infty$ the scattering solution $\psi_{\mathbf{v}}(x,t)$ is well approximated by the incoming asymptotic free solution $e^{-itH_0}\,\Phi_{\mathbf{v}}$, uniformly in time, up to an error that goes to zero as $1/v$. In this simple case it is not necessary to multiply the incoming asymptotic free solution by a phase. But, as we will see in the case of the Aharonov-Bohm effect [**5**], the phase appears and it is essential for the existence of this effect.

Let us see now how to use these results to prove the uniqueness of the inverse scattering problem and to give a formula for the reconstruction of the potential from the high-velocity limit of the scattering operator.

THEOREM 2.5. *Formula of Reconstruction.*
Suppose that $V^S \in \mathcal{V}_{SR}$ *and that (2.10) is satisfied. Then for all* $\Phi_{\mathbf{v}}$, $\Psi_{\mathbf{v}}$, *as in (1.19),*
(2.22)

$$v\left(i(S-I)\,\Phi_{\mathbf{v}},\,\Psi_{\mathbf{v}}\right) = \left(\int_{-\infty}^{+\infty} d\tau\, V^S(\mathbf{x}+\tau\hat{\mathbf{v}})\,\Phi_0,\,\Psi_0\right) + \begin{cases} o\left(v^{-\rho}\right), & \leq \rho < 1, \\ O\left(v^{-1}\right), & \rho = 1. \end{cases}$$

PROOF. The definition of the wave and scattering operators implies that

$$S - I = (W_+ - W_-)^*\, W_-.$$

Using Duhamel's formula and the intertwining relations $e^{-itH}\,W_\pm = W_\pm\, e^{-itH_0}$, it follows that

$$i(S-I)\,\Phi_{\mathbf{v}} = \int_{-\infty}^{+\infty} dt\, e^{itH_0}\, V^S\, W_-\, e^{-itH_0}\,\Phi_{\mathbf{v}}.$$

Observe that $W_-\, D(H_0) \subset D(H)$. Then, $W_-\, e^{-itH_0}\,\Phi_{\mathbf{v}} \in D(V^S)$. Hence,

$$v\left(i(S-I)\,\Phi_{\mathbf{v}},\,\Psi_{\mathbf{v}}\right) = \int_{-\infty}^{\infty} \ell_{\mathbf{v}}(\tau)\, d\tau + R(\mathbf{v}), \tag{2.23}$$

where $\tau = vt$ and the leading and remainder terms are, respectively,

$$\ell_{\mathbf{v}}(\tau) := \left(V^S(\mathbf{x})\, e^{-iH_0\,\tau/v}\,\Phi_{\mathbf{v}},\, e^{-iH_0\,\tau/v}\,\Psi_{\mathbf{v}}\right), \tag{2.24}$$

$$R(\mathbf{v}) := v\int_{-\infty}^{+\infty} dt\,\left((W_- - I)\, e^{-itH_0}\,\Phi_{\mathbf{v}},\, V^S\, e^{-itH_0}\,\Psi_{\mathbf{v}}\right). \tag{2.25}$$

It follows from Lemma 2.2 and Corollary 2.3 that

$$|R(\mathbf{v})| \leq C\int_{-\infty}^{+\infty} \left\|V^S\, e^{-itH_0}\,\Phi_{\mathbf{v}}\right\| dt \leq \frac{C}{v}\int_{-\infty}^{+\infty} h(|\tau|)\, d\tau. \tag{2.26}$$

By a simple calculation, recalling that $\Phi_{\mathbf{v}} = e^{im\mathbf{v}\cdot\mathbf{x}}\,\Phi_0$ and (2.1), 2.2) we obtain that

$$\ell_{\mathbf{v}}(\tau) = \left(V^S(\mathbf{x}+\tau\hat{\mathbf{v}})\, e^{-iH_0\,\tau/v}\,\Phi_0,\, e^{-iH_0\,\tau/v}\,\Psi_0\right). \tag{2.27}$$

Clearly, for each fixed τ we have that

$$\lim_{v\to\infty} \ell_{\mathbf{v}}(\tau) = (V^S(\mathbf{x}+\tau\hat{\mathbf{v}})\,\Phi_0,\,\Psi_0). \tag{2.28}$$

Furthermore, Lemma 2.2 gives the following integrable bound, uniformly in $\mathbf{v}$,

(2.29)
$$\begin{aligned}|\ell_{\mathbf{v}}(\tau)| \leq & C \left\| V^S(\mathbf{x})\, e^{-iH_0\tau/v}\, \Phi_{\mathbf{v}} \right\| \leq C \left\| V^S(\mathbf{x})\, e^{-iH_0\tau/v}\, f(\mathbf{p}-m\mathbf{v})\, (1+\mathbf{x}^2)^{-3/2} \right\| \\ \leq & C\, h(|\tau|).\end{aligned}$$

Then, by the dominated convergence theorem

$$\lim_{v\to\infty} \int_{-\infty}^{\infty} \ell_{\mathbf{v}}(\tau)\, d\tau = \int_{-\infty}^{\infty} (V^S(\mathbf{x}+\tau\hat{\mathbf{v}})\, \Phi_0,\ \Psi_0)\, d\tau. \tag{2.30}$$

The equation (2.22) with $\rho = 0$ follows from (2.26) and (2.30). For the proof in the case $\rho > 0$, see [**25**]. □

THEOREM 2.6. *The (high-velocity limit of) the scattering operator S uniquely determines the potential $V \in \mathcal{V}_{SR}$.*

PROOF. We identify any $\mathbf{z} = (z_1, z_2) \in \mathbf{R}^2$ with the vector $(z_1, z_2, 0, \ldots, 0) \in \mathbf{R}^n$ (or any other two–dimensional plane). Then for any Φ, Ψ with $\hat{\Phi}$, $\hat{\Psi} \in C_0^\infty(\mathbf{R}^n)$ let us designate by

$$\Phi(\mathbf{z}) = e^{-i\mathbf{p}\cdot\mathbf{z}}\, \Phi,\ \Psi(\mathbf{z}) = e^{-i\mathbf{p}\cdot\mathbf{z}}\, \Psi,$$

the states translated by $\mathbf{z}$, and

$$f(\mathbf{z}) := (V\, \Phi(\mathbf{z}),\ \Psi(\mathbf{z})). \tag{2.31}$$

Clearly $f(\mathbf{z})$ is a continuous and bounded function. Furthermore, taking $g(\mathbf{p})\,\Phi = \Phi$

(2.32)
$$\begin{aligned}|f(\mathbf{z})| \leq & \|V\, g(\mathbf{p})\, \Phi(\mathbf{z})\| \\ \leq & C \left\| V\, g(\mathbf{p})\, F(|\mathbf{x}| > |\mathbf{z}|/2) \right\| + \left\| V\, g(\mathbf{p}) \right\| \left\| F(|\mathbf{x}| < |\mathbf{z}|/2)\, e^{-i\mathbf{p}\cdot\mathbf{z}}\Phi \right\| \\ \in & L^2(\mathbf{R}^2, d\mathbf{z}).\end{aligned}$$

Here we used the short–range condition (1.7) and the rapid decay of Φ. For $\hat{\mathbf{v}}$ in the $\mathbf{z}$-plane, the Radon transform of $f(\mathbf{z})$ is given by

(2.33)
$$\begin{aligned}\tilde{f}(\hat{\mathbf{v}};\, \mathbf{z}) := & \int_{-\infty}^{+\infty} f(\mathbf{z}+\tau\hat{\mathbf{v}})\, d\tau = \int_{-\infty}^{+\infty} \Big(V(\mathbf{x}+\tau\hat{\mathbf{v}})\, \Phi(\mathbf{z}),\ \Psi(\mathbf{z}) \Big)\, d\tau \\ = & \lim_{v\to\infty} v \Big(i(S-I)\, \Phi_{\mathbf{v}}(\mathbf{z}),\ \Psi_{\mathbf{v}}(\mathbf{z}) \Big),\end{aligned}$$

where we used Theorem 2.5. Then, $\tilde{f}(\hat{\mathbf{v}};\, \mathbf{z})$ is uniquely determined by the (high-velocity limit of) the scattering operator S. As $f \in L^2(\mathbf{R}^2, d\mathbf{z})$ it follows from Theorem 2.17 in Chapter I of [**31**]) that $f(\mathbf{z})$ is uniquely determined by the (high-velocity limit of) the scattering operator S . In particular, $f(0) = (V(\mathbf{x})\, \Phi,\ \Psi)$ is uniquely determined for all Φ, Ψ with $\hat{\Phi}$, $\hat{\Psi} \in C_0^\infty(\mathbf{R}^n)$, and as this set is dense in $\mathcal{H}$ it follows that V is uniquely determined as an operator and as a function a.e. □

REMARK 2.7. The proof of Theorem 2.6 gives us a method for the unique reconstruction of a potential $V \in \mathcal{V}_{SR}$ from the high-velocity limit of the scattering operator. Using (2.22) we uniquely reconstruct the Radon transform of $f(\mathbf{z}) := (V\, \Phi(\mathbf{z}),\ \Psi(\mathbf{z}))$ from the high-velocity limit of the scattering operator. Inverting this Radon transform we uniquely reconstruct $f(\mathbf{z}) := (V\, \Phi(\mathbf{z}),\ \Psi(\mathbf{z}))$. As the set of Φ with $\hat{\Phi} \in C_0^\infty(\mathbf{R}^n)$ is dense in $\mathcal{H}$ we uniquely reconstruct V as an operator and in pointwise sense a.e. from $f(0) = (V\,\Phi,\ \Psi)$. Note that for the reconstruction of the potential it is enough to consider $f(\mathbf{z})$ only for $\mathbf{v}$ in a two-dimensional subspace. The problem is thus reduced to the inversion of the Radon transform of a bounded, continuous, square integrable function on $\mathbf{R}^2$.

For related literature on the inverse problem for potential scattering, see [**23**] and the references quoted there.

3. The Magnetic Aharonov-Bohm Effect

3.1. Introduction and Discussion of the Literature. As is well known, in classical physics the evolution of a charged particle under the action of a magnetic field is completely described by Newton's equation with the Lorentz force, $F = q\mathbf{v} \times B$, where B is the magnetic field, q is the charge of the particle, and $\mathbf{v}$ its velocity. It is a consequence of Newton's equation that the magnetic field acts locally, namely, if a particle propagates in a region of space where the magnetic field is zero, the Lorentz force is zero and the trajectory of the particle is a straight line. This implies that the dynamics of a classical particle is not affected by magnetic fields that are located outside of the region of space where the particle travels. There is no *action at a distance* of the magnetic fields on charged particles, and moreover, the relevant physical quantity is the magnetic field. The magnetic potentials have no physical meaning; they are just a mathematical tool.

The situation in quantum physics is fundamentally different. As quantum mechanics is a Hamiltonian theory it is necessary to introduce the magnetic potential to formulate the Schrödinger equation of a charged particle in the presence of a magnetic field. Moreover, in a region of space with non-trivial topology, such as the exterior of a torus, the magnetic potential has to be different from zero if there is a magnetic flux inside the torus, even if there is no magnetic field outside. There is a simple reason for this: Stokes' theorem implies that if the magnetic potential is zero outside the torus, the magnetic flux inside has to be zero.

Aharonov and Bohm pointed out in[**5**] that this implies that in quantum physics the magnetic flux inside the torus can *act at a distance* in a charged particle in a situation where the magnetic field is identically zero along the trajectory of the particle and, furthermore, that the action of the magnetic field is carried over by the magnetic potential, which gives it a real physical significance. Aharonov and Bohm [**5**] suggested an experiment to verify their theoretical prediction. They proposed the use of a thin straight solenoid with a magnetic field confined to its interior. A coherent electron wave packet is to be sent towards the solenoid, to be split in two parts, each one going through one side of the solenoid. Finally one has to bring both parts together behind the solenoid in order to create an interference pattern, due to the difference in phase in the wave function of each part, produced by the magnetic field inside the solenoid. In fact, the existence of this interference pattern was first predicted by Franz [**27**].

The predictions of Aharonov and Bohm [**5**] are a strong departure from the physical intuition of classical physics. They were, and still are, very controversial. For a review of the literature up to 1989 see [**47**] and [**49**]. In particular, in [**49**] there is a detailed discussion of the large controversy that involves over 300 papers. For a recent update see [**53, 57**].

Of course, from the experimental point of view it is not possible to have an infinite solenoid. It has to be finite, and the magnetic field has to leak outside.

In fact, if we assume that the magnetic field outside the finite solenoid can be neglected there is no Aharonov-Bohm effect at all because, in this case, the exterior of the finite solenoid is a simply connected region of space without a magnetic field where the magnetic potential can be gauged away to zero. To take this issue into account it was proposed to use a toroidal magnet that can contain a magnetic field inside without a leak. The experiments with toroidal magnets were carried over by Tonomura et al. [**48, 54, 55**]. In these remarkable experiments they split a coherent electron wave packet into two parts. One propagated inside the hole of the magnet and the other outside the hole. They brought both parts together behind the magnet and they measured the phase shift produced by the magnetic flux enclosed in the magnet. These experiments [**48, 54, 55**] gave strong evidence of the existence of the Aharonov-Bohm effect; actually, they are widely considered the only convincing experimental evidence of the existence of this effect.

After these experiments the controversy shifted into the interpretation of the results obtained by Tonomura et al. It was claimed that they could be explained by the action of a classical force acting on the electrons that travel through the hole of the magnet. For examples, see [**13, 30**] and the references quoted there. This force would accelerate the electron and it would produce a time delay. In a recent experiment Caprez et al. [**15**] observed that the time delay is zero, what excludes the explanation of the results of Tonomura et al. by the action of a force.

Aharonov and Bohm [**5**] proposed an Ansatz for the solution to the Schrödinger equation in simply connected regions of space where there are no electromagnetic fields which consists of multiplying the free evolution by the Dirac magnetic factor [**19**]. This Ansatz predicts both, the interference fringes observed by Tonomura et al. [**48, 54, 55**] and the absence of acceleration observed by Caprez et al. [**15**], because in the Aharonov-Bohm Ansatz the electron propagates following the free evolution, with the wave function multiplied by a phase.

After the experimental results of Tonomura et al. [**48, 54, 55**] and Caprez et al. [**15**], the controversy can be summarized in the following mathematical question: Is the Aharonov-Bohm Ansatz a good approximation to the exact solution to the Schrödinger equation for toroidal magnets and under the conditions of the experiments of Tonomura et al.? There have been many attempts to give an answer to this question. For a review of the literature up to 1989, see [**47**] and [**49**]. For a recent update, see [**9, 10, 11**]. Actually, the lack of any definite rigorous results on the validity of the Aharonov-Bohm Ansatz is perhaps the reason why this controversy lasted for so many years.

It is only very recently that this situation has changed. In [**10**] the authors give the first rigorous proof that the Aharonov-Bohm Ansatz is a good approximation to the exact solution of the Schrödinger equation. This paper gives, for the first time, a rigorous quantitative mathematical analysis of the Aharonov-Bohm effect with toroidal magnets under the conditions of the experiments of Tonomura et al. See [**48, 54, 55**]. In [**10**] it was assumed that the incoming free electron is represented by a Gaussian wave packet, what is physically reasonable. The authors obtained a rigorous, simple, quantitative error bound for the difference in norm between the exact solution and the approximate solution given by the Aharonov-Bohm Ansatz. The error bound is uniform in time. These results rigorously prove that quantum mechanics predicts the results observed by Tonomura et al. and by Caprez et al.

Moreover, in [**9, 11**], it is proven in a qualitative way that the Aharonov-Bohm Ansatz is a good approximation to the exact solution of the Schrödinger equation for high velocity for a very general class of magnets K and of incoming asymptotic states. This proves that the experimental results of Tonomura et al. [**48, 54, 55**] and of Caprez et al. [**15**] and the rigorous mathematical results of [**10**] hold in general and that they do not depend on the particular geometry of the magnets, on the velocities of the incoming electrons used in the experiments, and on the Gaussian shape of the wave packet.

Below we review the results of [**11**] and then those of [**10**].

3.2. The Magnet, the Magnetic Field and the Electric Potential. We consider the propagation of a non-relativistic particle, for example, an electron, outside a bounded magnet, K, in three dimensions, that is to say, the electron propagates in the exterior domain $\Lambda := \mathbf{R}^3 \setminus K$. We suppose that in the interior of K there is a magnetic field that produces a magnetic flux. We also assume that in Λ there are an electric potential V and a magnetic field B. This is a more general situation than in the case of the Aharonov-Bohm effect.

We assume that the magnet K is a compact submanifold of $\mathbf{R}^3$. Moreover, $K = \cup_{j=1}^{L} K_j$, where $K_j, 1 \leq j \leq L$ are the connected components of K. We suppose that the K_j are handle bodies. For a precise definition of handlebodies, see [**9**]. In intuitive terms, K is the union of a finite number of bodies diffeomorphic to tori or to balls. Some of them can be patched through the boundary. See Figure 1.

The magnetic field and the electric potential satisfy the following assumptions [**9**]. We denote by Δ the self-adjoint realization of the Laplacian in $L^2(\mathbf{R}^3)$ with domain $\mathcal{H}_2(\mathbf{R}^3)$. We assume that V is Δ-bounded with relative bound zero. By this we mean that the extension of V to $\mathbf{R}^3$ by zero is Δ-bounded with relative bound zero. Using an extension operator from $\mathcal{H}_2(\Lambda)$ to $H_2(\mathbf{R}^3)$ [**58**] it is easy to prove that this is equivalent to requiring that V is bounded from $\mathcal{H}_2(\Lambda)$ into $L^2(\Lambda)$ with relative bound zero. By $\|\cdot\|_{\mathcal{B}(\mathbf{R}^3)}$ we denote the operator norm in $L^2(\mathbf{R}^3)$.

ASSUMPTION 3.1. We suppose that the magnetic field, B, is a real-valued, bounded 2-form in $\overline{\Lambda}$, that is, continuous in a neighborhood of ∂K, and furthermore,

(1) B is closed : $dB|_{\Lambda} \equiv \operatorname{div} B = 0$.

(2) There are no magnetic monopoles in K:

$$\int_{\partial K_j} B = 0, \; j \in \{1, 2, \cdots, L\}. \tag{3.1}$$

(3)

$$|B(x)| \leq C(1+|x|)^{-\mu}, \text{ for some } \mu > 2. \tag{3.2}$$

(4) $d * B|_{\Lambda} \equiv \operatorname{curl} B$ is bounded and

$$|\operatorname{curl} B| \leq C(1+|x|)^{-\mu}. \tag{3.3}$$

(5) The electric potential, V, is a real-valued function, it is $\Delta-$bounded, with relative bound zero and

$$\left\| V(-\Delta+I)^{-1} F(|x| \geq r) \right\|_{\mathcal{B}(\mathbf{R}^3)} \leq C(1+r)^{-\alpha}, \text{ for some } \alpha > 1. \tag{3.4}$$

Condition (3.1) means that the total contribution of magnetic monopoles inside each component K_j of the magnet is 0. Formally we can use Stokes' theorem to prove that

$$\int_{\partial K_j} B = 0 \Longleftrightarrow \int_{K_j} \operatorname{div} B = 0, j \in \{1, 2, \cdots, L\}.$$

As $\operatorname{div} B$ is the density of magnetic charge, $\int_{\partial K_j} B$ is the total magnetic charge inside K_j. Hence, condition (3.1) means that the total magnetic charge inside K_j is zero, $j = 1, 2, \cdots, L$. Note that condition (3.4) is equivalent to the following assumption [**45**],

$$\left\|F(|x| \geq r)V(-\Delta + I)^{-1}\right\|_{\mathcal{B}(\mathbf{R}^3)} \leq C(1+r)^{-\alpha}, \text{ for some } \alpha > 1. \tag{3.5}$$

Equation (3.5) is a condition on the decay of V at infinity. However, (3.4) is technically more convenient.

We now introduce our class of magnetic potentials. Let $\{\hat{\gamma}_j\}_{j=1}^m$ be the closed curves defined in equation (2.6) of [**9**] (see Figure 1). It is proven in Corollary 2.4 of [**9**] that the equivalence classes of these curves are a basis of the first singular homology group of Λ. We now define a function that gives the magnetic flux across surfaces that have $\{\hat{\gamma}_j\}_{j=1}^m$ as their boundaries.

DEFINITION 3.2. The flux, Φ, is a function $\Phi : \{\hat{\gamma}_j\}_{j=1}^m \to \mathbf{R}$.

We proceed to introduce a class of magnetic potentials with a given flux modulo 2π.

DEFINITION 3.3. Let B be a closed 2-form that satisfies Assumption 3.1. We denote by $\mathcal{A}_{\Phi,2\pi}(B)$ the set of all continuous 1-forms, A, in $\overline{\Lambda}$ that satisfy.

(1)

$$|A(x)| \leq C\frac{1}{1+|x|}, \tag{3.6}$$

$$|A(x) \cdot \hat{x}| \leq C(1+|x|)^{-\beta_l}, \quad \beta_l > 1, \text{ where } \hat{x} := x/|x|. \tag{3.7}$$

(2)

$$\int_{\hat{\gamma}_j} \mathbf{A} = \Phi(\hat{\gamma}_j) + 2\pi n_j(A), n_j(A) \in \mathbb{Z}, \ j \in \{1, 2, \cdots, m\}. \tag{3.8}$$

(3)

$$dA|_\Lambda \equiv \ \operatorname{curl} A = B|_\Lambda. \tag{3.9}$$

Furthermore, we say that two potentials, $A, \tilde{A} \in \mathcal{A}_{\Phi,2\pi}(B)$ have the same fluxes if

$$\int_{\hat{\gamma}_j} A = \int_{\hat{\gamma}_j} \tilde{A}, \ j \in \{1, 2, \cdots, m\}. \tag{3.10}$$

Finally, we say that $A \in \mathcal{A}_{\Phi,2\pi}(B)$ is short range if

$$|A(x)| \leq C\frac{1}{(1+|x|)^\beta}, \quad \beta > 1. \tag{3.11}$$

We denote by $\mathcal{A}_{\Phi,2\pi,\mathrm{SR}}(B)$ the set of all potentials in $\mathcal{A}_{\Phi,2\pi}(B)$ that are short range.

The definition of the flux Φ depends on the particular choice of the curves $\{\hat{\gamma}_j\}_{j=1}^m$. However, the class $\mathcal{A}_{\Phi,2\pi}(B)$ is independent of this particular choice. In fact, it can be equivalently defined taking any other basis of the first singular homology group in Λ [**9**]. By Stokes' theorem the circulation $\int_{\hat{\gamma}_j} A$ of a potential $A \in \mathcal{A}_{\Phi,2\pi}(B)$ represents the flux of the magnetic field B in any surface whose boundary is $\hat{\gamma}_j, j = 1, 2, \cdots, m$.

It is proven in Lemma 3.8 of [**9**] that for any $A, \tilde{A} \in \mathcal{A}_{\Phi,2\pi}(B)$ with the same fluxes there is a gauge transformation between them. In other words, there is a C^1 0− form λ in $\overline{\Lambda}$ such that

$$\tilde{A} - A = d\lambda. \tag{3.12}$$

Furthermore, we can take $\lambda(x) := \int_{C(x_0,x)} (\tilde{A} - A)$, where x_0 is any fixed point in Λ and $C(x_0, x)$ is any curve from x_0 to x contained in Λ. Moreover, $\lambda_\infty(x) := \lim_{r\to\infty} \lambda(rx)$ exists and it is continuous in $\mathbb{R}^3 \setminus \{0\}$ and homogeneous of order zero, that is to say, $\lambda_\infty(rx) = \lambda_\infty(x), r > 0, x \in \mathbf{R}^3 \setminus \{0\}$.

3.3. The Schrödinger Equation and the Wave and Scattering Operators. We designate $\mathbf{p} := -i\nabla$. The Schrödinger equation for an electron in Λ with electric potential $\mathbf{V}$ and magnetic field $\mathbf{B}$ is given by

$$i\hbar \frac{\partial}{\partial t}\phi = \frac{1}{2M}(\mathbf{P} - \frac{q}{c}\mathbf{A})^2 + q\,\mathbf{V}, \tag{3.13}$$

where $\hbar$ is Planck's constant, $\mathbf{P} := \hbar\mathbf{p}$ is the momentum operator, c is the speed of light, M and q are, respectively, the mass and the charge of the electron, and $\mathbf{A}$ a magnetic potential with curl$\mathbf{A} = \mathbf{B}$. To simplify the notation we multiply both sides of (3.13) by $\frac{1}{\hbar}$ and we obtain

$$i\frac{\partial}{\partial t}\phi = \frac{1}{2m}(\mathbf{p} - A)^2\phi + V\phi, \tag{3.14}$$

with $m := M/\hbar, A = \frac{q}{\hbar c}\mathbf{A}$ and $V := \frac{q}{\hbar}\mathbf{V}$. We fix the flux modulo 2π by taking $A \in \mathcal{A}_{\Phi,2\pi}$, where $B := \frac{q}{\hbar c}\mathbf{B}$. Observe that this corresponds to fixing the circulations of $\mathbf{A}$ modulo $\frac{\hbar c}{q}2\pi$, or equivalently, to fixing the fluxes of the magnetic field $\mathbf{B}$ modulo $\frac{\hbar c}{q}2\pi$.

We define the Hamiltonian $H(A, V)$ with Dirichlet boundary condition at $\partial\Lambda$, i.e. $\psi = 0$ for $x \in \partial\Lambda$, because it is the standard boundary condition that corresponds to an impenetrable magnet K. See [**11**] for a discussion of this issue.

The free Hamiltonian is the self-adjoint operator

$$H_0 := \frac{1}{2m}\mathbf{p}^2, \tag{3.15}$$

with domain $\mathcal{H}_2(\mathbf{R}^3)$.

The wave operators are defined as in the introduction, but now that the free and the interacting spaces of states are different, we have to introduce an identification operator J. Let J be the operator from $L^2(\mathbf{R}^3)$ onto $L^2(\Lambda)$ given by multiplication by the characteristic function of Λ. Then, the wave operators are defined as

$$W_\pm(A, V) := \text{s-}\lim_{t\to\pm\infty} e^{itH(A,V)}\, J\, e^{-itH_0}. \tag{3.16}$$

We prove in [**9**] that if Assumption 3.1 holds, the wave operators exist and are partially isometric for every $A \in \mathcal{A}_{\Phi,2\pi}(B)$.

Moreover, if $A, \tilde{A} \in \mathcal{A}_{\Phi,2\pi}(B)$ and they have the same fluxes (for the case where the fluxes are not equal see [**9**])

$$W_\pm(\tilde{A}, V) = e^{i\lambda(x)}\, W_\pm(A, V)\, e^{-i\lambda_\infty(\pm\mathbf{p})}. \tag{3.17}$$

The scattering operator is defined as

$$S(A, V) := W_+^*(A, V)\, W_-(A, V). \tag{3.18}$$

If $A, \tilde{A} \in \mathcal{A}_{\Phi,2\pi}(B)$ with the same fluxes [**9**],

$$S(\tilde{A}, V) = e^{i\lambda_\infty(\mathbf{p})}\, S(A, V)\, e^{-i\lambda_\infty(-\mathbf{p})},\ \tilde{A}, A \in \mathcal{A}_{\Phi,2\pi}(B). \tag{3.19}$$

If, furthermore, $A, \tilde{A} \in \mathcal{A}_{\Phi,2\pi,\mathrm{SR}}(B)$, $S(\tilde{A}, V) = S(A, V)$.

3.4. The Aharonov-Bohm Effect. Qualitative Results. We will consider the case where the magnetic field outside K is zero but with a non-trivial magnetic flux, Φ, inside K. We designate by $\mathcal{A}_{\Phi,2\pi}(0)$ the set of potentials $A \in \mathcal{A}_{\Phi,2\pi}(B)$ in the case that $B = 0$. We proved in [**11**] that we can use the freedom of taking a gauge transformation to assume that $A \in \mathcal{A}_{\Phi,2\pi}(0) \cap C^1(\overline{\Lambda}, \mathbf{R}^3)$ and that support $A \subset B_R$, where $\Lambda \subset B_R$. This is what we will do from now on.

The Aharonov-Bohm Ansatz for the high-velocity solution is defined as follows [**11**].

DEFINITION 3.4 (Aharonov-Bohm Ansatz). Let A be a magnetic potential with curl $A = 0$, defined in a region $\mathcal{C}$ with trivial first group of singular homology. Let $A = \nabla\lambda(x)$ for some scalar function λ. Let

$$\psi_{\mathbf{v}}(x, t) := e^{-i\frac{t}{\hbar}H(A)} W_-(A, V)\, \varphi_{\mathbf{v}} \tag{3.20}$$

be the solution to the Schrödinger equation that behaves like $\psi_{\mathbf{v},0} := e^{-itH_0}\varphi_{\mathbf{v}}$ when time goes to minus infinite. We suppose that $\psi_{\mathbf{v}}$ is approximately localized at all times in $\mathcal{C}$. Then, the following change of gauge formula holds:

$$\psi_{\mathbf{v}} \approx \psi_{AB,\mathbf{v}}(x, t) := e^{i\lambda(x)} e^{-itH_0} \varphi_{\mathbf{v}}. \tag{3.21}$$

Observe that the Aharonov-Bohm Ansatz is the multiplication of the free solution by the Dirac magnetic factor $e^{i\lambda(x)}$ [**19**].

For the validity of the Aharonov-Bohm Ansatz it is necessary for the electron to stay in the simply connected region $\mathcal{C}$ disjoint from the magnet because if the electron hits K it will be reflected no matter how big the velocity is and it will not follow the free evolution multiplied by a phase, as is the case in the Aharonov-Bohm Ansatz. Hence, we need asymptotic states that have negligible interaction with K at all times. This is possible if the velocity is high enough.

For any $\mathbf{v} \neq 0$ we designate

$$\Lambda_{\hat{\mathbf{v}}} := \{x \in \Lambda : x + \tau\hat{\mathbf{v}} \in \Lambda,\ \forall \tau \in \mathbf{R}\}. \tag{3.22}$$

Let us consider asymptotic states (1.19) where φ has compact support contained in $\Lambda_{\hat{\mathbf{v}}}$. At the distance $z = vt$ the state is given by

$$e^{-i\frac{z}{v}H_0}\, \varphi_{\mathbf{v}} = e^{im\mathbf{v}\cdot x}\, e^{-i\frac{mzv}{2}}\, e^{-i\frac{z}{v}H_0}\, e^{-i\mathbf{p}\cdot z\hat{\mathbf{v}}}\varphi, \tag{3.23}$$

where we used (2.3). The term $e^{-i\mathbf{p}\cdot z\hat{\mathbf{v}}}$ is a translation in straight lines along the classical free evolution

$$\left(e^{-i\mathbf{p}\cdot z\hat{\mathbf{v}}}\varphi\right)(x) = \varphi(x - z\hat{\mathbf{v}}). \tag{3.24}$$

The term $e^{-i\frac{z}{v}H_0}$ gives the quantum-mechanical spreading of the wave packet. For high velocities it is one order of magnitude smaller than the classical translation. Neglecting it we get

$$(e^{-i\frac{z}{v}H_0}\varphi_{\mathbf{v}})(x) \approx e^{i\frac{mzv}{2}}\varphi_{\mathbf{v}}(x - z\hat{\mathbf{v}}), \text{ for large } v. \tag{3.25}$$

Under this approximation, for high velocities the asymptotic state evolves along the classical trajectory; modulo the unimportant global phase factor $e^{i\frac{mzv}{2}}$. The important point is that the support of our incoming wave packet remains in $\Lambda_{\mathbf{v}}$ for all distances, or for all times, and in consequence it has no interaction with K. We can expect that for high velocities, the exact solution, $\psi_{\mathbf{v}}$ (3.20), to the interacting Schrödinger equation (3.14) is close to the incoming wave packet $\psi_{\mathbf{v},0}$ and that it has negligible interaction with K. Our rigorous results justify this heuristic picture.

We need to make precise the statement that a trajectory in $\Lambda_{\hat{\mathbf{v}}}$ goes through the holes of K that are in $\Lambda_{\hat{\mathbf{v}}}$ or that it goes outside of the holes of K. In general, $\Lambda_{\hat{\mathbf{v}}}$ crosses several holes of K and if two electrons cross different holes of K there can be no simply connected region that contains both of them at all times. For this purpose we have to decompose $\Lambda_{\hat{\mathbf{v}}}$ on its components that cross the same holes of K. This was done in [**9**]. Take R so large that $K \subset B_R$. For any $x \in \mathbf{R}^3$ and any unit vector $\hat{\mathbf{v}}$ in $\mathbf{R}^3$, we denote

$$L(x,\hat{\mathbf{v}}) := x + \mathbf{R}\hat{\mathbf{v}}$$

and we give to $L(x,\hat{\mathbf{v}})$ the orientation of $\hat{\mathbf{v}}$.

Suppose that $L(x,\hat{\mathbf{v}}) \subset \Lambda$, and $L(x,\hat{\mathbf{v}}) \cap B_R \neq \emptyset$. We denote by $c(x,\hat{\mathbf{v}})$ the curve consisting of the segment $L(x,\hat{\mathbf{v}}) \cap \overline{B_R}$ and an arc on $\partial\overline{B_R}$ that connects the points $L(x,\hat{\mathbf{v}}) \cap \partial\overline{B_R}$. We orient $c(x,\hat{\mathbf{v}})$ in such a way that the segment of straight lines has the orientation of $\hat{\mathbf{v}}$.

DEFINITION 3.5. A line $L(x,\hat{\mathbf{v}}) \subset \Lambda$ goes through holes of K if $L(x,\hat{\mathbf{v}}) \cap B_R \neq \emptyset$ and $[c(x,\hat{\mathbf{v}})]_{H_1(\Lambda;\mathbf{R})} \neq 0$. Otherwise we say that $L(x,\hat{\mathbf{v}})$ does not go through holes of K.

Note that this characterization of lines that go or do not go through holes of K is independent of the R that was used in the definition. This follows from the homotopic invariance of homology. See Theorem 11.2, page 59 in [**28**].

In an intuitive sense $[c(x,\hat{\mathbf{v}})]_{H_1(\Lambda;\mathbf{R})} = 0$ means that $c(x,\hat{\mathbf{v}})$ is the boundary of a surface (actually of a chain) that is contained in Λ and then it can not go through holes of K. Obviously, as $K \subset B_R$, if $L(x,\hat{\mathbf{v}}) \cap B_R = \emptyset$ the line $L(x,\hat{\mathbf{v}})$ can not go through holes of K.

DEFINITION 3.6. Two lines $L(x,\hat{\mathbf{v}}), L(y,\hat{\mathbf{w}}) \subset \Lambda$ that go through holes of K go through the same holes if $[c(x,\hat{\mathbf{v}})]_{H_1(\Lambda;\mathbf{R})} = \pm[c(y,\hat{\mathbf{w}})]_{H_1(\Lambda;\mathbf{R})}$. Furthermore, we say that the lines go through the holes in the same direction if $[c(x,\hat{\mathbf{v}})]_{H_1(\Lambda;\mathbf{R})} = [c(y,\hat{\mathbf{w}})]_{H_1(\Lambda;\mathbf{R})}$.

DEFINITION 3.7. For any $\hat{\mathbf{v}} \in \mathbb{S}^2$ we denote by $\Lambda_{\hat{\mathbf{v}},\text{out}}$ the set of points $x \in \Lambda_{\hat{\mathbf{v}}}$ such that $L(x,\hat{\mathbf{v}})$ does not go through holes of K. We call this set the region without holes of $\Lambda_{\hat{\mathbf{v}}}$. The holes of $\Lambda_{\hat{\mathbf{v}}}$ is the set $\Lambda_{\hat{\mathbf{v}},\text{in}} := \Lambda_{\hat{\mathbf{v}}} \setminus \Lambda_{\hat{\mathbf{v}},\text{out}}$.

We now define an equivalence relation on $\Lambda_{\hat{\mathbf{v}},\text{in}}$. We say that $xR_{\hat{\mathbf{v}}}y$ if and only if $L(x,\hat{\mathbf{v}})$ and $L(y,\hat{\mathbf{v}})$ go through the same holes and in the same direction. By

$[x]$ we designate the classes of equivalence under $R_{\hat{\mathbf{v}}}$. We denote by $\{\Lambda_{\hat{\mathbf{v}},h}\}_{h\in\mathcal{I}}$ the partition of $\Lambda_{\hat{\mathbf{v}},\text{in}}$ given by this equivalence relation. It is defined as follows.

$$\mathcal{I} := \{[x]\}_{x\in\Lambda_{\hat{\mathbf{v}},\text{in}}}.$$

Given $h \in \mathcal{I}$ there is $x \in \Lambda_{\hat{\mathbf{v}},\text{in}}$ such that $h = [x]$. We denote

$$\Lambda_{\hat{\mathbf{v}},h} := \{y \in \Lambda_{\hat{\mathbf{v}},\text{in}} : yR_{\hat{\mathbf{v}}}x\}.$$

Then

$$\Lambda_{\hat{\mathbf{v}},\text{in}} = \cup_{h\in\mathcal{I}}\Lambda_{\hat{\mathbf{v}},h}, \quad \Lambda_{\hat{\mathbf{v}},h_1} \cap \Lambda_{\hat{\mathbf{v}},h_2} = \emptyset,\ h_1 \neq h_2.$$

We call $\Lambda_{\hat{\mathbf{v}},h}$ the subset of $\Lambda_{\hat{\mathbf{v}}}$ that goes through the holes h of K in the direction of $\hat{\mathbf{v}}$. Note that

$$\{\Lambda_{\hat{\mathbf{v}},h}\}_{h\in\mathcal{I}} \cup \{\Lambda_{\hat{\mathbf{v}},\text{out}}\} \tag{3.26}$$

is a disjoint open cover of $\Lambda_{\hat{\mathbf{v}}}$.

The dynamics of the electrons that travel through the holes of K in $\Lambda_{\hat{\mathbf{v}},h}$ can be described as follows. For large negative times the incoming electron wave packet is in Λ, far away from K. As time increases the electron travels towards K and it is inside B_R. At these times the electron has to be in $\Lambda_{\hat{\mathbf{v}},h}$ in order to cross B_R through the holes of K in $\Lambda_{\hat{\mathbf{v}},h}$. After crossing the holes it travels again away from K towards spatial infinity in Λ. This implies that the classical trajectories have to be in the following domain

$$\mathcal{C}_h := \left[\Lambda \setminus \left(\overline{B_R} \cup P_{\hat{\mathbf{v}}}\right)\right] \cup \left(\overline{B_R} \cap \Lambda_{\hat{\mathbf{v}},h}\right), \tag{3.27}$$

where $P_{\hat{\mathbf{v}}}$ is the plane orthogonal to $\hat{\mathbf{v}}$ that passes through zero

$$P_{\hat{\mathbf{v}}} := \left\{x \in \mathbf{R}^3 : x \cdot \hat{\mathbf{v}} = 0\right\}. \tag{3.28}$$

We take away from $\mathcal{C}_h$ the part of $P_{\hat{\mathbf{v}}}$ that does not intersect $\Lambda_{\mathbf{v},h}$ so that that the only way that the electron in $\mathcal{C}_h$ can classically cross the plane $P_{\hat{\mathbf{v}}}$ is through $\Lambda_{\hat{\mathbf{v}},h}$.

Furthermore, the classical trajectories of the electrons that do not cross any hole of K have to be on the set

$$\mathcal{C}_{\text{out}} := \left(\Lambda \setminus \overline{B_R}\right) \cup \left(\overline{B_R} \cap \Lambda_{\hat{\mathbf{v}},\text{out}}\right). \tag{3.29}$$

Let x_0 be a fixed point with $x_0 \cdot \hat{\mathbf{v}} < -R$. We define

(3.30)

$$\lambda_h(x) := \int_{C^h} A, \quad h \in \mathcal{I}, \text{ where } C^h \text{ is any differentiable path from } x_0 \text{ to } x \text{ in } \mathcal{C}_h,$$

and

(3.31)

$$\lambda_{\text{out}}(x) := \int_{C^{\text{out}}} A, \quad \text{where } C^{\text{out}} \text{ is any differentiable path from } x_0 \text{ to } x \text{ in } \mathcal{C}_{\text{out}}.$$

Since $H_1(\mathcal{C}_h; \mathbf{R}), h \in \mathcal{I}$ and $H_1(\mathcal{C}_{\text{out}}; \mathbf{R})$ are trivial **[11]**, $\lambda_h, h \in \mathcal{I}$ and λ_{out} do not depend on the particular curve form x_0 to x that we take, respectively, in $\mathcal{C}_h, h \in \mathcal{I}$ and $\mathcal{C}_{\text{out}}$. Furthermore, they are differentiable and $\nabla\lambda_h(x) = A(x), x \in \mathcal{C}_h, h \in \mathcal{I}$ and $\nabla\lambda_{\text{out}}(x) = A(x), x \in \mathcal{C}_{\text{out}}$.

We denote **[9]**

$$F_h := \int_{c(x,\hat{\mathbf{v}})} A, \quad x \in \Lambda_{\hat{\mathbf{v}},h}, h \in \mathcal{I}.$$

F_h is the magnetic flux over any surface (or a chain) in $\mathbf{R}^3$ whose boundary is $c(x,\hat{\mathbf{v}})$. That is to say, it is the flux associated to the holes of K in $\Lambda_{\hat{\mathbf{v}},h}$. F_h is constant for all $x \in \Lambda_{\hat{\mathbf{v}},h}$. Moreover, we have that

$$F_h = \lambda_h(x), \quad x \in C^0_+, \tag{3.32}$$

and

$$\lambda_{\text{out}}(x) = 0, \quad \text{for } x \in C^0_+ \tag{3.33}$$

where

$$C^0_+ := \left\{ x \in \Lambda \setminus \overline{B_R} : x \cdot \hat{\mathbf{v}} > 0 \right\}$$

is the region behind the magnet.

We now state our main results on the validity of the Aharonov-Bohm Ansatz [**11**].

THEOREM 3.8. *For any $1 > \delta > 0$ and any compact set $\Lambda_0 \subset \Lambda_{\hat{\mathbf{v}},h}, h \in \mathcal{I}$ there is a constant C_δ such that $\forall t \in \mathbf{R}$ and for all $\varphi \in \mathcal{H}_2(\mathbf{R}^3)$ with support in Λ_0*

$$\left\| e^{-itH(A,V)}\, W_-(A,V)\, \varphi_{\mathbf{v}} - e^{i\lambda_h}\, \chi_{\mathcal{C}_h}\, e^{-itH_0}\, \varphi_{\mathbf{v}} \right\|_{L^2(\mathbf{R}^3)} \leq \frac{C_\delta}{v^{1-\delta}} \|\varphi\|_{\mathcal{H}_2(\mathbf{R}^3)}. \tag{3.34}$$

Furthermore, for any $1 > \delta > 0$ and any compact set $\Lambda_0 \subset \Lambda_{\hat{\mathbf{v}},\text{out}}$ there is a constant C_δ such that $\forall t \in \mathbf{R}$ and for all $\varphi \in \mathcal{H}_2(\mathbf{R}^3)$ with support in Λ_0

$$\left\| e^{-itH(A,V)}\, W_-(A,V)\, \varphi_{\mathbf{v}} - e^{i\lambda_{\text{out}}} \chi_{\mathcal{C}_{\text{out}}}\, e^{-itH_0}\, \varphi_{\mathbf{v}} \right\|_{L^2(\mathbf{R}^3)} \leq \frac{C_\delta}{v^{1-\delta}} \|\varphi\|_{\mathcal{H}_2(\mathbf{R}^3)}. \tag{3.35}$$

Let us consider any $\varphi \in \mathcal{H}_2(\mathbf{R}^3)$ with compact support in $\Lambda_{\hat{\mathbf{v}}}$. As (3.26) is a disjoint open cover of $\Lambda_{\hat{\mathbf{v}}}$

$$\varphi = \sum_{h \in \mathcal{I}} \varphi_h + \varphi_{\text{out}}, \tag{3.36}$$

where $\varphi_h, \varphi_{\text{out}} \in \mathcal{H}_2(\mathbf{R}^3), \varphi_h$ has compact support in $\Lambda_{\hat{\mathbf{v}},h}, h \in \mathcal{I}$, and φ_{out} has compact support in $\Lambda_{\hat{\mathbf{v}},\text{out}}$. The sum is finite because φ has compact support. We denote

$$\varphi_{\mathbf{v}} := e^{im\mathbf{v}\cdot x}\, \varphi,\ \varphi_{\mathbf{v},h} := e^{im\mathbf{v}\cdot x}\, \varphi_h, h \in \mathcal{I}, \varphi_{\mathbf{v},\text{out}} := e^{im\mathbf{v}\cdot x}\, \varphi_{\text{out}}. \tag{3.37}$$

We denote

$$\psi_{AB,\mathbf{v},h} := \chi_{\mathcal{C}_h}\, e^{i\lambda_h}\, e^{-itH_0}\, \varphi_{\mathbf{v},h}, h \in \mathcal{I}, \quad \psi_{AB,\mathbf{v},\text{out}} := \chi_{\mathcal{C}_{\text{out}}}\, e^{i\lambda_{\text{out}}}\, e^{-itH_0}\, \varphi_{\mathbf{v},\text{out}}, \tag{3.38}$$

$$\psi_{AB,\mathbf{v}} := \sum_{h \in \mathcal{I}} \psi_{AB,\mathbf{v},h} + \psi_{AB,\mathbf{v},\text{out}}. \tag{3.39}$$

Equation (3.39) gives the Aharonov-Bohm Ansatz in the domain $\cup_{h\in\mathcal{I}}\mathcal{C}_h \cup \mathcal{C}_{\text{out}}$ that has non-trivial first group of singular homology as the sum of the Aharonov-Bohm Ansatz in each of the components, $\mathcal{C}_h, h \in \mathcal{I}, \mathcal{C}_{\text{out}}$ that have trivial first group of singular homology.

The next theorem proves that the Aharonov-Bohm Ansatz is actually valid on the biggest domain where it can be valid, namely in $\cup_{h\in\mathcal{I}}\mathcal{C}_h \cup \mathcal{C}_{\text{out}}$. This result provides an approximate solution at all times for every electron that does not hit the magnet.

THEOREM 3.9 (The Validity of the Aharonov-Bohm Ansatz [**11**]). *For any $1 > \delta > 0$ and any compact set $\Lambda_0 \subset \Lambda_{\hat{\mathbf{v}}}$ there is a constant C_δ such that $\forall t \in \mathbf{R}$ and for all $\varphi \in \mathcal{H}_2(\mathbf{R}^3)$ with support in Λ_0 the solution to the Schrödinger equation $e^{-itH(A,V)}\, W_-(A,V)\, \varphi_{\mathbf{v}}$ that behaves as $e^{-itH_0}\, \varphi_{\mathbf{v}}$ as $t \to -\infty$ is given at time t by the Aharonov-Bohm Ansatz, $\psi_{AB,\mathbf{v}}$, up to the following error*

$$\left\| e^{-itH(A,V)}\, W_-(A,V)\, \varphi_{\mathbf{v}} - \psi_{AB,\mathbf{v}} \right\|_{L^2(\mathbf{R}^3)} \leq \frac{C_\delta}{v^{1-\delta}} \|\varphi\|_{\mathcal{H}_2(\mathbf{R}^3)}. \tag{3.40}$$

According to 3.32 and 3.33, behind the magnet in C_+^0,

$$\psi_{AB,\mathbf{v},h} := \chi_{C_h}\, e^{iF_h}\, e^{-itH_0}\, \varphi_{\mathbf{v},h}, h \in \mathcal{I}, \quad x \in C_+^0 \tag{3.41}$$

and

$$\psi_{AB,\mathbf{v},\mathrm{out}} := \chi_{\mathcal{C}_{\mathrm{out}}}\, e^{-itH_0}\, \varphi_{\mathbf{v},\mathrm{out}}, \qquad x \in C_+^0. \tag{3.42}$$

The phase shifts e^{iF_h} were measured in the experiments of Tonomura et al. [**48, 54, 55**]. Because the Aharonov-Bohm Ansatz is free evolution, up to a phase, the electron is not accelerated. This explains the results of the experiment of Caprez et al. [**15**]. Then, Theorem 3.9 rigorously proves that quantum mechanics predicts the results of the experiments of Tonomura et al. and of Caprez et al.

3.5. The Aharonov-Bohm Effect. Quantitative Results. As we mentioned in the introduction, the experiments of Tonomura at al. [**48, 54, 55**] are widely considered to be the only experimental evidence of the physical existence of the Aharonov-Bohm effect. They constructed small toroidal magnets so that the magnetic field is practically zero outside them. In [**55**] the magnets are impenetrable and, furthermore, they are covered by super conductive layers that prevent the leakage of the magnetic field outside the magnets. We consider a magnet as in the experiments, $K := \{(x_1, x_2, x_3) \in \mathbf{R}^3 : 0 < \tilde{r}_1 \leq (x_1^2 + x_2^2)^{1/2} \leq \tilde{r}_2, |x_3| \leq \tilde{h}\}$. The inner radius is $\tilde{r}_1$, $\tilde{r}_2$ is the outer radius, and $2\tilde{h}$ is the height. As before we denote by $\Lambda := \mathbf{R}^3 \setminus K$ the exterior of the magnet. The magnetic field is zero for outside K.

In the experiments an electron wave packet was sent towards the magnet. It was superimposed behind it with a reference electron wave packet to produce the interference pattern. The experiments were prepared in such a way that the reference electron wave packet was not influenced by the magnet, and that the electron wave packet and the reference electron wave packet interfered only behind the magnet, forming interference patterns that provided strong evidence of the physical existence of the Aharonov-Bohm effect.

The electron wave packet was much larger than the magnet. It was 3 micrometers in size in the direction of the electron propagation and 20 micrometers in size in a plane perpendicular to the propagation direction [**56**]. The image that was produced behind the magnet clearly showed the shadow of the magnet and also the hole and the exterior of the magnet (see [**54, 55**]). Tonomura et al. [**54, 55**], pointed out that this indicates that there was no interference between the part of the electron wave packet that went through the hole and the one that either hit the magnet or traveled outside. This shows that it is reasonable to concentrate on the part of the wave packet that goes through the hole and to take it as the electron wave packet itself. It is either the part of the electron wave packet that goes through the hole or a smaller electron wave packet that really goes through the hole.

For the purpose of obtaining quantitative results that can be compared with the experimental observations, it is convenient to re-introduce the physical mass $M = m\hbar$ and to consider the momentum operator $\mathbf{P} := -i\hbar\nabla$ and the free Hamiltonian

$$H_0 := \frac{1}{2M}\mathbf{P}^2.$$

We consider that the emitted electron wave packet is given by the free evolution of a Gaussian wave function, $\varphi_{\mathbf{v}}$, with velocity $\mathbf{v}$. From the physical point of view, this is a reasonable assumption.

So, we define

$$\varphi_{\mathbf{v}} := e^{i\frac{M}{\hbar}\mathbf{v}\cdot x}\,\varphi, \text{ where } \quad \varphi := \frac{1}{(\sigma^2\pi)^{3/4}}e^{-\frac{x^2}{2\sigma^2}}, \tag{3.43}$$

with variance σ smaller than the inner radius of the magnet. We have chosen the variance transverse to the velocity of propagation, $\mathbf{v}$, equal to the longitudinal variance in the direction of propagation. In fact, the size of the longitudinal variance is not essential for our arguments and we have chosen it equal to the transversal variance only for simplicity.

Then, at the time of emission the electron is represented by the following Gaussian wave packet that is a solution to the free Schrödinger equation

$$\psi_{\mathbf{v},0}(x,t) := e^{-i\frac{t}{\hbar}H_0}\,\varphi_{\mathbf{v}}(x). \tag{3.44}$$

As before, the exact electron wave packet, $\psi_{\mathbf{v}}(x,t)$ is given by

$$\psi_{\mathbf{v}}(x,t) := e^{-i\frac{t}{\hbar}H(A)}\,W_-\,\varphi_{\mathbf{v}}.$$

To define the Aharonov-Bohm Ansatz we need to introduce a change of gauge formula from the zero vector potential; that, of course, has to be discontinuous somewhere. We take the surface of discontinuity as follows:

$$\mathcal{S} := \{(x_1,x_2,x_3) \in \mathbf{R}^3 : (x_1^2+x_2^2)^{1/2} > \tilde{r}_2, x_3 = 0\},$$

and we define the gauge transformation in the domain, $\mathcal{D}$, given by

$$\mathcal{D} := \Lambda \setminus \mathcal{S}.$$

We can assume that the support of A is contained on the convex hull of $\tilde{K}$ (see [**10**]). For every $x \in \mathcal{D}$, and a fixed point x_0 in $\mathcal{D}$ with vertical component less than $-\tilde{h}$, we define the gauge transformation as

$$\lambda_{A,0}(x) := \int_{x_0}^{x} A,$$

where the integral is over a path in $\mathcal{D}$. Note that for any $x \in \mathcal{D}$ with $x_3 > 0$ the integration contour has to go necessarily through the hole of the magnet.

With this gauge, the Aharonov-Bohm Ansatz is given by

$$\psi_{AB,\mathbf{v}}(x,t) = e^{i\lambda_{A,0}(x)}e^{-i\frac{t}{\hbar}H_0}\varphi_{\mathbf{v}}, \tag{3.45}$$

which, again, is the multiplication of the free incoming solution by the Dirac magnetic factor $e^{i\lambda_{A,0}(x)}$ [**19**].

3.6. Experimental Data. We consider the two different magnets with their dimensions given in Table I in [**48**]. We denote them by $K_j, j \in \{1,2\}$,

$$K_j := \{x = (x_1, x_2, x_3) \in \mathbf{R}^3 : \tilde{r}_{1,j} \leq \sqrt{x_1^2 + x_2^2} \leq \tilde{r}_{2,j}, |x_3| \leq \tilde{h}\}. \tag{3.46}$$

The height $\tilde{h}$ is $10^{-6} cm$ for both magnets and

$$\tilde{r}_{1,1} = 1.5 \times 10^{-4} cm,$$
$$\tilde{r}_{2,1} = 2.5 \times 10^{-4} cm,$$

$$\tilde{r}_{1,2} = 1.75 \times 10^{-4} cm,$$
$$\tilde{r}_{2,2} = 2.75 \times 10^{-4} cm.$$

In the experiments by Tonomura et al. [**55**] the electron has an energy of $150\, keV$. In these experiments they consider impenetrable magnets as we do in this paper. In the experiments [**54**] they consider penetrable magnets and energies of $80\, keV$, $100\, keV$, and $125\, keV$. Since our method also applies in the case of penetrable magnets, we will consider in our estimates below the two extreme energies and an intermediate energy, although the most important one is the one of $150\, keV$, which is the one used for the case of impenetrable magnets. Thus we consider the following energies.

$$E_1 = 150\, keV,$$
$$E_2 = 100\, keV,$$
$$E_3 = 80\, keV.$$

The velocities, and the velocities times m corresponding to the energies E_1, E_2, E_3 are given by:

$$v_1 = 2.2971 \times 10^{10} cm/s, \quad mv_1 = 1.9842 \times 10^{10} cm^{-1},$$
$$v_2 = 1.8755 \times 10^{10} cm/s, \quad mv_2 = 1.6201 \times 10^{10} cm^{-1},$$
$$v_3 = 1.6775 \times 10^{10} cm/s, \quad mv_3 = 1.4491 \times 10^{10} cm^{-1}.$$

In what follows we assume that the obstacle $\tilde{K}$ is either $\tilde{K}_1$ or $\tilde{K}_2$ and that the velocity v is either v_1, v_2, or v_3.

3.7. Quantitative Results on the Aharonov-Bohm Ansatz. We take $\mathbf{v} = (0, 0, v)$. Since under the free evolution the electron wave packet is concentrated along the classical trajectory, we can expect that if the velocity $\mathbf{v}$ is large enough, the exact electron wave packet will keep away, at all times, from the surface, $\mathcal{S}$, where the gauge transformation is discontinuous. Consequently, the Aharonov-Bohm Ansatz should be a good approximation to the exact solution. Theorem 3.10 proves that this is true under the conditions of the Tonomura et al. experiments [**48, 54, 55**], provided that the appropriate, quantitative, lower and upper bounds on the variance the σ of the Gaussian wave function are satisfied. The requirement for the variance σ to lie within the interval below ensures that the interaction of the electron with the magnet and the surface $\mathcal{S}$ is small. We give the error bounds below in terms of , $r_1 = \tilde{r}_1 - \varepsilon, r_2 = \tilde{r}_2 + \varepsilon$, with $\varepsilon := \tilde{r}_1/50$. This is convenient since in the proofs we replace the sharp cutoff J by a smooth cutoff. See [**10**] for details.

THEOREM 3.10 (The error bound [**10**]). *Suppose that the magnets and energies are the ones of the experiments of Tonomura et al. Then, for every Gaussian wave function,* φ, *with variance* $\sigma \in [\frac{4.5}{mv}, \frac{\tilde{r}_1}{2}]$ *and every* $t \in \mathbf{R}$, *the solution to the Schrödinger equation,* $e^{-i\frac{t}{\hbar}H(A)} W_- \varphi_{\mathbf{v}}$, *which behaves as* $e^{-i\frac{t}{\hbar}H_0} \varphi_{\mathbf{v}}$ *as* $t \to -\infty$ *is given at the time* t *by*

$$\psi_{AB,\mathbf{v}} := e^{i\lambda_{A,0}(x)} e^{-i\frac{t}{\hbar}H_0} \varphi_{\mathbf{v}}, \tag{3.47}$$

up to the following error,

$$\begin{aligned} &\| e^{-i\frac{t}{\hbar}H} W_-(A)\varphi_{\mathbf{v}} - e^{i\lambda_{A,0}} e^{-i\frac{t}{\hbar}H_0}\varphi_{\mathbf{v}} \| \\ &\leq 7 e^{-\frac{r_1^2}{2\sigma^2}} + 177 \times 10^3 e^{-\frac{33}{34}\frac{(\sigma m v)^2}{2}} + 10^{-100}, \end{aligned} \tag{3.48}$$

where, $m := M/\hbar$.

The main factors that produce the error bound are:

(1) The size of electron wave packet factor:

$$e^{-\frac{r_1^2}{2\sigma^2}}. \tag{3.49}$$

(2) The opening angle of the electron wave packet factor:

$$e^{-\frac{33}{34}\frac{(\sigma m v)^2}{2}}. \tag{3.50}$$

When the variance σ is close to the radius of the magnet, (3.49) is close to 1 and (3.50) is very small, because in this case $\sigma m v$ is big. Hence, when the electron is big compared to the inner radius, (3.49) is the important term. On the contrary, when the variance is very small, so that $\sigma m v$ is close to 1, the factor (3.50) is close to one and (3.49) is extremely small. In this case, the important factor is (3.50). Note that when the variance in position, σ, is small, the variance in momentum is big according to the Heisenberg uncertainly principle. Then, the component of the momentum transversal to the axis of the magnet is large. Consequently, the opening angle of the electron wave packet is large, and the electron spreads fast as it propagates.

Our uniform in time error bound applies to all experimental magnets and energies. Its behaviour is the same for the three energies and the two magnets, so there is no loss of generality if we select a magnet and an energy in our analysis. In the case of the biggest energy, E_1, and the second magnet, K_2, for $\sigma \in [6.7591 \times 10^{-6} r_1, \frac{r_1}{22}]$, or equivalently, $\sigma \in [\frac{23}{mv}, \frac{154678}{mv}]$, the variance sigma ranges over the interval $[1.1592 \times 10^{-9}, 7.7955 \times 10^{-6}]$. For these values of sigma the error bound in (3.48) is less than 10^{-99}. Furthermore, the probability of interaction of the electron with the magnet is less than 10^{-199}. See [**10**] for a detailed analysis of our error bound for all ranges of σ, with precise numerical values, as well as for more information on the physical interpretation of our results in terms of the radius of the electron and of its opening angle.

In conclusion, our results give a rigorous and quantitative proof that quantum mechanics predicts the results observed in the Tonomura et al. experiments. We found the time evolution of the electron up to an error bound that we provide explicitly. The approximate wave function of the electron is given by the Aharonov-Bohm Ansatz. Furthermore, in the Aharonov-Bohm Ansatz the electron is not accelerated; it propagates following the free evolution, with the wave function multiplied

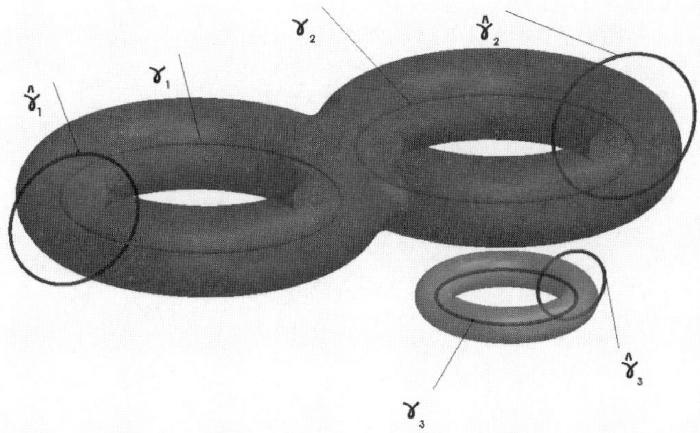

FIGURE 1. The magnet $K = \cup_{j=1}^{L} K_j \subset \mathbf{R}^3$, where K_j are handle-bodies, for every $j \in \{1, \cdots, L\}$. The exterior domain, $\Lambda := \mathbf{R}^3 \backslash K$. The curves $\gamma_k, k = 1, 2, \cdots m$ are a basis of the first singular homology group of K and the curves $\hat{\gamma}_k, k = 1, 2, \cdots m$ are a basis of the first singular homology group of Λ.

by a phase. As we prove that the Aharonov-Bohm Ansatz approximates the exact solution with an error bound uniform in time that can be smaller than 10^{-99} in norm, we rigorously prove that quantum mechanics predicts that no force acts on the electron, in agreement with the experimental results of Caprez et al. [**15**].

Summing up, the experiments of Tonomura et al. [**48, 54, 55**] give strong evidence of the existence of the interference fringes predicted by Franz [**27**] and by Aharonov and Bohm [**5**]. The experiment of Caprez et al. [**15**] verifies that the interference fringes are not due to a force acting on the electron, and the results of [**9, 10, 11**] rigorously prove that quantum mechanics theoretically predicts the observations of these experiments in a extremely precise way. This gives a firm experimental and theoretical basis to the existence of the Aharonov-Bohm effect, namely, that magnetic fields *act at a distance on charged particles* and that this action is carried by the circulation of the magnetic potential, which is what gives magnetic potentials a real physical significance.

References

[1] T. Adachi and K. Maehara, *On multidimensional inverse scattering for Stark Hamiltonians*, J. Math. Phys. **48** (2007), 042101. MR2319881 (2008b:81321)

[2] T. Adachi, T. Kamada, M. Kazuno, and K. Toratani, *On multidimensional inverse scattering in an external electric field asymptotically zero in time*, Inverse Problems **27** (2011), 065006. MR2805763 (2012h:81280)

[3] R. A. Adams and J. J. F. Fournier, *Sobolev Spaces*, Academic Press, Oxford, 2003. MR2424078 (2009e:46025)

[4] S. Agmon, *Lectures on Elliptic Boundary Value Problems*, D. Van Nostrand, Princeton, New Jersey, 1965. MR0178246 (31:2504)

[5] Y. Aharonov and D. Bohm, *Significance of electromagnetic potentials in the quantum theory*, Phys. Rev. **115** (1959), 485–491. MR0110458 (22:1336)

[6] S. Arians, *Geometric approach to inverse scattering for the Schrödinger equation with magnetic and electric potentials*, J. Math. Phys. **38** (1997), 2761–2773. MR1449530 (98d:81125)

[7] S. Arians, *Inverse Streutheorie für die Schrödinger Gleichung mit Magnet Felder*, Dissertation RWTH-Aachen, Berlin, Logos, 1997.

[8] S. Arians, *Geometric approach to inverse scattering for hydrogen-like systems in a homogeneous magnetic field*, J. Math. Phys. **39** (1998), 1730–1743. MR1614773 (99a:81190)

[9] M. Ballesteros and R. Weder, *High-velocity estimates for the scattering operator and Ahanorov-Bohm effect in three dimensions*, Commun. Math. Phys. **285** (2009), 345–398. MR2453598 (2010d:81115)

[10] M. Ballesteros and R. Weder, *The Aharonov-Bohm effect and Tonomura et al. experiments: Rigorous results*, J. Math. Phys. **50** (2009), 122108. MR2582580 (2010m:81122)

[11] M. Ballesteros and R. Weder, *Aharonov-Bohm Effect and High-Velocity Estimates of Solutions to the Schödinger Equation*, Commun. Math. Phys. **303** (2011), 175–211. MR2775119 (2012e:81084)

[12] H. Batelaan and A. Tonomura, *The Aharonov-Bohm effect: variations on a subtle theme*, Physics Today **62** (2009), 38–43.

[13] T. H. Boyer, *Darwin-Lagrangian analysis for the interaction of a point charge and a magnet: considerations related to the controversy regarding the Aharonov-Bohm and the Aharonov-Casher phase shifts*, J. Phys. A: Math. Gen. **39** (2006), 3455–3477. MR2214223 (2007a:78009)

[14] G. E. Bredon, *Topology and Geometry*, Springer-Verlag, New York, 1993. MR1224675 (94d:55001)

[15] A. Caprez, B. Barwick, H. Batelaan, *Macroscopic test of the Aharonov-Bohm effect*, Phys. Rev. Lett. **99** (2007), 210401, 4 pp.

[16] T. Daudè and F. Nicoleau, *Recovering the mass and the charge of a Reissner-Nordström black hole by an inverse scattering experiment*, Inverse Problems **24** (2008), 025017. See also Corrigendum: *Recovering the mass and the charge of a Reissner-Nordström black hole by an inverse scattering experiment*, Inverse Problems **25** (2009), 059801. MR2408554 (2009f:83042)

[17] T. Daudè and F. Nicoleau, *Inverse scattering in de Sitter-Reissner-Nordström black hole spacetimes*, Rev. Math. Phys. **22** (2010), 431–484. MR2647042 (2011g:83103)

[18] G. de Rham, *Differentiable Manifolds*, Springer-Verlag, Berlin, 1984. MR760450 (85m:58005)

[19] P. Dirac, *Quantized singularities in the electromagnetic field*, Proc. R. Soc. A **133** (1931), 60–72.

[20] V. Enss, *Propagation properties of quantum scattering states*, J. Func. Anal. **52** (1983), 219–251 . MR707205 (85b:81187)

[21] V. Enss, *A New Look at the Multidimensional Inverse Scattering Problem*, Understanding Physics (A.K. Richter, ed.), Copernicus Gesellschaft, Katlenburg-Lindau, 1998, 31–48.

[22] V. Enss and W. Jung, *Geometrical Approach to Inverse Scattering*, Proceedings of the First MaPhySto Workshop on Inverse Problems, April 22–24, 1999, Aarhus. MaPhySto Miscellanea, 1999, 5–21.

[23] V. Enss and R. Weder, *Inverse potential scattering: a geometrical approach.* Mathematical quantum theory. II. Schrödinger operators, Vancouver, BC, 1993. CRM Proc. Lecture Notes, 8, Amer. Math. Soc. Providence, RI, 1995, pp. 151–162. MR1332039 (96d:35103)

[24] V. Enss and R. Weder, *Uniqueness and reconstruction formulae for inverse N-particle scattering*, Differential Equations and Mathematical Physics, Proceedings of the International Conference, Univ. of Alabama at Birmingham, March 1994. Editor: I. Knowles. International Press Boston (ca. 1995), pp. 53–66. MR1703572 (2000k:81397)

[25] V. Enss and R. Weder, *The geometric approach to multidimensional inverse scattering*, J. Math. Phys. **36** (1995), 3902–3921. MR1341964 (96f:81128)

[26] V. Enss and R. Weder, *Inverse two-cluster scattering*, Inverse Problems **12** (1996), 409–418. MR1402100 (97d:81205)

[27] W. Franz, *Elektroneninterferenzen im Magnetfeld*, Verh. D. Phys. Ges. (3) **20** , No. 2 (1939), 65–66; Physikalische Berichte, **21** (1940), 686.

[28] M. J. Greenberg and J. R. Harper, *Algebraic Topology. A First Course*, Addison-Wesley, New York, 1981. MR643101 (83b:55001)

[29] A. Hatcher, *Algebraic Topology*, Cambridge University Press, Cambridge, 2002. MR1867354 (2002k:55001)

[30] G. C. Hegerfeldt and J. T. Neumann, *The Aharonov–Bohm effect: the role of tunneling and associated forces*, J. Phys. A: Math. Theor. **41** (2008), 155305, 11 pp. MR2449583 (2009j:81048)

[31] S. Helgason, *Groups and Geometric Analysis*, Academic Press, Orlando, 1984. MR754767 (86c:22017)

[32] B. Helffer, *Effet d'Aharonov-Bohm sur un état borné de l'équation de Schrödinger*, Comm. Math. Phys. **119** (1988), 315–329. MR968700 (90g:81030)

[33] H. T. Ito, *Inverse scattering problems for Dirac operator with time-dependent electromagnetic potentials* (Japanese) *Spectral-scattering theory and related topics* (Japanese), Kyoto, 1997. Sūrikaisekikenkyūsho Kōkyūroku **1047** (1998), 26–35. MR1668168

[34] H. T. Ito, *An inverse scattering problem for Dirac equations with time-dependent electromagnetic potentials*, Publ. RIMS, Kyoto Univ. **34** (1998), 355–381. MR1651391 (99h:35228)

[35] W. Jung, *Der Geometrische Ansatz zur Inversen Streutheorie bei der Dirac-Gleichung*, Diplomarbeit RWTH-Aachen, 1996.

[36] W. Jung, *Geometrical approach to inverse scattering for the Dirac equation*, J. Math. Phys. **38** (1997), 39–48. MR1424640 (98a:81252)

[37] W. Jung, *Gauge Transformations and Inverse Quantum Scattering with Medium-Range Magnetic Fields*, Math. Phys. Electron. J. **11** (2005), paper 5. MR2192547 (2006k:81405)

[38] T. Kato, *Perturbation Theory for Linear Operators, Second Edition*, Springer-Verlag, Berlin, 1976. MR0407617 (53:11389)

[39] F. Nicoleau, *An inverse scattering problem with the Aharonov-Bohm effect*, J. Math Phys. **41** (2000), 5223–5237. MR1770949 (2001g:81276)

[40] F. Nicoleau, *Inverse scattering for Stark Hamiltonians with short-range potentials*, Asymptot. Anal. **35** (2003), 349–359. MR2011794 (2004j:35221)

[41] F. Nicoleau, *An inverse scattering problem for short-range systems in a time-periodic electric field*, Math. Res. Lett. **12** (2005), 10001–10012. MR2189247 (2006m:81297)

[42] F. Nicoleau, *Inverse scattering for a Schrödinger operator with a repulsive potential*, Acta Math. Sin., Springer **22** (2006), 1485–1492 . MR2251408 (2008b:81322)

[43] F. Nicoleau, *An inverse scattering problem for the Schrödinger equation in a semiclassical process*, J. Math. Pures Appl. **86** (2006), 463–470. MR2281447 (2008j:35191)

[44] M. Reed and B. Simon, *Methods of Modern Mathematical Physics II. Fourier Analysis, Self-Adjointness*, Academic Press, New York, 1975. MR0493420 (58:12429b)

[45] M. Reed and B. Simon, *Methods of Modern Mathematical Physics III. Scattering Theory*, Academic Press, New York, 1979. MR529429 (80m:81085)

[46] M. Sandoval Romero and R. Weder, *The Initial Value Problem, Scattering and Inverse Scattering, for Non-Linear Schrödinger Equations with a Potential and a Non-Local Non-Linearity*, J. Phys. A **39** (2006), 11461–11478. MR2259648 (2008c:81038)

[47] S. Olariu and I. I. Popescu, *The quantum effects of electromagnetic fluxes*, Rev. Modern. Phys. **57** (1985), 339–436.

[48] N. Osakabe T. Matsuda, T. Kawasaki, J. Endo, A. Tonomura, S. Yano and H. Yamada, *Experimental confirmation of Aharonov-Bohm effect using a toroidal magnetic field confined by a superconductor*, Phys. Rev. A **34** (1986), 815–822.

[49] M. Peshkin and A. Tonomura, *The Aharonov-Bohm Effect*, Lecture Notes in Phys. **340**, Springer, Berlin, 1989. MR1016149 (91c:81004)

[50] Ph. Roux, *Scattering by a toroidal coil*, J. Phys. A: Math. Gen. **36** (2003), 5293–5304. MR1985319 (2004g:81301)

[51] Ph. Roux and D. R. Yafaev, *On the mathematical theory of the Aharonov-Bohm effect*, J. Phys. A: Math. Gen. **35** (2002), 7481–7492. MR1945524 (2003m:81085)

[52] S. N. M. Ruijsenaars, *The Aharonov-Bohm effect and scattering theory*, Ann. Phys. **146** (1983), 1–34. MR701261 (85e:81117)

[53] A. Tonomura, *Direct observation of thitherto unobservable quantum phenomena by using electrons*, Proc. Natl. Acad. Sci. U.S.A. **102** (2005), 14952–14959.
[54] A. Tonomura, T. Matsuda, R. Suzuki, A. Fukuhara, N. Osakabe, H. Umezaki, J. Endo, K. Shinagawa, Y. Sugita and H. Fujiwara, *Observation of Aharonov-Bohm effect by electron holography*, Phys. Rev. Lett. **48** (1982), 1443–1446.
[55] A. Tonomura, N. Osakabe, T. Matsuda, T. Kawasaki, J. Endo, S. Yano and H. Yamada, *Evidence for Aharonov-Bohm effect with magnetic field completely shielded from electron wave*, Phys. Rev. Lett. **56** (1986), 792–795.
[56] A. Tonomura, private communication, 2008.
[57] A. Tonomura and F. Nori, *Disturbance without the force*, Nature **452-20** (2008), 298–299.
[58] H. Triebel, *Interpolation Theory, Function Spaces, Differential Operators*, North-Holland, Amsterdam, 1978. MR503903 (80i:46032b)
[59] G. D. Valencia and R. Weder, *High-velocity estimates and inverse scattering for quantum N-body systems with Stark effect*, J. Math. Phys. **53** (2012), 102105, 30 pp. MR3050561
[60] F. W. Warner, *Foundations of Differentiable Manifolds*, Springer-Verlag, Berlin, 1983. MR722297 (84k:58001)
[61] R. Weder, *Multidimensional inverse scattering in an electric field*, J. Funct. Anal. **139** (1996), 441–465. MR1402772 (97f:81168)
[62] R. Weder, *Inverse scattering for N-body systems with time-dependent potentials*, Inverse problems of wave propagation and diffraction (Aix-les-Bains, 1996), Lecture Notes in Physics **486**, 1997, pp. 27–46. MR1601179 (98m:81207)
[63] R. Weder, *A Time–Dependent Method for Inverse Scattering*, Modern Mathematical Methods in Diffraction Theory and its Applications in Engineering. Editor: E. Meister, Peter Lang, Frankfurt, 1997, pp. 251–262. MR1482567
[64] R. Weder, *Inverse Scattering for the Nonlinear Schrödinger Equation*, Comm. Partial Differential Equations **22** (1997), 2089–2103. MR1629534 (99c:35232)
[65] R. Weder, *Inverse scattering for the nonlinear Schrödinger equation and $Lp - L\acute{p}$ estimates*, Spectral and scattering theory and related topics (Japanese) (Kyoto, 1999). Sūrikaisekikenkyūsho Kōkyūroku **1156** (2000), 157–168. MR1798804
[66] R. Weder, *$Lp - L\acute{p}$ estimates for the Schrödinger equation on the line and inverse scattering for the nonlinear Schrödinger equation with a potential*, J. Funct. Anal. **170** (2000), 37–68. MR1736195 (2001e:35135)
[67] R. Weder, *Uniqueness of inverse scattering for the nonlinear Schrödinger equation and reconstruction of the potential and the nonlinearity*, Mathematical and numerical aspects of wave propagation (Santiago de Compostela, 2000). Editors: A. Bermúdez, D. Gómez, C. Hazard, P. Joly and J. Roberts. SIAM Proceedings Series, SIAM, Philadelphia, 2000, pp. 631-634. MR1785968
[68] R. Weder, *Inverse scattering on the line for the nonlinear Klein-Gordon equation with a potential*, J. Math. Anal. Appl. **252** (2000), 102–123. MR1797847 (2001j:35272)
[69] R. Weder, *$L^p - L^{\acute{p}}$ Estimates for the Schrödinger Equation and Inverse Scattering*, Differential Equations and Mathematical Physics. Editors: R. Weikard and G. Weinstein. American Mathematical Society and International Press, U.S.A., 2000, pp. 435–448. MR1764771 (2001b:35238)
[70] R. Weder, *Inverse scattering for the non-linear Schrödinger equation: reconstruction of the potential and the non-linearity*, Math. Meth. Appl. Sci. **24** (2001), 245–254. MR1814412 (2002b:35191)
[71] R. Weder, *Inverse scattering for the nonlinear Schrödinger equation. II. Reconstruction of the potential and the nonlinearity in the multidimensional case*, Proc. Amer. Math. Soc. **129** (2001), 3637–3645. MR1860498 (2002h:35343)
[72] R. Weder, *Direct and inverse scattering for the nonlinear Schrödinger equation with a potential*, VI Seminar on Free Boundary Value Problems and their Applications, Part 1 (Spanish) (Rosario, 1998). MAT Ser. A, **3**, 2001, pp. 13–20. MR1941699 (2003k:35200)
[73] R. Weder, *Time-Dependent Approach to Inverse Scattering. Partial Differential Equations and Spectral Theory*, Editors: M. Demuth and B.-W. Schultz. Operator Theory Advances and Applications **126**. Birkhäuser, Berlin, 2001, pp. 321–328. MR1946685 (2003k:35251)
[74] R. Weder, *Multidimensional inverse scattering for the nonlinear Klein-Gordon equation with a potential*, J. Differential Equations **184** (2002), 62–77. MR1929146 (2004d:35178)

[75] R. Weder, *The Aharonov-Bohm effect and time-dependent inverse scattering theory*, Inverse Problems **18** (2002), 1041–1056. MR1929281 (2003i:81238)

[76] R. Weder, *The Time-Dependent Approach to Inverse Scattering*, Advances in Differential Equations and Mathematical Physics. Editors: Y. Karpeshina, G. Stolz, R. Weikard, and Y. Zeng. Contemporary Mathematics, vol. **327**, Amer. Math. Soc., Providence, RI, 2003, pp. 359–377. MR1991554 (2004g:35173)

[77] R. Weder, *The $Lp-L\acute{p}$ estimate for the Schrödinger equation on the half-line*, J. Math. Anal. Appl. **281** (2003), 233–243. MR1980087 (2003m:35200)

[78] R. Weder, *Time–Dependent Methods in Inverse Scattering Theory*, New Analytic and Geometrical Methods in Inverse Problems. Editors: K. Bingham, Ya. V. Kurylev and E. Somersalo. Springer, Berlin, 2004, pp. 367–381. MR2053429 (2005c:35226)

[79] R. Weder, *Scattering for the forced non-linear Schrödinger equation with a potential on the half-line*, Math. Meth. Appl. Sci. **28** (2005), 1219–1236. MR2142269 (2005m:35289)

[80] R. Weder, *The forced non-linear Schrödinger equation with a potential on the half-line*, Math. Meth. Appl. Sci. **28** (2005), 1237–1255. MR2142270 (2006b:35308)

[81] R. Weder, *The electric Aharonov-Bohm effect*, J. Math. Phys. **52** (2011), 052109, 17 pp. MR2839065 (2012f:81106)

[82] R. Weder, *Topological effects in quantum mechanics and high-velocity estimates*, Int. J. Modern Phys. D **20** (2011), 951–961. MR2801524 (2012f:81294)

[83] D. R. Yafaev, *Scattering matrix for magnetic potentials with Coulomb decay at infinity*, Integral Equations Operator Theory **47** (2003), 217–249. MR2002666 (2004f:47018)

[84] D. R. Yafaev, *Scattering by magnetic fields*, St. Petersburg Math. J. **17** (2006), 875–895. MR2241430 (2007c:81186)

Fellow, Sistema Nacional de Investigadores, Departamento de Física Matemática, Instituto de Investigaciones en Matemáticas Aplicadas y en Sistemas, Universidad Nacional Autónoma de México, Apartado Postal 20-726, México DF 01000, México

E-mail address: weder@unam.mx

Selected Published Titles in This Subseries

66 **E. Khruslov, L. Pastur, and D. Shepelsky, Editors,** Spectral Theory and Differential Equations, 2014 (TRANS2/233)

65 **Michael Levitin and Dmitri Vassiliev, Editors,** Operator Theory and Its Applications, 2010 (TRANS2/231)

64 **Arina A. Arkhipova and Alexander I. Nazarov, Editors,** Nonlinear Partial Differential Equations and Related Topics, 2010 (TRANS2/229)

63 **Alexei Alexandrov, Anton Baranov, and Sergey Kislyakov, Editors,** Linear and Complex Analysis, 2009 (TRANS2/226)

62 **T. Suslina and D. Yafaev, Editors,** Spectral Theory of Differential Operators, 2008 (TRANS2/225)

61 **V. M. Buchstaber and I. M. Krichever, Editors,** Geometry, Topology, and Mathematical Physics, 2008 (TRANS2/224)

60 **Yu. Neretin, M. Olshanetsky, and A. Rosly, Editors,** Moscow Seminar on Mathematical Physics, II, 2007 (TRANS2/221)

59 **M. S. Birman and N. N. Uraltseva, Editors,** Nonlinear Equations and Spectral Theory, 2007 (TRANS2/220)

58 **V. Kaimanovich and A. Lodkin, Editors,** Representation Theory, Dynamical Systems, and Asymptotic Combinatorics, 2006 (TRANS2/217)

57 **M. V. Karasev, Editor,** Quantum Algebras and Poisson Geometry in Mathematical Physics, 2005 (TRANS2/216)

56 **Ernest Vinberg, Editor,** Lie Groups and Invariant Theory, 2005 (TRANS2/213)

55 **V. M. Buchstaber and I. M. Krichever, Editors,** Geometry, Topology, and Mathematical Physics, 2004 (TRANS2/212)

54 **S. G. Gindikin, Editor,** Lie Groups and Symmetric Spaces, 2003 (TRANS2/210)

53 **M. V. Karasev, Editor,** Asymptotic Methods for Wave and Quantum Problems, 2003 (TRANS2/208)

52 **Yu. M. Suhov, Editor,** Analytic Methods in Applied Probability, 2002 (TRANS2/207)

51 **M. S. Agranovich and M. A. Shubin, Editors,** Partial Differential Equations, 2002 (TRANS2/206)

50 **V. Turaev and A. Vershik, Editors,** Topology, Ergodic Theory, Real Algebraic Geometry, 2001 (TRANS2/202)

49 **Michael Semenov-Tian-Shansky, Editor,** L. D. Faddeev's Seminar on Mathematical Physics, 2000 (TRANS2/201)

48 **L. Lerman, G. Polotovskiĭ, and L. Shilnikov, Editors,** Methods of Qualitative Theory of Differential Equations and Related Topics, 2000 (TRANS2/200)

47 **R. A. Minlos, Senya Shlosman, and Yu. M. Suhov, Editors,** On Dobrushin's Way. From Probability Theory to Statistical Physics, 2000 (TRANS2/198)

46 **Vladimir Arnold, Maxim Kontsevich, and Anton Zorich, Editors,** Pseudoperiodic Topology, 1999 (TRANS2/197)

45 **Ya. Eliashberg, D. Fuchs, T. Ratiu, and A. Weinstein, Editors,** Northern California Symplectic Geometry Seminar, 1999 (TRANS2/196)

44 **Alexander Astashkevich and Serge Tabachnikov, Editors,** Differential Topology, Infinite-Dimensional Lie Algebras, and Applications, 1999 (TRANS2/194)

43 **A. Yu. Morozov and M. A. Olshanetsky, Editors,** Moscow Seminar in Mathematical Physics, 1999 (TRANS2/191)

42 **S. Tabachnikov, Editor,** Differential and Symplectic Topology of Knots and Curves, 1999 (TRANS2/190)

For a complete list of titles in this series, visit the
AMS Bookstore at **www.ams.org/bookstore/trans2series/**.